The Historiography of the First Russian Antarctic Expedition, 1819–21

Rip Bulkeley

The Historiography of the First Russian Antarctic Expedition, 1819–21

palgrave
macmillan

Rip Bulkeley
Oxford, UK

ISBN 978-3-030-59545-6   ISBN 978-3-030-59546-3   (eBook)
https://doi.org/10.1007/978-3-030-59546-3

Cover illustration: A Russian warship at the Admiralty Yard in 1818 or 1819, detail from the
Panorama of St Petersburg by Angelo Toselli (1820)

This Palgrave Macmillan imprint is published by the registered company Springer Nature
Switzerland AG.
The registered company address is: Gewerbestrasse 11, 6330 Cham, Switzerland

Forgetfulness, and even historical error, are essential for the formation of a nation, from which it follows that the achievements of historical research often pose a threat to nationality.
Ernest Renan

The past always changes; only the future remains the same.
Soviet-era Polish quip

CORRESPONDENCE

Any reader wishing to contact the author about aspects of this book, or perhaps to obtain his 2016 summary of (a) some small mistakes in his previous book about the Bellingshausen expedition, and (b) further evidence supporting conclusions reached in that book which came to hand after publication, is invited to write to him at rbalkli@protonmail.com

ABSTRACT

—

This book looks at the different ways in which Russian historians and other authors have thought about their country's first Antarctic expedition (1819–1821) over the past 200 years, and the effects which that process has had on Russia's Antarctic policy and may yet have on Antarctica itself. The focus is on the Soviet decision in 1949, in line with the cultural policies of late Stalinism, to revise the traditional view of the expedition in order to claim that the Antarctic mainland was first sighted by Russian seamen in January 1820, a claim which remains the official position in Russia today. The book shows, however, that the case for such a claim has never been established, and that the work of Russian historians was marred by their unsuccessful attempts to prove it. A revision of the 1949 revision is therefore overdue.

Keywords Antarctica; Antarctic Treaty; Bellingshausen; Cold War; Discovery; Prestige; Russia; Soviet Union; Stalin; *Zhdanovshhina*

ACKNOWLEDGEMENTS

The author is sincerely grateful to the Russian Naval Archives and their director Valentin Smirnov for the prompt and reliable supply of requested documents and for consenting to the reproduction of a detail from Bellingshausen's track chart in Figs. 3.1 and 3.2; to Sheila Bransfield for early drafts of her book about her ancestor and other help; to Aleksandr Chekulaev in St Petersburg for locating and providing an abundance of Soviet and post-Soviet books about the Antarctic; to Prof. Aant Elzinga in Göteborg for explaining Lomonosov's Swedish article about icebergs; to his friends Judith Holland and Christopher Kirwan, for their rigorous scrutiny of the text at various stages; to Irina Lukka, a senior librarian at the National Library of Finland, for her generous assistance with research in 2011; to Dr Aleksandr Ovlashhenko in Riga for digital copies of his recent books; to Timo Palo, a participant in the recent Estonian memorial cruise to the Antarctic, for permission to publish the photograph in Fig. 6; to the late Dr Ian R. Stone, formerly the outstanding editor of *Polar Record*, for his support and friendship throughout this long project; and to Dr Erki Tammiksaar in Tartu, Estonia, for information about or digital copies of rare works, for his contributions to this book, and for discussions of these matters over several years.

CONTENTS

Abbreviations[1]

AANII	Arctic and Antarctic Research Institute
ANII	Arctic Research Institute
ATS	Antarctic Treaty System
AWS	Automatic Weather Station
CE	Common Era
CHM	Common Heritage of Mankind
GLONASS	Global Navigation Satellite System
GPS	Global Positioning System
I. / Is.	Island/Islands
IGY	International Geophysical Year (1957–1958)
IPY	International Polar Year
IRN	Imperial Russian Navy
LHS	Leningrad Historical Section (of the Soviet Geographical Society}
MHS	Moscow Historical Section (of the same)
n.d.	no date
NSR	Northern Sea Route
NSRA	Northern Sea Route Administration
NWbW	North-West by West—example of compass bearing
O.S.	Old Style or Julian calendar
PEP	Protocol on Environmental Protection
RAE	Russian Antarctic Expeditions
RN	Royal Navy
SAE	Soviet Antarctic Expeditions

---

[1] (translated where appropriate)

SCAR	Scientific Committee on Antarctic Research, International Science Council
s.d.	ship's date
SMP	South Magnetic Pole
UN	United Nations
USP	Unique Selling Proposition

LIST OF FIGURES

List of Tables

CHAPTER 1

Preliminaries

BELLINGSHAUSEN

Captain Faddej Faddeevich Bellingshausen of the Imperial Russian Navy (1778–1852; Fig. 1.1) was never one to make a fuss. When his father died young and his family, already in financial difficulties, could not afford the expensive army career provided for two older brothers, he entered the Naval Cadet College, which was in part a charitable institution for boys in such circumstances. When the Navy required him to replace his German forenames with Russian ones, he did so, reserving the original 'Fabian Gottlieb Benjamin' for domestic use thereafter. And when he lost the chance to command a second polar expedition because it sailed a few weeks before he completed the book about his first one, he calmly reverted to a more orthodox naval career. Twenty years later he had risen to the top of his profession as a full Admiral and member of the Admiralty Council.[1]

Setting aside his achievements as an Antarctic explorer, other aspects of Bellingshausen's still rather obscure life suggest exceptional qualities.[2] Although he graduated from the Cadet Corps halfway down his class list,

---

[1] The original texts of translated quotations, indicated by numbers in square brackets, are provided at the end of each chapter. For the system of romanization used, the rendering of dates, and a few other matters, please consult the Apparatus at the end of the book.

[2] Reportedly, more is known about him in Estonia than has yet been published internationally (personal communication from E. Tammiksaar, 10 June 2020).

© The Author(s), under exclusive license to Springer Nature Switzerland AG 2021
R. Bulkeley, *The Historiography of the First Russian Antarctic Expedition, 1819–21*,
https://doi.org/10.1007/978-3-030-59546-3_1

Fig. 1.1 Statue of Admiral F.F. Bellingshausen by Ivan Nikolaevich Schröder, 1870. (Photograph by the author)

his mathematical abilities were so considerable that five years later he was singled out by Captain Adam Krusenstern (1770–1846) for the post of navigational officer on the first Russian circumnavigation, from 1803 to 1806, a landmark voyage in Russian naval history. He also contributed much to the cartography of that expedition.

Bellingshausen's family were Baltic Germans living on the island of Ösel (Saaremaa), then part of the Russian province of Livland and now in Estonia. He was raised as a Lutheran and spoke no Russian until he went to the Cadet Corps at the age of ten. But in 1826, aged 48, he married an Orthodox Russian woman 30 years his junior from the landed aristocracy of Pskov province, even though such interfaith marriages, while frequent for ethnically Russian men, were uncommon for their sisters. Also, few Russian naval officers published technical manuals on aspects of their profession, as Bellingshausen did with a treatise on naval gunnery in 1839.

The fact that Bellingshausen never claimed to have discovered an Antarctic continent, despite having sighted the first land south of the Antarctic Circle in January 1821, was not solely due to his temperament. For one thing he knew that vessels from other countries were already in the far south when he reached it in December 1819. And for another he may have been handicapped by an authoritative theory about the Antarctic ice cap, promulgated by the great naturalist Georges-Louis Leclerc de Buffon (1707–1788) in 1778, according to which a vast land-like and permanent accumulation of ice was sure to be encountered between 70°S and the South Pole, whether or not it covered any bodies of land. But land, not ice, was what Bellingshausen's orders instructed him to seek. Buffon provided a map in which his conjectured southern ice cap extended up to and in places north of the Antarctic Circle (Buffon 1778: after p. 615), and by the time of Bellingshausen's voyage the theory was shown as probable fact in at least one of the best available atlases.

But that said, temperament was also a factor. According to his son-in-law, Bellingshausen took a very matter-of-fact view of his Antarctic voyage:

> Bellingshausen never considered his achievements [as an explorer] to have been anything out of the ordinary. In his own opinion, he had simply fulfilled his duty and added something useful to knowledge [1]. (G[ershau] 1892: 382)

DECEPTION

This book tells the story of an unusual, but also very successful, historical deception, which over the past 70 years has transformed a fine explorer who by his own account was *not* the first to sight and so in popular parlance 'discover' the continent of Antarctica 200 years ago, into a national hero whom many Russians now 'know' did precisely that. It presents a detailed analysis of that metamorphosis with a view to understanding why and how it happened, and offers some thoughts about its possible relevance to the political and economic future of the region.

The deception was unusual in having been carried through by people who were for the most part honest, and who tried to investigate the matter carefully, or to relay the findings of others who had done so. Perhaps by the highest standards those researchers were insufficiently critical; they were certainly insufficiently self-critical. But they lived in times that were inimical to dispassionate historical enquiry, and even in more favourable circumstances, such as those enjoyed by the present author, historians must always walk in peril of their own subjectivity. Most of the protagonists are now dead, and while their mistakes should be documented for the sake of truth and of the brave expeditioners who died before them, their achievements and their occasional dissent from the line imposed by their rulers will also be respected in these pages.

PANFILOVTSY

Seeking the mandate of precedent, governments in many times and places have made free with history, either directly or through their sponsored agencies, and authors have sometimes endorsed such crooked dealings, especially in societies where tales of the past amounted to little more than Homeric praise-singing. For example, in 1644 the Dutch West India Company concocted a Dutch presence since 1598, along part of the north-eastern seaboard of North America, out of nothing more than ink and paper, and 11 years later they were backed up by the propagandist Adriaen van der Donck (Hunter 2009: 283–287). It is less common, nowadays, to find professional historians engaging in such dirty work. Perhaps something of the sort took place in the Soviet Union with regard to the Bellingshausen expedition, but to suppose that we are dealing here with a simple exercise in misinformation would be to make a serious mistake.

The relationship between the modern Russian state and historical truth is more complicated than that. Take the fascinating case of the 28 *Panfilovtsy*. According to Soviet press reports in November 1941, they were a group of soldiers in the 8th Guards Panfilov Division who were killed to a man while mounting a heroic resistance against a column of 54 German tanks near Volokolamsk. The editor of one newspaper boldly transplanted a line attributed to the commander of Napoleon's Old Guard at Waterloo, 'The Guard dies, but does not surrender', into the mouth of one of the leaders of these supposed heroes, despite its being perfectly well known in Russia in its original form, and despite the fact that the soldiers in question were not Guards at the time of their alleged sacrifice (Statiev 2012: 781).

The tale of the *Panfilovtsy* proved hard to confirm, despite which, and despite the arrest in May 1942 of a surviving member of the 'fatalities', it was vigorously upheld by the authorities both during and after the war, while other incontestable feats of heroism performed by Soviet units on the same front received less publicity. In 1947, however, the arrest of a second survivor from the group of 28 supposed martyrs prompted an investigation by military prosecutor Lieutenant-General Nikolaj Afanas'ev. Afanas'ev's report, completed in 1948, concluded that the *Panfilovtsy* episode had been fabricated by the journalists who first reported it; he also discovered four more survivors of the group.

When the report was placed before the Central Committee of the Soviet Communist Party, however, they decided to ignore it. The problem was that they had already endorsed the story before the whole country in 1942 by bestowing the title of Hero of the Soviet Union on 28 infantrymen, whose names had in reality been chosen at random from inaccurate casualty lists after Mikhail Ivanovich Kalinin (1875–1946), chairman of the Supreme Soviet, demanded to know who they were. One survivor lived on until 1982, and the *Panfilovtsy* grave at Nelidovo Village, although inscribed with 22 of the 28 names (omitting, presumably, the six known to be alive in 1948), contains only six bodies.

Fast forward to 2016, and Minister of Culture Vladimir Rostislavovich Medinskij removes Sergej Vladimirovich Mironenko from his post as director of the National Archives of Russia after the latter has first declared the tale of the *Panfilovtsy* to be a myth, and then responded to criticism from Medinskij by publishing the Afanas'ev report on the Archives website. Medinskij blustered that the *Panfilovtsy* were 'civil saints' of 'blessed memory', and the tale of their heroism was 'a legendary story that has become true' [2].[3] But as so often the affair was more complex than at first appears. On the one hand further evidence has recently been found which suggests there may have been an episode of heroic self-sacrifice resembling, if not quite on the scale, or with the prowess, of the legendary version, during that battle. The makers of the crowdfunded blockbuster

---

[3] https://www.currenttime.tv/a/panfilovcy-28-war-voina-podvig-fight-medinsky/29634354.html.

film *The Panfilov 28* (2016), to which the Ministry of Culture had contributed 30 million roubles in 2014, will doubtless have been relieved to hear that. On the other hand the Minister's fitness to pass judgement on such matters has been called into question with a damning academic critique of his 2002 doctoral thesis, which found the work to be riddled with crude historical errors, dependent on using only sources favourable to the author's argument, and uncritical even of those (Kozlyakov and Others 2016).

The point of mentioning the *Panfilovtsy* here, however, is not to contribute to the historical debate, but rather to highlight features of the Mironenko affair which illustrate important aspects of the relationship between the Russian state and the Russian past. First, at no point from 1948 to the present day was any attempt made to suppress the Afanas'ev report. Next, it had already been published in full by the magazine *Novyj Mir*, 18 years before Mironenko put it online, so that presumably every self-respecting university library in Russia has a copy (Petrov and Èdel'man 1997). Mironenko had breached no censorship, merely offended certain conventions (a very 'Russian' thing to do) by shoving some unwelcome counterevidence in the face of the film's directors, Kim Druzhinin and Andrej Shalopa, who then called on the Minister of Culture, and ultimately President Putin, to support them. Furthermore, Mironenko was not dismissed, merely relieved of the burdens of the directorship and left to pursue his work as a senior researcher at the Archives—hardly a punitive measure for any true historian. Last but not least, five years after Mironenko posted it on 8 July 2015 the report remains online at the National Archives, together with Afanas'ev's letter presenting it to Politburo member Andrej Zhdanov (of whom more later).[4]

All of which shows that when Sky News reporter John Sparks claimed that he needed to bribe his way into the Archives in order to set eyes on the Afanas'ev report, he was simply staging a melodrama for propaganda purposes, instead of accessing the report online (or in print) like anyone else (Sparks 2016). More important, it also illustrates the abiding duality of official Russian attitudes to history. On the one hand there is vigilant concern, in respect of what might be called 'public history', that nothing thought detrimental to Mother Russia should be openly acknowledged, so that a consistently positive image of the national past can be curated and upheld. The state-owned TV channel *History* plays a large role, as does the

---

[4] http://statearchive.ru/607.

unified history syllabus for schools that was set in motion by President Putin in 2013. (Unfortunately, to judge from a few examples of related teaching materials seen by the present author, the standard expected from contributors is even lower than that reportedly set for Medinskij's D.Hist.)

On the other hand, however, Russia has an extensive and treasured system of national and provincial archives with hundreds, if not thousands, of busy and respected scholars hard at work in them every day. To cite another example, the present author encountered no difficulty a few years ago, when he published an article, in Russian and in the country's leading journal for the history of science, which cast doubt on the reliability of the only document which supports the Russian claim to priority in Antarctica (Balkli 2013). Unsurprisingly, the article has been generally ignored, but the author's access to Russian archives has not been affected in any way. The tradition of respect for history and its practitioners in Russia owes something to Marxism, to which lip-service was paid by the authorities for several decades, but more perhaps to the cultural continuity of Russian scholarship from pre-revolutionary times down to the present, despite much grievous personal and intellectual suffering along the way. That rational and humane heritage is associated with such names as Vladimir Vernadskij (1863–1945), Sergej Vavilov (1891–1951), Nikolaj Gorbunov (1892–1938), Petr Kapitsa (1894–1984), Mikhail Bakhtin (1895–1975), and Nikolaj Semyonov (1896–1986), amongst many others. It is still alive and well in Russia, and the archives and other institutions of science, social science, and history in which it thrives wield considerable soft power to counterbalance occasional interventions from misguided politicians.

There may be points in this book where the reader is tempted to dismiss Russia as a land without respect for culture in general and history in particular. It is true that intellectuals, including historians, have suffered greatly in the past when they stepped out of line with the requirements of the state, and are still sometimes harassed for so doing nowadays. But the real lesson of the *Panfilovtsy* is that history matters enormously to Russians, as it often does in a country which is not easy in its skin, as the French say. They have what is sometimes called a love-hate relationship with past realities, and that uneasy confusion, not the headline episodes of official bullying, is the key to understanding much of what follows.

Original Texts of Quotations

[1] Беллингсгаузенъ [так] никогда не смотрѣлъ на совершенный имъ подвигъ как на нѣчто выходящее изъ ряда обыхновенныхъ событій, а считалъ его простымъ исполненіемъ своего долга, принеснимъ нѣкоторую полъзу наукѣ.

[2] … свѣтлой памяти … этих гражданских святых … это легендарная история, которая стала историей истинной.

References[5]

Balkli, Rip. 2013. Pervye nablyudeniya materikovoj chasti Antarktidy: popytka kriticheskogo analiza [First Sightings of the Mainland of Antarctica: Towards a Critical Analysis]. *Voprosy istorii estestvoznaniya i tekhniki* 4: 41–56.

de Buffon, M. le Comte. 1778. *Histoire naturelle générale et particulière. Supplément.* Vol. 5. Paris: Imprimerie Royale.

G[ershau], P. 1892. Admiral″ Fhaddej Fhaddeevich″ Bellinsgauzen″ [Admiral Faddej Faddeevich Bellingshausen]. *Russkaya starina* 75: 373–395.

Hunter, Douglas. 2009. *Half Moon: Henry Hudson and the Voyage that Redrew the Map of the New World.* London: Bloomsbury Press.

Kozlyakov, V.N., and Others. 2016. Zayavlenie o lishenii Vladimira Rostislavovicha Medinskogo uchenoj stepeni doktora istoricheskikh nauk [Statement About Withdrawing the Degree of Doctor of History from Vladimir Rostislavovich Medinskij]. Ministry of Education, Russian Federation. April 25. http://wiki.dissernet.org/tools/vsyakosyak/MedinskyVR_ZoLUS.pdf.

Petrov, N., and O. Èdel′man. 1997. Novoe o sovetskikh geroyakh [A New Perspective on Soviet Heroes]. *Novyj Mir* 6: 140–151.

Sparks, John. 2016. How Russia is Engaged in a Battle for its Own History. *Sky News*, December 11. https://news.sky.com/story/how-russia-isengaged-in-a-battle-for-its-own-history-10691897.

Statiev, Alexander. 2012. "La Garde meurt mais ne se rend pas!": once again on the 28 Panfilov Heroes. *Kritika: Explorations in Russian and Eurasian History* 13: 769–798.

---

[5] Key to archival references: SARN = State Archives of the Russian Navy, F = Fond, S = Series [*Opis′*], P = Piece [*Delo*], fo./fos = folio/s [*list/y*], v = verso [*oborotnoe*]. (The latter is necessary because verso pages were usually unnumbered.) Dates on manuscript documents are given in Universal Time (CE).

CHAPTER 2

Introductions

THE PROBLEM

In February 1937 the Australian geographer Frank Debenham (1883–1965), a former geologist with Robert Scott's fateful *Terra Nova* expedition and the first director of the Scott Polar Research Institute at the University of Cambridge, sent a draft of his translation of Bellingshausen's narrative of his voyage to the British Antarctic historian Hugh Robert Mill (1861–1950). In an accompanying letter Debenham drew Mill's attention to Bellingshausen's 'virtual discovery' of part of the coast of East Antarctica in early 1820 'as far as seeing is concerned, even tho' not believing' (Debenham 1937). Two years later, he stated publicly that Bellingshausen had seen 'the first undoubted land of the main mass of the continent' on 17 February 1820, but without recognizing it as such (Debenham 1939: 1). In 1982 the British historian A.G.E. Jones (1914–2002) pointed out the many difficulties of interpreting the evidence, but felt able to conclude that the expedition had probably sighted an ice coast of the Antarctic mainland

R. Bulkeley, *The Historiography of the First Russian Antarctic Expedition, 1819–21*,
https://doi.org/10.1007/978-3-030-59546-3_2

on 28 January 1820, a date which Debenham had also considered but then rejected (Jones 1982: 89–94).[1,2]

Neither investigation was entirely satisfactory. Debenham had no Russian and worked only with Bellingshausen's book (referred to here as *Two Seasons* from an English version of the title), which he was translating in the sense of producing a fair copy from versions by other people. Jones used the third Russian edition of *Two Seasons*, published in 1960, and also consulted other documents, but in most cases only as briefly excerpted in another important source, the facsimile edition of Bellingshausen's track chart (Bellinsgauzen" 1821d) published by the historian Mikhail Ivanovich Belov (1916–1981) in 1963. Crucially, both Debenham and Jones based their assertions about a sighting of land in February 1820 on mistranslations of a Russian expression meaning 'iceberg', which they turned into a reference to mountains (Bulkeley 2019—see also Chap. 3).[3] Neither appears to have consulted linguistic experts or nineteenth-century dictionaries—Debenham in particular seems to have assumed that any living Russian speaker would be familiar with the nuances of the language as used by sea officers more than a century earlier.

Instead of proceeding at third hand, by discussing evaluations of the expedition reached by non-Russian historians, this book describes and analyses the findings of Russian authors on the subject from 1821, when the expedition returned to Russia, down to the present day, with only occasional references to foreign authors. Far from being a topic of merely academic interest, Russian versions of what the expedition achieved became embedded in Soviet Antarctic policy during the Cold War, and they remain fundamental for Russian Antarctic policy today.

In 1819 the Russian Emperor Alexander I sent a naval expedition, commanded by Junior Captain Bellingshausen, to search for land in the Southern Ice Ocean. The expedition was half of a 'double polar venture' (Barratt 1988: 235), for which two parallel, equally (and generously)

---

[1] The original texts of translated quotations, indicated by numbers in square brackets, are provided at the end of each chapter. For the system of romanization used, the rendering of dates, and a few other matters, please consult the Apparatus at the end of the book.

[2] Although published some years later, Debenham's first thoughts about the matter came in notes to his translation that were probably written before his letter to Mill and certainly before his 1939 statement. At that point he considered two of the expedition's southerly probes, on 28 January and 17 February 1820, as possible sightings, but expressed much greater confidence about the second (Bellingshausen 1945 1: 117, 128).

[3] Debenham also mentioned the possibility of 'land ice' for 28 January.

equipped squadrons were dispatched, one under Captain Lieutenant Mikhail Nikolaevich Vasil'ev (1770–1847) to the Arctic approaches of the North Pacific, and the other under Bellingshausen to the Antarctic. But whereas previous Russian expeditions, notably those led by Bering and Billings, had achieved significant results in the region assigned to Vasil'ev, Bellingshausen's expedition was a first for Russia and only the second in world history, coming 47 years after the first, which formed part of the second circumnavigation, in 1772–1776, commanded by Captain James Cook, RN (1728–1779), but had found no land south of the Antarctic Circle.

The Bellingshausen voyage lasted for just over two years and the people responsible for reporting its results were Bellingshausen and the expedition's astronomer Ivan Mikhailovich Simonov (1794–1855), from the University of Kazan. According to them the most significant achievement had been the discovery in January 1821, during their second Antarctic season, of the first land south of the Antarctic Circle, namely Peter I Island and the nearby Alexander I Coast. (The latter could not be made out, at the time, as between an island or part of a more extensive mainland.) That verdict was accepted in Russia and abroad for more than 120 years. The only significant change was that, as more land was discovered in the far south, both nearby and in other widely separated longitudes, commentators grew more confident that the Alexander I Coast would turn out to be part of the mainland. As for their first Antarctic season in 1820, Bellingshausen and Simonov recorded separately that, despite crossing the Antarctic Circle several times, the expedition saw no land between leaving the sub-Antarctic South Sandwich Is. on 17 January and making landfall at the southern end of Tasmania on 5 April.

Those established findings were radically revised on 10 February 1949, when the All-Union Geographical Society, meeting in Leningrad, passed a resolution affirming that: 'The Russian navigators Bellingshausen and Lazarev … passed around the Antarctic continent and were *the first* to approach its shores …' [1] (Anon 1949—emphasis added). That was a bold contention, because it was known by then that British and American sealers had sighted the mainland of Antarctica in January and November 1820 respectively, in the first case under charter to the Royal Navy and commanded by one of its officers. The resolution added that Russian priority entitled the Soviet Union, as the successor state, to take part in international negotiations about the future of Antarctica, a proposition which

could have been supported on stronger grounds than a questionable interpretation of history, but was not.

Bellingshausen's final, considered account of 28 January 1820, the date
required for priority, reads as follows:

> ... at noon, in latitude 69°21′28″, longitude 2°14′50″, we met ice which
> appeared to us, through falling snow, like white clouds. The wind, from NE,
> was dropping, with a heavy swell from NW. Because of the snow we could
> not see far. I gave orders to proceed SE, close-hauled. After proceeding for
> two [nautical] miles [3.7km] in that direction we saw that continuous ice
> extended from East through South to West.[4] Our course led directly into
> that ice field, scattered with hillocks [2]. (Bellinsgauzen″ 1831 1: 171–172)

Because it will be mentioned repeatedly here, it is appropriate also to
provide Bellingshausen's account of the probe made on 17 February
1820, the one which Debenham thought was a more plausible sighting of
the coast. That was also mentioned at the 1949 meeting, but it was not at
first given the same attention as 28 January, presumably because it did not
confer priority.

> Eventually, at a quarter past three after noon, we saw a quantity of large, flat,
> high ice islands, beset with light floes some of which overlay one another in
> places. The ice formations to SSW join together into hilly, solid standing
> ice[5]; its edges were perpendicular, forming coves, and its surface rose away
> to the South, to a distance whose limits we could not make out from the
> [main] cross-trees. ...
>   Seeing that the ice islands had similar surfaces and edges to the afore-
> mentioned large ice formation, which lay before us, we concluded that those
> huge ice masses and all similar formations get separated from the main coast
> by reason of their own weight or from other physical causes and, carried by
> the winds, drift out into the expanse of the Southern Ice Ocean [3].
> (Bellinsgauzen″ 1831 1: 188–189; see Fig. 3.1)

For fuller translations of these passages, which for convenience will be
repeated in later chapters, see Bulkeley (2019).

---

[4] Bellingshausen used the nautical, or in his words 'Italian' mile, and distances given by or
reported from him are converted to kilometres accordingly. Few Russian commentators,
unfortunately, have explained which sort of mile they meant when in their own voice.

[5] Bellingshausen's language strongly implied, but did not quite say, that the massive ice
formation was grounded on the ocean floor.

In an accompanying lecture the president of the Society, Lev Semyonovich Berg (1876–1950), reinforced the cautious wording of the resolution by stating that '*Nowadays* it is clear that the hill[ock]y ice described by Bellingshausen as extending from east to west in fact comprised the margin of the Antarctic continent' [4] (Anon 1949—emphasis added). Members of the Society duly hailed the expedition as the first discoverers, the 'Columbuses', of Antarctica. But as we shall see in Chaps. 6 and 8, they did so with varying degrees of doubt and confusion.

On 7 June 1950 the Soviet Union sent a memorandum to Western governments which referred to the Geographical Society meeting and elaborated on its findings:

> ... the Russian seamen Bellingshausen and Lazarev ... were the first to reach the shores of Antarctica, circumnavigated the continent, and thus demonstrated that the then widely held view, that there was no land beyond the Antarctic Circle, was mistaken [5]. (Anon 1950)

That achievement, the Note continued, meant that the Soviet Union had just as much interest in determining the future of Antarctica as states with ongoing expeditions in the region. Together with Soviet participation in the Antarctic research programmes of the International Geophysical Year of 1957–1958 (IGY), that historical thesis became the grounds on which, in its own eyes, the Soviet Union joined 11 Western countries as a founding member of the Antarctic Treaty in 1959.

An analysis of the Russian historiography of the expedition therefore needs to explain the rational and non-rational factors which contributed to the upgrading of its achievements in 1949, to trace the acceptance of and resistance to that revision in the Soviet Union, and to re-evaluate the Soviet claim that Antarctica was first discovered by Russian explorers, a view which is still generally held in Russia today. But those tasks should not be attempted without first explaining how the original assessment of the expedition was formed and then maintained for more than a century (Chaps. 3 and 4) before coming under various pressures to change (Chap. 5), and then undergoing a radical transformation after World War II (Chap. 6).

Although geographical exploration has commonly played a role in political events, the historiography of an individual expedition has seldom become a state-sponsored political process in its own right. It will be shown here that the Bellingshausen expedition was so treated during the Cold War and remains a cherished political meme in modern Russia. Since

1949, however, Russian historians treating the expedition have from time to time suppressed evidence, misstated facts, neglected sources, and overlooked the cultural context of Bellingshausen's day. The only way to put forward such a controversial thesis, so that even Russian readers might at least examine it, is by laying out the whole story and scrutinizing every important link in the long chain of commentary and inference which stretches from 1821 down to the present day. That should reveal how much of it can still be relied on, and, if possible, what went wrong in treatments of the expedition that are not to be relied on. The original texts of all translated passages must also be provided, so that Russian readers can check them for themselves. Such a comprehensive account will also, in the normal way, enable other scholars to offer counterarguments to those put forward here.[6]

Throughout, it will be important to remember that most of the Russian authors who are criticized here were working both without complete intellectual freedom, and without the powerful technologies which can now be deployed to find and search relevant texts from previous centuries.

It is 70 years since Terence Armstrong (1920–1996) suggested that:

> … it seems important, in view of the political significance attached by the Russians to [Bellingshausen's] explorations, that the sequence of events should be made clear. (Armstrong 1950: 477)

The only amendment from the present author would be to insert 'and arguments' after 'events'.

Another way to describe the subject of this book is to say that when there is international conflict in the Arctic, where all territory falls under recognized national sovereignty and regional agreements are less restrictive (Anderson 2009; Sale and Potapov 2010), it tends to occur quite openly. In the Antarctic, by contrast, states may engage in conflict indirectly because of the strength of the Antarctic Treaty System (ATS). In the case of Russia, the expression of national interests in the Antarctic has since 1949 placed considerable emphasis on a debatable version of past events.

---

[6] Besides the Antarctic historiography of the expedition, to be analysed here, there has been a second, tropical historiography comprising accounts of the expedition's exploration of New Zealand and parts of the South Pacific, which lasted four months, and the disposition and interpretation of the biological and ethnographic collections that were assembled during that part of the voyage. Much excellent work has been published on that topic by Professor Glynn Barratt of the University of British Columbia, amongst others, and no attempt will be made to repeat it here.

How This Book Is Organized

After outlining the problem to be addressed and other important matters in this chapter, the survey begins in Chap. 3 by describing the main primary sources for the Bellingshausen expedition. Where possible thereafter the book focuses on work by Russian historians in the Thucydidean sense of the word, namely people seeking to acquire new knowledge about or to reach a better understanding of some aspect of the recorded human past by discovering new or re-examining old sources, and by reconsidering the findings of previous research historians. Regrettably, few Russian scholars of the first rank have ever studied the expedition closely. Much that was published about it in the Soviet Union, especially during the boom years of 1949–1971, was not history, in the sense of original discovery or critical analysis, even when written by historians who had published on other subjects. Most of it was past-writing which summarized and communicated the findings of a small number of research historians to the Soviet public, which had a keen appetite for such informative and entertaining material and the numerous periodicals, including *Around the World* and *Knowledge Is Power*, which provided it. A survey of that tertiary literature might be useful for a more sociological historiography of the expedition, but as a survey and analysis of secondary authors, the historians proper, this essay can be more selective with tertiary commentators. That is just as well, since an attempt at a complete historiography, of the Soviet period only, has so far required three books to cover the 44 years from 1917 to 1960 (Ovlashhenko 2013, 2014, 2016). It should also be mentioned, because it is something that strikes a foreigner as very unusual, that throughout the 200 years to be described in this book Russian historians and other commentators on the expedition seldom discussed or even referred to each other's work, although, by mentioning most of the cases in which they did, this survey may not give a true picture of that aspect of their culture.[7]

There is a slight problem with access to Russian sources for anyone working outside the boundaries of the former Soviet Union. British libraries have limited holdings of the magazine *Around the World*, for example. The author believes that he has seen and studied all the published and

---

[7] From time to time during the period in question, the vast polity ruled from Moscow or St Petersburg has changed its name, constitution, and territorial boundaries. But there will be points in this book where it becomes necessary to mention or suggest very different periods in its history in a single paragraph, or even sentence. The terms 'Russia' and 'Russian' will be used in such passages.

most of the unpublished primary and almost all of the secondary sources, so that he is both aware of minor secondary items to which he has not gained access, and confident that all major secondary items, referred to by other historians and listed in bibliographies, have been reviewed.

Periodization of the Russian historiography of the expedition is hampered by the scarcity of original historical work both in the early years and in the present century. At several points the periods employed must overlap because different authors and their circumstances did so. The testimony of the expeditioners themselves (primary sources) is summarized in Chap. 3, together with some analysis of their ice terminology in order to understand them properly. Chapter 4 then surveys the secondary Russian sources until 1928, which together with primary sources constituted the original assessment of the expedition. A Norwegian landing on Peter I I. (one of Bellingshausen's discoveries) in February 1929, followed by an official declaration of Norwegian sovereignty two years later, initiated a period of two decades in which factors prompting a Soviet reconsideration of the expedition came into play (Chap. 5). The Soviet government's new claim to priority, and with it special rights, in Antarctica, put forward in 1949, is described in Chap. 6. At that point the chronology is interrupted briefly in order to analyse a letter written by a senior member of the expedition in 1821 (Chap. 7).[8] From the 1950s to the early 1970s (Chaps. 8 and 9) Soviet research and publication on the expedition went through various phases, starting with a focus on the original sources. The mid-1950s saw a thaw in East-West scientific relations which brought a calmer tone to Soviet studies of the Bellingshausen expedition and to the history of geography in general. In the early 1960s some of the best Soviet work on the expedition focused on fresh analyses of the sources. The later Soviet period, from 1966 to 1991, is discussed separately (Chap. 10), before moving on to the post-Soviet period, from 1992 onwards, in Chap. 11. Overall, this periodization reflects the fact that the question of whether and how the expedition discovered Antarctica was the primary concern of most of the secondary Russian authors surveyed and has been the main issue contested within the wider world of learning. It also reflects the author's aspiration to explain the Russian historiography, if possible, by invoking the broader historical context.

---

[8] Although the letter is an important primary source, it will be discussed out of chronological sequence because it was unknown until 1918 and was overlooked by Soviet historians until after the claim to absolute priority in Antarctica was launched in 1949.

No comprehensive bibliography of Russian literature on the expedition has ever been published, but that compiled by N.S. Bogatkina, and included by Admiral Evgenij Evgen'evich Shvede (1890–1977) in his important second edition of Bellingshausen's narrative (Bellinsgauzen 1949: 354–357), is an essential starting point which can now be consulted in online copies of that book. *The Historiography of the Natural Sciences in Russia* (1956), by the historian of science Vasilij Pavlovich Zubov (1900–1963), covers most of the nineteenth century and adds a few items from the periodical literature to Bogatkina. For the Soviet period, the *Bibliography of the History of Science* series, published by the Academy of Sciences, reached the year 1988 in 1997 with its last volume incomplete and the series apparently abandoned. From then on into the post-Soviet period, other sources must be consulted. However, the three-volume bibliography *Historiography in the USSR* by Militsa Vasil'evna Nechkina (1901–1985) and others is disappointing. Although its second volume—1917 to 1967—has a brief listing for Arctic historiography, the work largely ignores the historiography of geography, exploration, and navigation, and indeed that of the natural sciences in general. A short but useful bibliography with items from the post-Soviet period can be found on Bellingshausen's page at the *Maritime Encyclopedia* website (Anon n.d.-a).

Ovlashhenko's recent books (above) contain bibliography for additional tertiary items published between 1917 and 1960, as well as for important sources such as Vyshinskij's memorandum to Stalin. A few works were found with the help of successive drafts of an article by Dr Tammiksaar that were generously supplied by their author (Tammiksaar 2016). Lastly Barratt's historiographic remarks at several points in his *Russia and the South Pacific* series are by no means confined to Pacific aspects of the expedition, and offer an essential checklist for work on this topic.

THE PROGRESS OF EXPLORATION

Later assessments of the expedition were naturally influenced by the progress of empirical knowledge of the Antarctic. That process began with the discovery of the South Shetland Is. by the British trader and seal hunter William Smith (1790–1847) in 1819. Those islands and parts of the Antarctic Peninsula were then explored by British and American seamen, including Edward Bransfield (about 1782–1852)[9] in January and Nathaniel Palmer (1799–1877) in November 1820.[10] The Russian expedition made important discoveries far to the west and south of the Peninsula in January 1821, and then carried out its own brief survey of the South Shetlands. For two decades after that other explorers discovered islands or potentially coastal pieces of land in widely dispersed meridians but at similar latitudes to Peter I I. and the Alexander I Coast. Then in 1840 the hypothesis that a rocky continent existed near the South Pole was boosted by the announcement that a US expedition, commanded by Charles Wilkes (1798–1877), had discovered 2780 km of coastline between latitudes 64° and 67°S (Anon 1840).[11] Confusion followed, however, when the British explorer James Clark Ross (1800–1862) reported that he had sailed over part of the land claimed by Wilkes. Ross cautioned that even after the promising Antarctic discoveries of the mid-nineteenth century the existence of a south polar continent had not yet been established (Ross 1847 1: 280, 275). Indeed the idea was not generally accepted until the turn of the twentieth century, after much further exploration (Bulkeley 2016).

One reason for the delay was that after the Ross Antarctic expedition, the last of the age of sail, the attention of polar explorers became focused on the Arctic, partly for ideological motives arising from the disappearance of the Franklin Expedition in the late 1840s, and partly because Arctic exploration might prove commercially beneficial through finally discovering the long sought-after sea route between Europe and the Far East which is known as the North-West Passage. (The most profitable enterprise in the far south, seal hunting, had collapsed in the 1830s after almost annihilating the animal populations on which it depended.) Interest in the Antarctic revived in the 1870s when the Royal Navy's oceanographic *Challenger* expeditions discovered deposits of erratic rocks on the ocean floor in meridians where no land had yet been sighted in the far south,

[9] Bransfield's date of birth was estimated from Admiralty records rather than his death certificate, which places it three years later (personal communication from Sheila Bransfield, 31 July 2018).

[10] Bransfield's sighting of the Antarctic Peninsula on 30 January 1820 is the only serious alternative to the Russian claim to have sighted the mainland first on 28 January (Campbell 2000: 131).

[11] The distance was converted from 1500 nautical miles, the generally accepted figure for this claim. Wilkes himself reported his discovery as positions from which a supposed coast was observed.

because the most likely source for such deposits was the melting of icebergs calved from glaciers which had travelled over land (Shokal'skij 1898: 490). The sealing industry experienced a temporary revival, and whaling, modernized with steam-powered propulsion and winches as well as grenade harpoons, turned its attention to the Antarctic Ocean, where operations finally began in the 1890s. There was also a strong ideological impulse, voiced at international geographical meetings and heightened by the shortage of Antarctic stations for the International Polar Year of 1882–1883, towards solving one of the last great geographical questions, whether or not there was an Antarctic continent, and if so what its dimensions were.[12] Between 1890 and 1912 expeditions led by Borchgrevinck, de Gerlache, von Drygalski, Nordenskjöld, Scott, Charcot, Shackleton, and Amundsen confirmed the existence of a continent, delineated much of its outline, and began to explore the interior (Baughman 1994; Bulkeley 2016; Martin 1996). In doing so some of them benefitted greatly from knowledge about how to survive in polar conditions that had been acquired from indigenous peoples during previous decades of Arctic exploration.

It took longer to determine the exact nature of the mountainous coasts opposite the south-eastern shores of the South Shetland Is., first sighted in the 1820s, including Trinity Land, Palmer Land, and Graham Land, as well as the Alexander I Coast which lay some 1000 km to the south and west. The British Graham Land Expedition (BGLE) in the 1930s were the first to conclude from surface exploration that Trinity Land and others make up a long peninsula which forms part of the mainland of Antarctica, now known as the Antarctic Peninsula, and that Bellingshausen's Alexander I Coast is in fact an island (Rymill 1936), although no one would question its status as part of Antarctica in the broad sense, in which continents are deemed to include the islands rising from their continental shelves.[13]

---

[12] In Britain, especially, imperialist ambitions also played a part, as nourished by Sir Clements Markham (1830–1916), president of the Royal Geographical Society.

[13] Mikhail Lazarev was probably the first person to refer to the Alexander I Coast as an island, in his letter to Shestakov in 1821, followed by Krusenstern in an atlas published in 1824, and by Bellingshausen himself in the last paragraph of his book (below), which was completed in the same year. Having conferred with Bellingshausen on *Vostok* immediately after the voyage (personal communication from E. Tammiksaar, 10 June 2020), Krusenstern probably followed his protégé's work on the book quite closely, if only to obtain the latest information for his own work. After John Biscoe's expedition (1830–1833), however, he changed his mind and became one of the first people to refer to the Alexander I Coast as a

Before 1936 the hypothesis first advanced by the French explorer Jules Dumont d'Urville (1770–1842), that Trinity Land and others were islands separated from the continent by one or more straits, had been widely accepted (Yelverton 2004). It was endorsed by the Royal Navy (Hydrographic Department 1930: 72), and shown in British, Russian, and other maps (Shokal'skij 1898; Marks" 1916; Odhams about 1935: 128; S. Grigor'ev 1937: 8–9).

The historiographic significance of the BGLE was that its discoveries indicated that British and American sightings of the Peninsula in the 1820s were sightings of the continental mainland, as some of those present had surmised, and that the Russian sighting of the Alexander I Coast in January 1821 was not. This abrupt reversal of the previous status of those two early discoveries coincided with an important paradigm shift in which, as Antarctica became better understood, explorers and geographers began to treat the major and in those pre-climate change days 'permanent' ice shelves and ice tongues as integral parts of the continent, instead of restricting that title to rock alone. In the 1930s Frank Debenham was the first person to reconsider Bellingshausen's achievements in the light of this new conception of the Antarctic mainland. Both developments, the discoveries of the BGLE and the incorporation of major ice shelves into the mainland of Antarctica, were to influence mid-twentieth-century assessments of the Russian expedition by putting a British sighting of the Peninsula and a supposed Russian sighting of an ice coast in East Antarctica, a mere two days apart in late January 1820, into contention as candidates for the first ever sighting of the Antarctic mainland. However, as we shall see, the Cold War context made it virtually impossible for the issue to be discussed in a rational and open-minded way.

Original Texts of Quotations

[1] Русские мореплаватели Беллинсгаузен и Лазарев … обошли вокруг Антарктического материка, *впервые* подошли к его берегам … [разрядка наша].

[2] … въ полдень въ широтѣ 69°, 21′, 28″, долготѣ 2°, 14′, 50″, мы встрѣтили льды, которые представились намъ сквозь шедшій тогда снѣгъ, въ видѣ бѣлыхъ облаковъ. Вѣтръ былъ отъ NO умѣренный, при большой зыби отъ NW, по причинѣ снѣга, зрѣніе

'Land', probably connected to Graham Land (Krusenstern 1833, 1836: 30–31). His revised assessment was then taken up by Novosil'skij (1854c: 16–17).

наше не далеко простиралось; я привелъ въ бейдевинтъ на SO, и пройдя симъ напрвавленіемъ двѣ мили, мы увидели, что сплошные льды простираются отъ Востока чрезъ Югъ на Западъ; путь нашъ велъ прямо въ сіе льдяное поле, усѣянное буграми.

[3] наконецъ въ четверть четвертаго часа по полудни, увдѣли множество большихъ, плоскихъ, высокихъ льдяныхъ острововъ, затертыхъ плавающими мелкими льдами, и мѣстами одинъ на другомъ лежащими. Льды къ SSW примыкаются къ льду гористому, твердо стоящему; закраины онаго были перпендикулярны и образовали заливы, а поверхность возвышалась отлого къ Югу, на растояніе, предѣловъ котораго мы не могли видѣть съ салинга. ...

Видя льдяные острова, поверхностью и краями сходные съ поверхностью и краями большаго вышеупомянутаго льда, предъ нами находящагося, мы заключили, что сіи льдяныя громады и всѣ подобные льды, отъ собственной своей тяжести, или другихъ физическихъ причинъ, отдѣлились отъ матераго берега, вѣтрами отнесенные, плаваютъ по пространству Ледовишаго Южнаго Океана; ...

[4] *Теперь* ясно, что описываемые Беллинсгаузеномъ бугристые льды, простирашіеся с востока на запад, представляли собою именно окраину антарктического материка ... [разрядка наша].

[5] ... русские мореплаватели Беллинсгаузен и Лазарев ... впервые достигли берегов Антарктики, обошли этот материк кругом и тем самым доказали ошибочность распространенного в то время взгляда, будто за южным полярным кругом нет земли.

References[14]

Anderson, Alun. 2009. *After the Ice: Life, Death and Politics in the New Arctic*. London: Virgin.

Anon. 1840. Discovery of the Antarctic Continent. *Sydney Herald*, March 13: 2.

———. 1949. Russkie otkrytiya v Antarktike [Russian Discoveries in the Antarctic]. *Izvestiya*, February 11: 3.

---

[14] Key to archival references: SARN = State Archives of the Russian Navy, F = Fond, S = Series [*Opis'*], P = Piece [*Delo*], fo./fos = folio/s [*list/y*], v = verso [*oborotnoe*]. (The latter is necessary because verso pages were usually unnumbered.) Dates on manuscript documents are given in Universal Time (CE).

————. 1950. Memorandum Sovetskogo Pravitel'stva po voprosu o rezhime Antarktiki [The Soviet Government's Memorandum on the Question of an Antarctic Regime]. *Izvestiya*, June 10: 2.

————. n.d.-a. *Bellinsgauzen Faddej Faddeevich [Bellingshausen, Faddej Faddeevich]*. Morskaya Èntsiklopediya Website. http://95.31.135.131/card/view/24.

Armstrong, Terence. 1950. Recent Soviet Interest in Bellingshausen's Antarctic Voyage of 1819–21. *Polar Record* 5 (39): 475–478.

Barratt, Glynn. 1988. *The Russians and Australia*. Vancouver: University of British Columbia Press.

Baughman, T.H. 1994. *Before the Heroes Came: Antarctica in the 1990s*. Lincoln, NE: University of Nebraska Press.

Bellingshausen, Captain. 1945. *The Voyage of Captain Bellingshausen to the Antarctic Seas, 1819–1821*, ed. Frank Debenham, 2 vols. London: Hakluyt Society.

Bellinsgauzen, F.F. 1949. *Dvukratnye izyskaniya v yuzhnom ledovitom Okeane i plavanie vokrug sveta v prodolzhenie 1819, 1820, i 1821 godov*, ed. E.E. Shvede. Moscow: Geografgiz.

Bellinsgauzen", Kapitan". 1821d. Karta Plavaniya Shlyupov" Vostoka i Mirnago vokrug" Yuzhnago polyusa v" 1819, 1820 i 1821 godakh" pod" Nachal'stvom" Kapitana Billensgauzena [Chart of the Voyage of Sloops *Vostok* and *Mirnyj* around the South Pole in 1819, 1820 and 1821 under the Command of Captain Bellingshausen]. St Petersburg: SARN—F–1331, S–4, P–536, fos 5–19.

Bellinsgauzen", Kapitan". 1831. *Dvukratnyya izyskaniya v" yuzhnom" ledovitom" okeanye i plavanie vokrug" svyeta v" prodolzhenii 1819, 20 i 21 godov" [Two Seasons of Exploration in the Southern Ice Ocean and a Voyage around the World, During the Years 1819, 1820 and 1821]*, ed. L.I. Golenishhev"-Kutuzov", 2 vols. plus *Atlas*. St Petersburg: Glazunovs.

Bulkeley, Rip. 2016. Naming Antarctica. *Polar Record* 52 (1): 2–15.

————. 2019. Bellingshausen's 'Mountains': The 1820 Russian Sighting of Antarctica and Bellingshausen's Theory of the South Polar Ice Cap. *Polar Record* 55 (6): 392–401.

Campbell, R.J. 2000. *The Discovery of the South Shetland Islands*. London: Hakluyt Society.

Debenham, F. 1937. Letter to H.R. Mill, 13 February 1937. Cambridge: Archives of the Scott Polar Research Institute: MS/100/23/48.

————. 1939. Foreword. *Polar Record* 3 (17): 1–2.

Grigor'ev, S.G. 1937. *Vokrug yuzhnogo polyusa*. 3rd ed. Moscow: Textbook Press.

Hydrographic Department. 1930. *The Antarctic Pilot*. London: HMSO.

Jones, A.G.E. 1982. *Antarctica Observed*. Whitby: Caedmon.

von Krusenstern, Vice-Admiral. 1833. Über die Entdeckung des südlichen Continents [On the Discovery of the Southern Continent]. *Annalen der Erd-, Völker- und Staatenkunde* 8 (4): 95–96.

von Krusenshtern″, Vice-Admiral. 1836. *Dopolnenie k″ izdannym″ v″ 1826 i 1827 ob″yasneniyam″ osnovanij … Atlasa Yuzhnago Morya [Supplement to the Explanatory Principles of the Atlas of the Pacific Ocean, 1826 and 1827]*. St Petersburg: Scientific Committee, Ministry of the Navy.

Marks″, A.F. 1916. *Ustrojstvo poverkhnosti i rastitel'nyj pokrov″ Zemnogo shara [The Distribution of Land Surfaces and Vegetation across the Earth]*. Map. St Petersburg: Marks″. Source. https://www.davidrumsey.com/luna/servlet/detail/RUMSEY~8~1~255643~5519943.

Martin, Stephen. 1996. *A History of Antarctica*. Sydney: State Library of New South Wales Press.

Novosil'skij, P.M. 1854c. *Shestoj kontinent″*. 3rd ed. St Petersburg: Imperial Academy of Sciences.

Odhams. 1935. *The New Pictorial Atlas of the World*. London: Odhams.

Ovlashhenko, Aleksandr. 2013. *Materik l'da: pervaya russkaya antarkticheskaya èkspeditsiya i eyo otrazhenie v sovetskoj istoriografii (1920-e–1940-e gody) [The Continent of Ice: The First Russian Antarctic Expedition and its Footprint in Soviet Historiography (1920s to 1940s)]*. Saarbrücken: Palmarium.

———. 2014. *Antarkticheskij rubikon: tema otkrytiya Antarktidy v sovetskikh istochnikakh nachala 50-kh godov [Antarctic Rubicon: The Discovery of Antarctica as a Theme in Soviet Sources from 1950]*. Saarbrücken: Palmarium.

———. 2016. *Antarkticheskij renessans: provedenie pervykh kompleksnykh antarkticheskikh èkspeditsij i problema otkrytiya Antarktidy [Antarctic Renaissance: The Arrival of the first Combined Antarctic Expeditions and the Problem of the Discovery of Antarctica]*. Saarbrücken: Palmarium.

Ross, Captain Sir James Clark. 1847. *A Voyage of Discovery and Research in the Southern and Antarctic Regions During the Years 1839–43*. 2 vols. London: John Murray.

Rymill, J. R. 1936. An Antarctic Illusion: The Coasts of Grahamland. *The Times*, December 12: 13.

Sale, Richard, and Eugene Potapov. 2010. *The Scramble for the Arctic: Ownership, Exploitation and Conflict in the Far North*. London: Lincoln.

Shokal'skij, Yu.M. 1898. Polyarnyya strany Yuzhnago polushariya [Polar Countries of the Southern Hemisphere]. In *Èntsiklopedicheskij slovar'*, ed. K.K. Arsen'ev″ and F.F. Petrushevskij, vol. 24, 489–495. St Petersburg: Brokgauz and Efron.

Tammiksaar, E. 2016. The Russian Antarctic Expedition under the Command of Fabian Gottlieb von Bellingshausen and its Reception in Russia and the World. *Polar Record* 52 (5): 578–600.

Yelverton, David E. 2004. *The Quest for a Phantom Strait*. Guildford: Polar Publishing.

CHAPTER 3

What the Explorers Actually Said (1821–1855)

PRIMARY SOURCES

None of the logbooks or officers' journals of the voyage have survived, and neither Bellingshausen's own journal nor the original manuscript of his book, completed in October 1824, has ever been found. As the reader will soon discover, clarification of exactly what happened when is often hampered by the loss of such authoritative records. Setting to one side documents relating to the conception and preparation of the expedition, which have been preserved in considerable numbers, the list of primary sources for its execution and results begins with 16 reports and two tables of discoveries which Bellingshausen submitted during the voyage, usually to the Russian Minister of Marine, the Marquis de Traversay (1754–1831), but sometimes to the Admiralty College. Some of those documents were published in Russia, usually with cuts, between 1949 and 1952; some were first published in full online in 2019, albeit in challenging facsimile.[1] But the two documents which constitute Bellingshausen's final report have never been published in Russia, even in part (Bulkeley 2014a: 102–123). Some of the maps prepared by Bellingshausen and his officers have also survived, including two relating to the first season that were sent

---

[1] A 'virtual exhibition' about the Bellingshausen expedition, posted on the website of the Russian Naval Archives in 2019, offers more than 80 documents, many previously unknown (Kondakova 2019).

© The Author(s), under exclusive license to Springer Nature   25
Switzerland AG 2021
R. Bulkeley, *The Historiography of the First Russian Antarctic Expedition, 1819–21*,
https://doi.org/10.1007/978-3-030-59546-3_3

from Sydney in 1820, and a 15-sheet chart of the expedition's track throughout their two Antarctic seasons (Bellinsgauzen" 1821d; Belov 1963b).[2] In Belov's estimation Bellingshausen prepared the track chart from logbooks and journals during the homeward voyage (Belov 1963a: 32). It is a more immediate source than *Two Seasons*, and almost as comprehensive (Fig. 3.1).[3]

A less complete account of the voyage is made up of more than 200 sketches and paintings by the expedition's artist Pavel Nikolaevich Mikhajlov (1786–1840), many of which have recently been published (Petrova and Others 2012). Another shipboard record is a diary kept by the astronomer Simonov (Simonov" 1822a; Bulkeley 2014a: 144–158). Only 48 entries were published from a period of 106 days in the expedition's first Antarctic season; there is no way to tell whether other entries were not selected or were never written, and, if the former, who made the selection—Simonov or the editor of the magazine. Other documents surviving from the voyage itself include a seaman's diary, which as another immediate source from *Vostok* is sometimes useful on things like dates and damage to the ships (Kisilev" 1819–1821),[4] and extracts from Simonov's letters to his patron Mikhail Leont'evich Magnitskij (1778–1844), rector of the University of Kazan, which are not pertinent to this investigation.

Another important document is the private letter written two months after returning to Russia by Junior Captain Mikhail Petrovich Lazarev (1788–1851), who had just been promoted after commanding the expedition's second ship *Mirnyj* as a lieutenant, to his friend Aleksej Antipovich Shestakov (1786–1856). The letter was not discovered and published until almost a century later (Lazarev" 1821, 1918; Bulkeley 2014a: 166–173). As the only source which describes a sighting that might qualify as the discovery of Antarctica, it has often been upgraded by wishful-thinking but inexpert commentators into Lazarev's shipboard 'journal', but that is just nonsense.[5] The letter will be analysed in detail in Chap. 7.

[2] More accurately, the track of *Vostok*, Bellingshausen's command ship.

[3] The original texts of translated quotations, indicated by numbers in square brackets, are provided at the end of each chapter. For the system of romanization used, the rendering of dates, and a few other matters, please consult the Apparatus at the end of the book.

[4] This diary will not be referred to here because Russian commentators, apparently uninterested in the humble details it relates, hardly ever used it after it was discovered and published in 1949 and republished in 1951 (Ostrovskij 1949a; Andreev 1949c).

[5] For a recent example hosted by those bywords for accuracy, the BBC, see (Krechetnikov 2020).

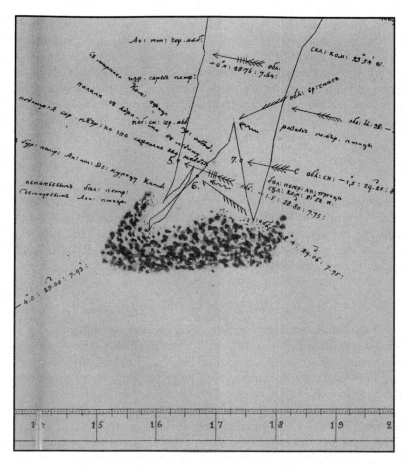

Fig. 3.1 Detail for 5(17) to 7(19) February 1820 from Sheet 3 of Bellingshausen's 1821 track chart. Scanned from Belov (1963b) and published here with the consent of the Russian Naval Archives

Simonov published a popular pamphlet about the voyage in 1822 which was translated into German and French (Simonov" 1822b), as well as scientific papers with little bearing on the discovery question. He also wrote, but never completed, a fuller account of the voyage which was not published until 1951 (below). After being rather poorly edited, Bellingshausen's narrative of the voyage was published in 1831, together with an *Atlas* with further maps by himself as well as lithographs of coastal profiles and studies of fauna, flora, and local people taken from Mikhajlov's sketches (Bellinsgauzen" 1831). The limited number of copies, just 600, became one of several hindrances to knowledge of the expedition. Less than ten years later Carl Friedrich Gauss (1777–1855), one of the most famous and best connected scientists in Europe, was unable to get hold of a copy. The last primary source to appear in the nineteenth century was a short biography of Bellingshausen by his son-in-law, which contains information obtained directly from its subject (G[ershau] 1892).[6]

ICE TALK

Before looking briefly at what the explorers said about their Antarctic discoveries, we need to know something about the language in which they said it and the ideas which may have influenced them. That is especially true when dealing with their putative sightings of mainland coasts in 1820, because those were probably confined to ice.

Mains and Continents

It is important to understand, first, that Bellingshausen may not have accepted what had until recently been the unanimous view of European mariners and geographers, including Behrens, Lomonosov, Phipps, and Cook in the eighteenth century, that large and lofty ice masses encountered in polar regions must have formed near or on land, as opposed to flatter and lower ice floes and ice sheets which could be formed at sea (Behrens 1738: 50–51; Lomonosow 1763; Phipps 1774: 72; Cook 1777 2: 230, 240–243).[7] One reason that people believed this was that, once whalers started visiting the Svalbard archipelago after its discovery by Willem Barentsz in 1596,[8] an example became widely known. The 'Seven Icebergs of Spitzbergen' were short glaciers ending in sea cliffs about 60 m high, from which they were observed to release 'ice islands' into the

---

[6] For Novosil'skij's retrospective treatments of the voyage, first published in 1853, see Chap. 4.

[7] Behrens did not originate the theory, which dates from at least the late sixteenth century (Davis 1595: 26–30).

[8] Reports of earlier discovery by Norse or Russian seamen have not yet been confirmed, but continue under active investigation (Okhuizen 2005).

sea below them (Martens 1675: 19; Scoresby 1820: 101–109).[9] Greenland was another case in point. It is said to have been christened 'mother of ice' by the Vikings, and the phrase was so well known by the seventeenth century that it occurred in a report from the Venetian Ambassador in London in 1622.

Bellingshausen disagreed with that consensus. In his book he explained that sea ice could unite and grow into very large and mountainous bodies (from which parts could also become separated as large icebergs), but he never suggested that land was needed for that to happen. In what reads like an encounter with such a vast marine ice cap he described it as 'hilly, solid standing ice' (below). One such immovable body of ice had been established at the South Pole, he thought, probably with some shoals and islands, two of which he had discovered, around its fringe (Bellinsgauzen" 1831 2: 243–251; for a translation, see Bulkeley 2019). Because his theoretical discussion disagrees with the older theory, that ice in quantities sufficient to produce very large icebergs could only be accumulated along coastlines or on land, it seems possible that Bellingshausen was influenced by a new theory proposed in 1778 by the French naturalist Buffon, whom he cited indirectly (via Forster) in this passage, and whose *Histoire Naturelle* was taken on the expedition (Anon 1819: 348; Bulkeley 2021). Buffon had applied the expression *calotte de glace* to polar ice caps[10] in a supplement at the end of his multi-volume work; had maintained that temperatures were so low in polar regions that permanent ice caps would have formed whether or not any land lay beneath them; and had predicted that because the southern hemisphere was colder than the northern and largely taken up by oceans, the southern ice cap should be much larger than the northern. Whether or not the Antarctic ice cap had engulfed any land was immaterial (*indifférent*), because it would have formed directly on the surface of the sea without it and in any case such land could not be surveyed (Buffon 1778: 601–613).[11]

---

[9] Martens' book showed that a new vocabulary was replacing the old. In it, both glaciers and ice islands were referred to as 'Eisberge'. See below.

[10] The 1809 edition of J.F. Rolland's *Nouveau Vocabulaire ou Dictionnaire Portatif* defined 'calotte' as a cap which covers only the crown of the head, in other words a skull cap.

[11] To the modern reader Buffon's armchair methods may seem unscientific. His style certainly had the usual overconfidence of the Enlightenment. But he worked with such empirical findings as there were, and before explorers had surveyed the Antarctic trying to understand it was rather like trying to understand the first nanoseconds of our universe, or the physics of parallel universes, today. Buffon contrasts favourably with John Cleves Symmes,

It is hard to date Buffon's new theory precisely.[12] In the 1778 *Supplément* to an edition of the *Histoire Naturelle* which commenced publication in 1769, he tells us that he felt his theory was *confirmed* by the narrative of Cook's voyage, published in 1777, which implies that he had formed it before reading that book. He may however have been influenced by early reports, in about 1775, that Cook had found no land in the Antarctic.

Before that Buffon accepted the traditional theory, according to which icebergs originated from large masses of ice which could only accumulate near land. Thus at the beginning of his great work he had conjectured, on the basis of sightings of massive icebergs in the Southern Ocean, that there might be a continent the size of Europe, Asia, and Africa combined near the South Pole (Buffon and Daubenton 1749: 212–216). He repeated that at the beginning of the 1769 edition, so that it was perhaps unfortunate for Bellingshausen that Buffon changed his mind nine years later to declare that polar ice caps could have formed in open ocean, and that this was more likely to have occurred in the southern hemisphere than in the northern. He died in 1788 without resolving the contradiction between the two theories, which continued to appear in successive editions of his work.

The hypothesis that Bellingshausen was influenced by Buffon is also consistent with the similarity between the two men's visual representations of dense pack ice (Fig. 3.2).[13]

The notion of a massive, possibly marine, Antarctic ice cap was taken up by Bernardin de Saint-Pierre, who also included a map of it (1796: 82–104, opp. p.87). The idea remained in print in editions and translations of his and Buffon's works far into the nineteenth century. It was endorsed by Nicolas Desmarest in the *Encyclopédie Méthodique* (Desmarest 1803: 653–655), and a Buffonesque ice cap was depicted in an 1817 atlas (Thomson 1817: 3, 7), though not extending as far north as Buffon had proposed.

---

who published an outright polar fantasy based on Edmond Halley's Hollow Earth conjecture in the year in which *Vostok*'s keel was laid down (Seaborn 1818).

[12] The author was previously misled by an incorrect dating on the internet into supposing that Buffon had published this idea earlier, in 1769 (Bulkeley 2014a: 57). That source should have been dated to 1800.

[13] Buffon was very particular about his polar projection maps, and asked for the Antarctic ice cap to be shaded more and more heavily towards the Pole in order to emphasize 'the vast extent' (*vaste étendue*) of the planet which he thought it covered (Buffon 1860: 52–53). The publisher did not entirely comply.

a

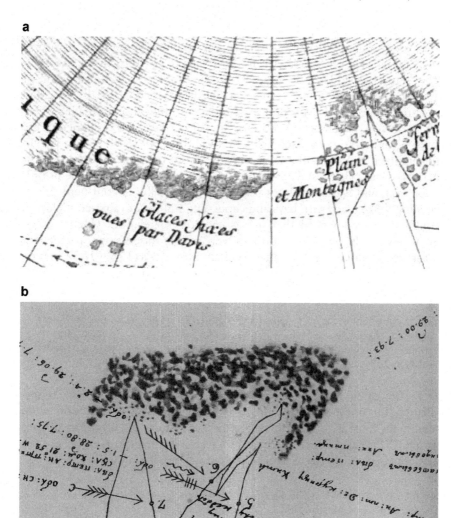

b

Fig. 3.2 Representations of Continuous Ice. (a) Buffon 1778. (b) Bellingshausen 1821. In (b) a detail from Sheet 3 of the track chart has been inverted to facilitate comparison. (See Fig. 3.1, published with the consent of the Russian Naval Archives)

It seems possible, therefore, that when Bellingshausen explained that 'The enormous ice formations which rear up into sloping hills the nearer they are to the South Pole, I call *main* ...' [1] (Bellinsgauzen" 1831 2: 250—emphasis added), he was alerting the reader to a technical term for a (probably) marine polar ice cap, which in practice he hardly ever had occasion to use. The expression resembled an English one, 'main ice' (or 'main of ice'), which had been in use for 200 years and was defined in the latest polar literature (Bulkeley 2014a: 59; John Ross 1819: lxiv–lxvi).[14] In the eighteenth century the expression began to be modernized as 'continent de glace' (Prévost 1759: 165—another author on the expedition's book list), and was sometimes used very graphically: 'a continent of ice, very high, like land, with bays and capes, and, till they examined it closely, could not be convinced that it was a mere congelation' (Henry 1774 Supp.: 111), although such authors were not yet discussing a permanent polar ice cap along Buffon's lines. William Scoresby's (1789–1857) usage, however, much discussed at around the time of Bellingshausen's visit to London, was closer to Buffon:

> ... a continent of ice mountains may exist in regions near the Pole, yet unexplored, the nucleus of which may be as ancient as the earth itself, and its increase derived from the sea and atmosphere combined. (Scoresby 1818: 294)[15]

In short, Bellingshausen's rare use of the phrase 'main of ice' in a report (below), but not in his book, and in a language which still lacked the word 'continent', needs to be understood in the context of early glaciological thought (Bulkeley 2014a: 83). But Scoresby and Bellingshausen were not academic scholars, accustomed to citing their sources, so that any

---

[14] John Ross's conduct as commander of a recent Arctic expedition was a hot topic in London periodicals and naval circles. His narrative appeared only a few months before Bellingshausen visited the city in August 1819 (Ross 1819), and the draft of a note from the Marquis de Traversay to the ambassador in London, urging the latter to purchase copies before the two exploring squadrons should arrive in England, is preserved in the Russian Naval Archives (de Traversay 1819).

[15] Various authors also rendered Scoresby's concept as 'ice/icy continent' (O'Reilly 1818: throughout; Anon 1818: 292). In O'Reilly's case the phrase received much publicity through being mocked in the London press. While in London, Bellingshausen and his officers stayed at the Hungerford Coffee House near Covent Garden, which was well supplied with periodicals and frequented by naval officers. They needed only to pick up the latest issue of *The Quarterly Review* to encounter the idea.

connection they may have had with Buffon can only be guessed at from resemblances of thought and opportunities for the transfer of ideas.

For some further linguistic background, at the time of the expedition the usual Russian expression for 'continent' was *chast' sveta*, literally 'part of the world',[16] of which the most authoritative dictionary considered there were four (Imperial Russian Academy 1822: 1243).[17] The same dictionary gave two senses for *materik*: first, a solid layer of ground (bedrock), by contrast with looser, covering layers (regolith, drift); and second, in geography, a large expanse of land, by contrast with islands (Imperial Russian Academy 1814: 712). A modern translation of the second sense would be 'mainland', but in English from the sixteenth to the nineteenth centuries 'main' was often used instead, as in 'the Spanish Main', and that is how *materik* will usually be translated here. In the 1850s the loan word *kontinent* began to be used in Russian by well-informed geographers like Pavel Mikhajlovich Novosil'skij (1800–1862) and Arkadij Il'ich Pavlovskij (1828–1889), but it did not appear in dictionaries until the late nineteenth century. Lastly *materik* (main) and its adjective, *materoj/-yj*, were very versatile, taking their meanings from the context. Dal' gave 'main channel' (*materik" ryeki*)[18] as an example of the noun (Dal' 1881: 311), and for the adjective, 'main forest' (*materoj lyes*) occurred in a collection of eighteenth-century legislation (Imperial Chancellery 1830: 30). As we have just seen, Bellingshausen's only recorded use of the phrase *materoj lyod* (main ice) was somewhat roundabout (literally 'The enormous ice formations, which ..., I call main, considering that ...' [2]) (Bellinsgauzen" 1831 2: 250). This adjectival version appeared in print by mid-century in the sense of a bank or shore of solid ice next to water (N. Smirnov" 1853: 128). When it began to appear in dictionaries, however, it was given other meanings, such as permafrost (Dal' 1863: 32),[19] or *Grundeis* (I. Pavlovskij 1879: 458) (a German word which came to mean permafrost in the twentieth century,

---

[16] A translation of the German word *Weldtheil*.

[17] Africa, America, Asia, and Europe, with Australia discounted as an island.

[18] Literally 'main of the river'.

[19] This was the first example of 'main ice' in the work of the great lexicographer Vladimir Ivanovich Dal' (1801–1872). Dal' neatly bridged the two senses of *materoj* when explaining *badaran'*, a well sunk to tap water from the permafrost layer, for which he used the expression 'main ice'. In the 1830s Dal' had spent about eight years as a naturalist and folklorist exploring the province of Orenburg in the southern Urals, and then became one of the 17 founding members of the Russian Geographical Society in 1845.

but in the nineteenth century meant ice that formed, or might possibly form, on the bed of a river before rising to the surface (Kant 1801: 10–14; Aycke 1836)). None of those later Russian authors, then, thought of *materoj lyod* as a vast polar ice cap. In 1820 Bellingshausen may well have been the first person ever to use the other, noun form of the phrase *materik l'da* [3] (main of ice), in his report on their first Antarctic season (Bellinsgauzen" 1820a: fo. 242v).

It was no accident that the fictional Captain Nemo took the *Nautilus* to the South Pole, by passing under the sort of vast floating ice cap first posited by Buffon and then explained by Bellingshausen, during the second half of the nineteenth century, while the existence of a rocky Antarctic continent remained in doubt (Verne 1871).

Icebergs

Besides 'main of ice', we need to understand Bellingshausen's expressions for icebergs. In the eighteenth and early nineteenth centuries the commonest term for an iceberg, in English, French, and Russian, was 'ice island' (*ledyanyj/-oj ostrov*). In the eighteenth century, however, a German word for 'glacier', *Eisberg*, was imported into English directly, and into Russian by translation as *ledyanaya gora* (literally, ice hill/mountain) (Anon 1804a: 1090–1097). Simultaneously *Eisberg* was acquiring a second, marine, in other words 'iceberg' sense in German (de Pagès 1783: 229),[20] as were the equivalents in French and English. At much the same time the Russian word for an ice-house, *lednik*, began to be used for glaciers, a newly popular subject with travel writers, and that left 'ice hill' (*ledyanaya gora*) free to mean 'iceberg' in Russian too, alongside the older expression 'ice island'.

In 1763 the great savant Mikhail Vasil'evich Lomonosov (1711–1765) may have been the first to use the new expression, taking care as he did so to point out when he meant 'glacier' and when he meant 'iceberg'. But according to the editor of the second edition, that book was not discovered and published until 1847, well after Bellingshausen wrote his reports and subsequent book (Lomonosov" 1847, 1854: I).[21] Even without support from Lomonosov, however, 'ice hill' was being used to mean

---

[20] This was a German translation of a French book taken on the expedition.

[21] Lomonosov's other work on the subject was in Swedish (Lomonosow 1763), and so less likely to influence Russian usage.

'iceberg' before Bellingshausen even learned Russian (Maksimovich" 1788: 11, 313). In that example, the phrase referred without any explanation to icebergs encountered on a polar voyage, exactly as it would be used by Bellingshausen about 35 years later. In the early nineteenth century the expression was still sometimes treated as a novelty (Stojkovich" 1813: 106); in one case the English translation 'Ice-bergs, Ice-islands' was provided (Berkh" 1823: 147). But the 1788 example shows that, in the context of polar exploration, such explanations were not always given. For whatever reason, Bellingshausen (like Cook) generally preferred the older *ledyanoj ostrov* (ice island), but he did use *ledyanaya gora* (ice hill) a few times, when the icebergs in question were surrounded by a 'terrain' of pack ice and so looked most like real hills.[22] Eventually the phrase became the normal expression for icebergs, and was used in turn to explain the German word when it was finally imported into Russian, as *ajsberg*, in the 1930s (Pinegin 1934: 99).

Another word which poses certain challenges is *bugor"*, which was used by Bellingshausen to refer to ice hillocks, a feature of the scene observed on 28 January 1820 (above). There was as yet no systematic ice vocabulary in Russian, although one in English was offered by Scoresby just after Bellingshausen passed through London (Scoresby 1820: 226–229). In foreign-language dictionaries *bugor"* was sometimes treated as equivalent to 'hill' (Anon. 1813: 85; Alexandrow 1904: 38), but an authoritative Russian dictionary explained that it meant a topographic feature smaller than a proper hill (*kholm"*) (Dal' 1863: 119). Given that Bellingshausen had originally described the scene, in his first report from Sydney, as *icebergs* surrounded by an ice field, or pack ice (Bellinsgauzen" 1820a: fo. 242v), he may have decided to replace that in his book by stressing that they were *small* icebergs, perhaps something like bergy bits, an expression once defined as 'about the size of a small cottage' (Hydrographic Office 1952: 4). If so, his usage would have resembled that of Lomonosov, who used the same word in 1763 to describe the summer break-up of hummocky ice-sheets (Lomonosov" 1854: 56).

In his translation Frank Debenham rendered *bugry* as 'hillocks' (Bellingshausen 1945 1: 117), and the present author agrees with his choice; 'bergy bits' would be an anachronism. It is worth adding that the smallest Russian 'ice hills' in the author's collection of texts were described

---

[22] Another document from the expedition, which used the expression more generally, may have been written by *Mirnyj*'s surgeon Nikolaj Galkin (Anon. 1829: 47).

as 3–5 m high (Anon 1922), thus well within the scope of the 'hillocks' with which Bellingshausen replaced 'icebergs' in his book.

THE DISCOVERY QUESTION

The meagre primary sources have relatively little to say about the most controversial historical issue, namely whether or not a substantive sighting of the Antarctic coast was achieved in January or February 1820. Unfortunately Mikhajlov took no sketches during the relevant period, or at least none that survive. For his part, Simonov failed to confirm Lazarev's later recollection that '*main ice of extraordinary height*' was sighted in January 1820 (Chap. 7). In his diary entry for 19 January 1820 Simonov had mentioned the likely existence of 'impenetrable domes [of ice] which cover the surface of the sea like crusts around both Poles', [4] just the sort of thing Lazarev later reported, but Simonov's entry for 28 January could be paraphrased as 'yet another ice field' (Bulkeley 2014a: 152), showing that to his mind the marine ice cap he was hoping to see was not encountered that day. Frustratingly however, there is no entry in his published diary for 17 February 1820, the other most likely date for an early sighting of an ice coast. He did mention both dates in a scientific paper a few years later, but only as falling within a period when the expedition was 'In open water below the Antarctic Circle' [5] (Simonov" 1825: 115).[23] Because of these various silences and contradictions, historians are largely dependent on Bellingshausen's reports, book, and navigational chart for the details of the voyage.

Bellingshausen's first report from Sydney was dated 20 April 1820 and took a year to reach St Petersburg (Bellinsgauzen" 1820a). In it he stated that after discovering the sub-Antarctic Traversay Is. near South Georgia in January 1820 he had found no signs of land in the far south that season. In mid-February, however:

> … *between the 5th and 6th* [O.S.] *I reached latitude S 69°7′30″, longitude E 16°15′. There, beyond ice fields comprising small ice and [ice] islands, a main*

[23] The article records that noon and midnight temperatures were one or two degrees (Réaumur) higher between 28 January and 8 February than they were between 16 and 25 February 1820. If, as Debenham and the present author have proposed, the expedition is more likely to have approached and sighted an ice-coast of the Antarctic mainland during the second period than the first, did Simonov's thermometers perhaps record indirect evidence for that interpretation?

of ice was sighted, the edges of which had broken away perpendicularly, and which stretched as far as we could see, rising to the south like land. The flat ice islands that are located close to this main are evidently nothing but detached fragments of this main, since they have edges and upper surfaces which resemble the main [6]. (Bellinsgauzen" 1820a: fo. 242v)[24]

A summary of that report was published early in May 1821 and the full text two years later (Bellinsgauzen" 1821b, 1823). For security reasons his report from Rio de Janeiro, on the homeward journey, described only briefly the discoveries of Peter I I. and the Alexander I Coast in January 1821, the first land ever found south of the Antarctic Circle (Bellinsgauzen" 1821a). Fuller information was provided in his final report, with an accompanying table of discoveries, which was signed and dated on 5 August 1821, the day the expedition dropped anchor off their home port of Kronstadt (Bellinsgauzen" 1821e). Once again a press report appeared within weeks, giving the names and coordinates of the Antarctic discoveries (Anon 1821a).

To repeat the quotation in Chap. 2, Bellingshausen's final account of 28 January 1820, the date required for priority, reads as follows:

… at noon, in latitude 69°21′28″, longitude 2°14′50″, we met ice which appeared to us, through falling snow, like white clouds. The wind, from NE, was dropping, with a heavy swell from NW. Because of the snow we could not see far. I gave orders to proceed SE, close-hauled. After proceeding for two [nautical] miles [3.7km] in that direction we saw that continuous ice extended from East through South to West. Our course led directly into that ice field, scattered with hillocks [7]. (Bellinsgauzen" 1831 1: 171–172)

The passage in the book corresponding to his report about the probe made on 17 February 1820 (above) also bears repeating because it was so different:

Eventually, at a quarter past three after noon, we saw a quantity of large, flat, high ice islands, beset with light floes some of which overlay one another in places. The ice formations to SSW join together into hilly, solid standing ice; its edges were perpendicular, forming coves, and its surface rose away to the

---

[24] Two Russian depictions of icescapes and ships may have been inspired by this historic event (Bulkeley 2014a: Fig. 6).

South, to a distance whose limits we could not make out from the [main] cross-trees. …

Seeing that the ice islands had similar surfaces and edges to the aforementioned large ice formation, which lay before us, we concluded that those huge ice masses and all similar formations get separated from the main coast by reason of their own weight or from other physical causes and, carried by the winds, drift out into the expanse of the Southern Ice Ocean … [8]. (Bellinsgauzen″ 1831 1: 188–189)

The contrast between an 'ice field, scattered with hillocks' and 'hilly, solid standing ice [which] … rose away to the South, to a [limitless] distance' is very clear. Only the latter, not the former, might qualify as the great Antarctic ice cap for which Bellingshausen once used the expression 'main ice', and which he suggested, in a theoretical essay towards the end of his book, must be immobile and stretch 'beyond the Pole' from where he observed it in February 1820 (Bellinsgauzen″ 1831 2: 250), in short for a distance of roughly 4600 km from one side of the world to the other. The difference between the two passages explains why Debenham concluded that the second was more plausible as a sighting of the coast.[25]

Whatever he thought about the Antarctic ice cap, or we think now, Bellingshausen restricted his discovery claims below the Antarctic Circle to the two sightings of land that were achieved in January 1821. He was supported in this by Simonov, whose early account of the voyage included a partial endorsement of Cook's negative conclusions:

In his voyage the famous navigator Cook says that: '*I had now made the circuit of the southern ocean in a high latitude, and traversed it in such a manner as to leave not the least possibility of there being a continent,[26] unless near the Pole, and out of the reach of navigation.*' We had probed further than that navigator in many regions, and remained below the Antarctic Circle for much longer than he did, but, unless the Alexander Coast is the edge of such a land, we find ourselves obliged to confirm his findings by stating that we saw no signs of the supposed southern continent, unless [it lay] further than could be seen from the point at which eternal ice presented an impassable boundary for seamen [9]. (Simonov″ 1822b: 55)

[25] For fuller translations and explanations of these passages from Bellingshausen's book, see Bulkeley (2019).

[26] Because the word *kontinent* had not yet entered Russian, Simonov translated it from Cook as *materoj zemli*, literally 'mainland'. See 'Ice Talk' above.

Simonov's main scientific report from the expedition was published before *Two Seasons* (Simonov" 1828). However it said nothing about the expedition's work in the Antarctic because it was solely concerned with astronomical determinations of the geographical coordinates of places where Simonov had established an observatory on shore, all located in the southern temperate or tropical zones. The sequel indicated by its sub-title, *Part One*, never appeared, but would perhaps have been a record of his observations, now lost, of selected fixed stars in the southern hemisphere.[27] Apart from some rather vague remarks by Belov (below) no Russian commentator has ever drawn attention either to Simonov's missing star catalogue, or to the fact that he published latitudes for Rio, Sydney, Queen Charlotte Sound (NZ), and Tahiti, but longitudes only for Tenerife, Rio, and Sydney (Simonov" 1828). In his fullest but unfinished account of the voyage, which once again failed to cover their second Antarctic season,[28] Simonov added little to his previous description of the first (1820) season except to say that it had been Lazarev who first proposed that they should break off that year's probing of the ice fields, something which had already been mentioned by Bellingshausen (Simonov 1990b: 134; Bellingsgauzen" 1831 1: 192).

Apart from recalculating some of his navigational coordinates, Bellingshausen maintained his and Simonov's initial assessments in *Two Seasons*, but made one small change. In his account of the first season Bellingshausen added the promising speculation that some birds seen between 19 and 25 February 1820 might have been from land-based species (Bellinsgauzen" 1831 1: 193–200; 2: 250). Debenham later suggested that he could have been misled by Arctic terns, which migrate across the Southern Ocean at great distances from land (Bellingshausen 1945 1: 131).

As mentioned above, Lazarev's letter to Shestakov, the only other primary source for the discovery question, will be discussed in Chap. 7 in order to reflect its place in the historiographic sequence.

*　*　*

[27] Belov recorded the 'Astronomical and physical observations of Prof. Simonov' as if it were a separate work, also published in 1828 (Belov 1963a: 8). It was indeed so recorded in the card catalogue of the National Library in St Petersburg, but today their electronic catalogue makes it clear that that was a series title, of which only this first part, that on geographical coordinates, ever appeared. The book itself confirms this.

[28] The narrative stops on 10 October 1820, at the point when the expedition returned to the ice after visiting Macquarie I.

It will be all too easy, in the chapters that follow, to forget the great skill and courage of the seamen who gathered the information that we shall be discussing, some of the first ever obtained from Antarctica itself. To take just one incident, in the small hours of 21 January 1820 *Mirnyj* collided head on with a low ice floe and suffered major damage to her forestem, the massive upright timber at the 'sharp end'.[29] Had she been moving faster or met the floe to one side of the forestem, then with *Vostok* out of sight ahead she would probably have gone down with all hands. Bellingshausen would doubtless have persisted with his appointed task, but the possible injection of even the slightest additional caution into his decisions might have deprived the venture of its crowning achievements, the sighting of a mainland ice coast a month later and the discovery of the first land south of the Antarctic Circle in 1821.

ORIGINAL TEXTS OF QUOTATIONS

[1] Огромные льды, которые по мѣрѣ близости къ Южному полюсу подымаются въ отлогія горы, называю я матерыми …

[2] Огромные льды, которые … называю я матерыми …

[3] *Материкъ льда …*

[4] … техъ не проницаемыхъ сводовъ, кои на подобіе коры покрываютъ морскую поверхность у обоихъ полюсовъ, …

[5] Въ открытомъ морѣ за Южнымъ Полярнымъ кругомъ …

[6] *… съ 5 на 6 число [ст.ст.] дошелъ до широты S. 69°, 7′, 30″, долготы O. 16°, 15′. Здесь за льдеными [так] полями мелкаго льда, и Островами виденъ Материкъ льда. коего края отъ ломаны перпендикулярно, и Которой [так] продолжался по Мерѣ на-шего зренія, возвышаясь къ Югу подобно берегу. Плоскіе льдяные острова, близь сего материка находящіися [так], ясно показываютъ, что онѣ суть отломки сего материка; ибо имѣютъ края и верьхную поверьхность подобную материку.*

---

[29] By an unfortunate error Debenham's translation gives 'stern' instead of 'stem' (Bellingshausen 1945 1: 119). 'Forestem' was clear enough in the original, but matters cannot have been helped at Cambridge by Bellingshausen's use of what seems to have been an extremely rare variant, *gref*, of the word *grep*, meaning 'cutwater', which is itself too rare for most dictionaries (Bellinsgauzen″ 1831 1: 175; Larionov 1963: 129).

[7]   ... въ полденъ въ широтѣ 69°, 21′, 28″, долготѣ 2°, 14′, 50″, мы встрѣтили льды, которые представились намъ сквозъ шедшій тогда снѣгъ, въ видѣ бѣлыхъ облаковъ. Вѣтръ былъ отъ NO умѣренный, при большой зыби отъ NW, по причинѣ снѣга, зрѣніе наше не далеко простиралось; я привелъ въ бейдевинтъ на SO, и проидя симъ напрвавленіемъ двѣ мили, мы увидели, что сплошные льды простираются отъ Востока чрезъ Югъ на Западъ; путъ нашъ велъ прямо въ сіе льдяное поле, усѣянное буграми.

[8]   ... наконецъ въ четверть четвертаго часа по полудни, увдѣли множество большихъ, плоскихъ, высокихъ льдяныхъ острововъ, затертыхъ плавающими мелкими льдами, и мѣстами одинъ на другомъ лежащими. Льды къ SSW примыкаются къ льду гористому, твердо стоящему; закраины онаго были перпендикулярны и образовали заливы, а поверхность возвышалась отлого къ Югу, на разстояніе, предѣловъ котораго мы не могли видѣть съ салинга. ...
Видя льдяные острова, поверхностью и краями сходные съ поверхностью и краями большаго вышеупомянутаго льда, предъ нами находящагося, мы заключили, что сіи льдяныя громады и всѣ подобные льды, отъ собственной своей тяжести, или другихъ физическихъ причинъ, отдѣлились отъ матераго берега, вѣтрами отнесенные, плаваютъ по пространству Ледовишаго Южнаго Океана; ...

[9]   Знаменитый мореходецъ Кукъ, въ путешествіи своемъ говоритъ: я обошелъ во кругъ Южнаго полушарія въ большей широтѣ такимъ образомъ, что неоспоримо доказалъ, что нетъ въ ономъ никакой матерой земли, развѣ въ окрестностяхъ полюса, куда не возможно достиунуть. Мы углублялись во многихъ мѣстахъ далѣе сего мореходца, оставались не въ примѣръ долѣе его за полярнымъ кругомъ, и ежели если берегъ Александра не есть оконечность земли сей, то принужденными найдемся подтвердить слова его, должны будемъ сказать, что мы не видѣли никакихъ признаковъ предполагаемаго южнаго материка, развѣ за предѣлами зрѣнія отъ тѣхъ мѣстъ, гдѣ вѣчные льды положили непроницаемую межу для плавателей.

References[30]

Alexandrow, A. (pseud.). 1904. *A Complete Russian-English Dictionary*. 3rd ed. London: Nutt.

Andreev, A.I., ed. 1949c. *Plavanie shlyupov «Vostok» i «Mirnyj» v Antarktiku v 1819, 1820 i 1821 godakh [The Antarctic Voyage of the Sloops Vostok and Mirnyj in 1819, 1820 and 1821]*. Moscow: Geografgiz.

Anon., ed. 1804a. *Slovar' geograficheskij Rossijskago Gosudartsva [Geographical Dictionary of the Russian Empire]*. Vol. 3. (K–M). Moscow: University Press.

———. ed. 1813. *Nouveau dictionnaire Russe-Français-Allemand [A New Russian, French, and German Dictionary]*. Vol. 1. St Petersburg: Glasounov.

———. 1818. Foreign Articles. *Niles Weekly Register* 15 (381). Baltimore.

———. 1819. n.t. SARN—F–166, S–1, P–660a, fos 347–348.

———. 1821a. Dal'nyejshiya svyedeniya o plavanii otryada, sostoyashhago iz" shlyupov" Vostoka i Mirnago, pod" komandoyu Kapitana Bellinsgauzena [Further Information about the Voyage of the Squadron Comprising Sloops Vostok and Mirnyj, Commanded by Captain Bellingshausen]. *Russkij Invalid* 196: 786–788.

———. 1829. Pis'mo k * [Letter to *]. *Kazanskij vyestnik"* 35 (1): 46–56.

———. 1922. Ledyanyya gory [Icebergs]. *Segodniya* 33: 3. Riga.

Aycke, J.C. 1836. Bemerkungen über Grundeis [Remarks on Grundeis]. *Annalen der Physik und Chemie* 2 (9): 122–129.

Behrens, C.F. 1738. *Der wohlversuchte Südländer: Reise um die Welt [The Much Sought After Southerner: A Circumnavigation]*. Leipzig: Self.

Bellingshausen, Captain. 1945. *The Voyage of Captain Bellingshausen to the Antarctic Seas, 1819–1821*, ed. Frank Debenham. 2 vols. London: Hakluyt Society.

Bellinsgauzen", Kap. 1823. Donesenie Kapitana 2 ranga Bellinsgauzena iz" Porta Zhaksona, o svoem" plavanii [The Report of Junior Captain Bellingshausen from Port Jackson, About His Voyage]. *Zapiski izdavaemyya Gosudarstvennym" Admiraltejskim" Departamentom"* 5: 201–219.

Bellinsgauzen", Kapitan". 1821b. Vypiska iz" doneseniya Kapitana 2 ranga Bellinsgauzena k" Morskomu Ministru ot" 8 Aprelya 1820 goda [O.S.] iz" Porta Zhaksona [Extract from the Report of Junior Captain Bellingshausen from Port Jackson to the Minister of Marine, April 8 1820 [O.S.]]. *Syn" Otechestva*, April 23 1821 (O.S.), 69: 133–135.

———. 1821d. Karta Plavaniya Shlyupov" Vostoka i Mirnago vokrug" Yuzhnago polyusa v" 1819, 1820 i 1821 godakh" pod" Nachal'stvom" Kapitana

[30] Key to archival references: SARN = State Archives of the Russian Navy, F = Fond, S = Series [*Opis'*], P = Piece [*Delo*], fo./fos = folio/s [*list/y*], v = verso [*oborotnoe*]. (The latter is necessary because verso pages were usually unnumbered.) Dates on manuscript documents are given in Universal Time (CE).

Billensgauzena [Chart of the Voyage of Sloops *Vostok* and *Mirnyj* around the South Pole in 1819, 1820 and 1821 under the Command of Captain Bellingshausen]. St Petersburg: SARN—F–1331, S–4, P–536, fos 5–19.

Bellinsgauzen″, Kapitan″. 1831. *Dvukratnyya izyskaniya v″ yuzhnom″ ledovitom″ okeanye i plavanie vokrug″ svyeta v″ prodolzhenii 1819, 20 i 21 godov″ [Two Seasons of Exploration in the Southern Ice Ocean and a Voyage around the World, During the Years 1819, 1820 and 1821]*, ed. L.I. Golenishhev″-Kutuzov″, 2 vols. plus *Atlas*. St Petersburg: Glazunovs.

Bellinsgauzen″, F.F. 1820a. Report to the Marquis de Traversay from Sydney, April 20 1820. St Petersburg: SARN—F–166, S–1, P–660b, fos 239–245v.

———. 1820b. Personal letter to the Marquis de Traversay from Sydney, April 20 1820. St Petersburg: SARN—F–166, S–1, P–660b, fos 246–249v.

Bellinsgauzen″, F.F. 1821e. Report to the Marquis de Traversay from Kronstadt, 5 August 1821. St Petersburg: SARN—F–203, S–1, P–826, fos 1–18v.

Bellinsgauzen″, F.F, Kap. 1821a. Report to the Marquis de Traversay from Rio de Janeiro, March 17 1821. St Petersburg: SARN—F–166, S–1, P–660b, fos 352–353v.

Belov, M.I. 1963a. O kartakh pervoj russkoj antarkticheskoj èkspeditsii 1819–1821 gg. [The Maps of the First Russian Antarctic Expedition 1819–1821]. In *Pervaya russkaya antarkticheskaya èkspeditsiya 1819–1821 gg. i eyo otchyotnaya navigatsionnaya karta*, ed. M.I. Belov, 5–56. Leningrad: Morskoj Transport.

———., ed. 1963b. *Pervaya russkaya antarkticheskaya èkspeditsiya 1819–1821 gg. i eyo otchyotnaya navigatsionnaya karta [The First Russian Antarctic Expedition 1819–1821 and its Official Navigational Chart]*. Leningrad: Morskoj Transport.

[Berkh″, V.N.], ed. 1823. *Khronologicheskaya istoriya vsyekh″ puteshestvii v syevernyya polyarnyya strany [Chronological History of All Journeys Into Arctic Regions]*. 2nd ed. St Petersburg: Imperial Academy of Sciences.

de Buffon, M. le Comte. 1778. *Histoire naturelle générale et particulière. Supplément*. Vol. 5. Paris: Imprimerie Royale.

———. 1860. *Correspondance inédite de Buffon [Unpublished Correspondance]*. Vol. 2. Paris: Hachette.

de Buffon, M. le Comte, and L.-J.-M. Daubenton. 1749. *Histoire naturelle générale et particulière*. Vol. 1. Paris: Imprimerie Royale.

Bulkeley, Rip. 2014a. *Bellingshausen and the Russian Antarctic Expedition, 1819–21*. Basingstoke: Palgrave.

———. 2019. Bellingshausen's 'Mountains': The 1820 Russian Sighting of Antarctica and Bellingshausen's Theory of the South Polar Ice Cap. *Polar Record* 55 (6): 392–401.

———. 2021. Bellingshausen in Britain: Supplying the Russian Antarctic expedition, 1819. *The Mariner's Mirror* 107 (1): 40–53.

Cook, James. 1777. *A Voyage Towards the South Pole, and Round the World; Performed in His Majesty's Ships the Resolution and Adventure, in the Years 1772, 3, 4, and 5*. 2 vols. London: Strahan and Cadell.

Dal', V.I. 1863. *Tolkovyj slovar' zhivago velikorusskago yazyka [Reference Dictionary of the Greater Russian Language]*. Vol. 1. Moscow: Semyon.

Dal', Vladimir. 1881. *Tolkovyj slovar' zhivago velikorusskago yazyka [Reference dictionary of the Greater Russian language]*. Vol. 2. St Petersburg–Moscow: Vol'f.

Davis, J. 1595. *The Worlds Hydrographicall Discription*. London: Self.

Desmarest, [Nicolas]. 1803. *Encyclopédie Méthodique: Géographie-Physique [Systematic Encyclopedia: Physical Geography]*. Vol. 2. Paris: Agasse.

G[ershau], P. 1892. Admiral'' Fhaddej Fhaddeevich'' Bellinsgauzen'' [Admiral Faddej Faddeevich Bellingshausen]. *Russkaya starina* 75: 373–395.

Henry, David. 1774. Journal of a Voyage to Discover the North East Passage; Under the Command of the Hon. Commodore Phipps, and Capt. Skiffington Lutwych, ... In *An Historical Account of all the Voyages Round the World, Performed by English Navigators*... Vol. 4. Supplement: 29–118. London: Newbery.

Hydrographic Office. 1952. *A Functional Glossary of Ice Terminology*. Washington, DC: U.S. Navy Hydrographic Office.

Imperial Chancellery. 1830. *Polnoe sobranie zakonov'' Rossijskoj Imperii [Complete Collection of the Laws of the Russian Empire]*. Vol. 18, 1767–1769. St Petersburg: Imperial Chancellery.

Imperial Russian Academy. 1814. *Slovar' Akademii Rossijskoj [Dictionary of the Russian Academy]*. Pt 3 K–N. St Petersburg: Imperial Academy of Sciences.

———. 1822. *Slovar' Akademii Rossijskoj [Dictionary of the Russian Academy]*. Pt 6 S–end. St Petersburg: Imperial Russian Academy.

Kant, I. 1801. *Physische Geographie [Physical Geography]*. Part 1. Mainz: Vollmer.

Kisilev'', Egor'. 1819–1821. *Pamit'nik'' prinadlezhit'' matrozu 1j stat'ej Egoru Kisilevu [The Notebook of Seaman 1st Class Egor Kisilev]*. Manuscripts Division, Russian State Library, Moscow: Fond 178, MS 10897.8.

Kondakova, O.N. 2019. *K 200-letiyu otkrytiya Antarktidy Èkspeditsiej F.F. Bellinsgauzena i M.P. Lazareva na shlyupakh "Vostok" i "Mirnyj" [For the Bicentenary of the Discovery of Antarctica by F.F. Bellingshausen and M.P. Lazarev on Sloops Vostok and Mirnyj]*. St Petersburg: Russian Naval Archives. https://rgavmf.ru/virtualnye-vystavki/k-200-letiu-otrritiya-antarktidi.

Krechetnikov, A. 2020. Kak rossiyane i britantsy odnovremenno otkryli Antarktidu [How Russians and Britons Simultaneously Discovered Antarctica]. BBC Russian Service, January 28. https://www.bbc.com/russian/features-51264899.

Larionov, A.L. 1963. Korabli Pervoj russkoj antarkticheskoj èkspeditsii—shlyupi «Vostok» i «Mirnyj» [The Ships of the First Russian Antarctic Expedition—Sloops *Vostok* and *Mirnyj*]. In *Pervaya russkaya antarkticheskaya èkspeditsiya 1819–1821 gg. i eyo otchyotnaya navigatsionnaya karta*, ed. M.I. Belov, 128–142. Leningrad: Morskoj Transport.

Lazarev", M.P. 1821. Pis'mo Mikhaila Petrovicha Lazareva k" Aleksyeyu Antonovichu Shestakovu, 24 sentyabrya 1821 [A letter from Mikhail Petrovich Lazarev to Aleksyej Antonovich Shestakov, September 24 1821 [O.S]]. St Petersburg: SARN—F–315, S–1, P–775, fos 1–6v.

———. 1918. Letter 1 in: Pis'ma Mikhaila Petrovicha Lazareva k" Aleksyeyu Antonovichu Shestakovu v" g. Krasnyj Smolenskoj gubernii [Letters from Mikhail Petrovich Lazarev to Aleksyej Antonovich Shestakov at Krasnyj in the Smolensk Gubernorate]. *Morskoj sbornik"* 403 (1): 51–66.

Lomonosov", M.V. 1847. Kratkoe opisanie raznykh" puteshestvij po syevernym" moryam", i pokazanie vozmozhnago prokhodu Sibirskim" okeanom" v" Vostochnuyu Indiyu [A Short Description of Various Journeys Through Northern Seas and An Exposition of a Possible Route to the East Indies Via the Siberian Ocean]. In *Sochineniya Lomonosova*, ed. Aleksandr" Smirdin". St Petersburg: Imperial Chancellery—not seen.

———. 1854. Kratkoe opisanie raznykh" puteshestvij po syevernym" moryam", i pokazanie vozmozhnago prokhodu Sibirskim" okeanom" v" Vostochnuyu Indiyu [A Short Description of Various Journeys Through Northern Seas and An Exposition of a Possible Route to the East Indies Via the Siberian Ocean]. In *Proekt" Lomonosova i èkspeditsiya Chichagova*, ed. A. Sokolov", 3–141. St Petersburg: Hydrographic Department.

Lomonosow, Michael. 1763. Tankar, om Is-bergens ursprung uti de Nordiska Hafven [Thoughts on the Origin of Icebergs in Northern Seas]. *Kongl. Vetenskaps Academiens Handlingar* 24: 34–40.

Maksimovich", L.M. 1788. *Novyj i polnyj geograficheskij slovar' Rossijskago gosudarstva [New Complete Geographical Dictionary of the Russian Empire]*. Vol. 3. Moscow: University Press and Novikov.

Martens, Friderich. 1675. *Spitzbergische oder Groenlandische Reisebeschreibung gethan im Jahr 1671 [Account of a Voyage to Spitsbergen and Greenland in 1671]*. Hamburg: Schultzen.

O'Reilly, Bernard. 1818. *Greenland, the Adjacent Seas, and the North-West Passage to the Pacific Ocean*. London: Baldwin, Cradock, and Joy.

Okhuizen, Edwin. 2005. Dutch Pre-Barentsz Maps and the Pomor Thesis About the Discovery of Spitsbergen. *Acta Borealia* 22 (1): 21–41.

Ostrovskij, B.G. 1949a. Novoe ob istoricheskom pokhode Bellinsgauzena–Lazareva v Antarktiku [A New Source for Bellingshausen and Lazarev's Historic Voyage to the Antarctic]. *Zvezda* 2: 96–99.

de Pagès, [P.M.F.]. 1783. Tagebuch einer Seefahrt gegen den Nordpol [Journal of a voyage towards the North Pole]. Extract translated from the French original. *Der Teutsche Merkur*. 3rd Viertelj. 193–242.

Pavlovskij, I.Ya. 1879. *Russko–nyemetskij slovar' [Russian–German Dictionary]*. 2nd ed. Riga: Kummel.

Petrova, Evgenia, and Others, ed. 2012. *Pavel Mikhailov 1786–1840: Voyages to the South Pole*. St Petersburg: Palace.

Phipps, C.J. 1774. *A Voyage towards the North Pole*. London: Nourse.

Pinegin, Nikolaj. 1934. *Sem' desyat dnej bor' by za zhizn' [Seventy Days in a Fight for Life]*. Arkhangel'sk: Northern Regions Press.

Prévost d'Exiles, Antoine-François. 1759. *Histoire générale des voyages [A General History of Voyages]*. Vol. 15. Paris: Didot.

Ross, Captain John. 1819. *A Voyage of Discovery, Made Under the Orders of the Admiralty, in His Majesty's Ships Isabella and Alexander*. Vol. 1. London: Longman and Others.

de Saint-Pierre, Bernardin. 1796. *Etudes de la Nature [Studies of Nature]*. Vol. 1. rev. ed. London: Spilsbury.

Scoresby, William, Jr. 1818. On the Greenland or Polar Ice. *Memoirs of the Wernerian Society* 2: 261–388.

———. 1820. *An Account of the Arctic Regions*. Vol. 1. Edinburgh: Constable.

Seaborn, Captain Adam. (pseudonym of J.C. Symmes). 1818. *Symzonia: A Voyage of Discovery*. New York: Seymour.

Simonov", I. 1828. *Opredyelenie geograficheskogo polozheniya myest" yakornago stoyaniya shlyupov" VOSTOKA i MIRNAGO i.t.d. [A Determination of the Geographical Location of the Anchorages of Sloops Vostok and Mirnyj etc.]*. St Petersburg: Department of Education.

Simonov, I.M. 1990b. «Vostok» i «Mirnyj» [*Vostok* and *Mirnyj*]. In *Dva plavaniya vokrug Antarktidy*, ed. T.Ya. Sharipova, 45–248. Kazan: Kazan University Press.

Simonov", Prof. 1825. O raznosti temperatury v" Yuzhnom" i Syevernom" polushariyakh" [On the Difference in Temperature between the Southern and Northern Hemispheres]. *Kazanskij vyestnik"* 14: 99–119.

Simonov", Prof. i Kav. 1822a. Plavanie shlyupa *Vostoka* v" Yuzhnom" Ledovitom" Morye [The Voyage of the Sloop *Vostok* in the Southern Ice Ocean]. *Kazanskij vyestnik"* 4 (3): 156–165, 4 (4): 211–216, 5 (5): 38–42, 5 (7): 174–181, 6 (10): 107–116, 6 (12): 226–232.

———. 1822b. *Slovo o uspyekhakh" plavaniya shlyupov" Vostoka i Mirnago okolo svyeta i osobenno v" Yuzhnom" Ledovitom" morye, v" 1819, 1820 i 1821 godakh" [An Address about the Results from the Voyage of the Sloops Vostok and Mirnyj around the World and Especially in the Southern Ice Ocean, in 1819, 1820 and 1821]*. Kazan: Kazan University Press.

Smirnov", N.M. 1853. *Sobranie russkikh" voennykh" razskazov" [A Collection of Russian War Stories]*. St Petersburg: Military Education Department.

Stojkovich", Athanasij. 1813. *Nachal'nyya osnovaniya fizicheskoj geografii [Preliminary Elements of Physical Geography]*. Kharkov: University Press.

Thomson, John. 1817. *A New General Atlas*. Edinburgh: Self.

de Traversay, Marquis. 1819. Memo, June 23 1819. St Petersburg: SARN—F–166, S–1, P–660a, fo. 42.

Verne, Jules. 1871. *Vingt Mille Lieues sous les Mers [20,000 Leagues under the Sea]*. Paris: Hetzel.

CHAPTER 4

General Acceptance (1833–1928)

The first and longest period in the Russian historiography of the Bellingshausen expedition elapsed between their triumphant return in 1821 and a commemorative article published in 1928 by the oceanographer Yulij Mikhajlovich Shokal'skij (1856–1940), who did much to spread knowledge of the expedition both at home and abroad. During that time there was widespread admiration for, and little substantial disagreement about, what had been achieved.[1]

EARLY RESPONSES

Initial Russian enthusiasm for the expedition, in the 1820s, was motivated, at least in part, by the customary adulation for the reigning monarch which dominated public life in the capital (see for example Khvostov" 1825). However one passing mention, also in the year that Emperor Alexander died, said nothing about Bellingshausen having discovered anything (Kornilovich" 1825: 335). In the following reign, the publicist N.A. Polevoj (1796–1846) appears from a few remarks he let fall in 1833 to have been an admirer, but in the passage quoted by V. Zubov (1956:

---

[1] The original texts of translated quotations, indicated by numbers in square brackets, are provided at the end of each chapter. For the system of romanization used, the rendering of dates, and a few other matters, please consult the Apparatus at the end of the book.

© The Author(s), under exclusive license to Springer Nature Switzerland AG 2021
R. Bulkeley, *The Historiography of the First Russian Antarctic Expedition, 1819–21*,
https://doi.org/10.1007/978-3-030-59546-3_4

320–321) he said nothing of any substance *about* the expedition (Polevoj 1833: 80).[2] Also in 1833 Krusenstern boosted his protégé to Alexander von Humboldt. Encouraged by John Biscoe's recent discovery of Enderby Land in East Antarctica, Krusenstern upgraded Bellingshausen's discoveries of Peter I Island and the Alexander I Coast to the status of 'Lands' and surmised that, when taken together with Enderby and Graham Lands and the South Shetland and South Orkney Islands, they meant that an Antarctic continent had finally been discovered (Krusenstern 1833).

Others were less enthusiastic. A Russian survey of geographical discoveries in the first three decades of the century appeared in 1841. Alongside numerous foreign explorers such as Flinders and Lapérouse it listed the Russians Krusenstern, Lisyanskij, Lazarev (from an earlier voyage), Kotzebue, Lütke, and Bellingshausen, in that order. The latter had 'discovered Alexander and Peter Islands' [1], but had been prevented by ice from getting closer than eight leagues (about 44 km). With ominous carelessness, a mere twenty years after the expedition and ten years after Bellingshausen's book appeared, the expedition was misdated by five years (Anon 1841: 10). A year later, after recapitulating the discoveries of previous explorers including Bellingshausen, the expeditions of Wilkes and Ross were reported as 'Discovery of a *new* mainland in the southern hemisphere' [2] (Anon 1842—emphasis added).

Another cool response came from the Russian Geographical Society. According to the meteorologist Ludwig Friedrich Kämtz (1801–1867) at the University of Dorpat (Tartu), the expedition's work in the Antarctic was insignificant compared to its discoveries in the South Pacific (Kemtts" 1848: 41–42). Count Fyodor Petrovich Lütke (1797–1882), the Society's influential first president, had a low opinion of the Bellingshausen expedition (Tammiksaar 2016: 582).

One early and positive Russian treatment of the expedition came from someone with no connection to it. In 1844 the painter and lithographer Ignatij Stepanovich Shhedrovskij (1815–1871) created a small tribute to Bellingshausen at Kronstadt. Bellingshausen had joined the Admiralty Council that year after being promoted to full admiral in 1843, either of which events might have prompted such a self-advertising response from Shhedrovskij. A colour lithograph on the front cover appears to show a tropical or temperate foreshore facing an Antarctic island (Peter I?) on the horizon. That juxtaposition may reflect the quality of the contents, another

---

[2] Polevoj's periodical *Moskovskij Telegraf* is still unavailable on the internet.

indicator being the mistaken designation of *Mirnyj* as the squadron's command ship in the legend below the picture (Shhedrovskij 1844).[3]

The headline achievements of the expedition were recapitulated in obituaries of Bellingshausen, Simonov, and Lazarev in the 1850s, and few commentators, foreign or Russian, queried or amended their findings over the next 70 years (V.I. 1881: 350). The expedition was widely admired outside Russia, especially after the early translations of works by Simonov were joined by a competent German summary of *Two Seasons* (Lowe 1842).

MISTAKES

Two early misapprehensions, sometimes involving foreign commentators, should be mentioned. Bellingshausen's furthest south was 69°53′ on 21 January 1821; Cook's had been 71°10′ on 30 January 1774. After months of dogged, very dangerous, but finally unsuccessful, attempts to equal or surpass his predecessor, Bellingshausen would have been mortified to see claims that he had in fact done so. Yet such reports were quite frequent; thus: 'Captain Bellingshausen, who pushed further towards the South Pole than earlier mariners ...' [3] (Anon 1830: 842), or: 'Bellingshausen ... sought the field which had disheartened Cook, and reached a point two degrees further south than he had done' (Anon 1868: 142). The mistake perhaps arose from Simonov's statement that: 'We had probed further than that navigator in many regions, and remained below the Antarctic Circle for much longer than he did ...' [4], which was literally accurate but might be condensed into an *absolute* further south by a careless reader. In the 1860s Petermann clarified Simonov to the effect that Bellingshausen had gone further south than Cook in many longitudes, but never further than Cook's furthest (Petermann 1863: 409). But by then the damage was done.

A variation was the inaccurate claim that Bellingshausen had at least reached 70°S. That took root earlier (Anon 1824), and was more persistent: '... in the second [season], he pushed as far as 70°, ...' [5] (Anon 1842: 152); 'Bellingshausen ... crossed the latitude of 70°S in 1°30′W'

---

[3] This privately produced work was a Russian technological landmark, since it was created a year before the artist published his groundbreaking album *Our Folks!* (Suris 1957: 4). Probably very few copies were made; research carried out on behalf of the author suggests that none survive in public ownership. Indeed the booklet is so rare that it is not even listed in Russian catalogues of rarities. The Russian Naval Museum in St Petersburg holds only a photograph of the cover (Lyalin and Others 2007: 182).

(Bruce 1894: 59). The misconception perhaps arose from a carelessly ordinal treatment of space, for which anything between 69°00′01″ and 70°00′00″S lies 'in' or 'under' the 70th degree of latitude. Thus Simonov told his audience in 1822 that the Emperor's flag had fluttered 'in 70° south latitude' [6] (1822b: 4); and Lt. Arkadij Sergeevich Leskov (1797–after 1858), one of Bellingshausen's officers, made a similar claim in a letter to Admiral Moller (1764–1848), chief of staff of the Imperial Navy: *'Russian navigators ... were the first to settle this important question by discovering land under the 70th degree south'* [7] (Leskov″ 1823; see also Krasheninnikov″ 1853: 242; Men'shikov″ 1891: 389). Simonov and Leskov were using expressions vague enough to be true in the ordinal sense. But geographical degrees were conventionally subdivided into 3600 seconds so that scientists and navigators could measure and report positions with errors that were hopefully smaller than a clear horizon. By choosing flattery instead of accuracy in his published and translated lecture, Simonov caused others to misapprehend that the expedition had reached or even passed the absolute latitude of 70°00′00″S.[4]

A more puzzling blunder was the assertion, apparently launched by the Scottish geographer James Laurie, that Bellingshausen, rather than Bransfield or Palmer, had discovered Trinity Land, at the tip of what we now call the Antarctic Peninsula (Laurie 1842: 1018). Neither Bellingshausen nor Simonov ever claimed such a discovery, leave alone naming it after a British maritime institution (Trinity House). Laurie's French source said nothing of the kind, nor was it responsible for his grossly inaccurate date of 1829 for the discoveries of Peter I I. and Alexander I Coast in the same passage. The Trinity Land error was repeated in the mid-nineteenth century, but not widely.

Novosil'skij

The most significant Russian contributions to knowledge of the expedition in the nineteenth century, after those of Bellingshausen and Simonov, were two pieces first published by Novosil'skij in 1853. A former officer with the expedition on board HIMS *Mirnyj*, Novosil'skij wrote an account of the voyage and then a sequel in which he surveyed Antarctic

---

[4] That does not excuse one spectacularly inaccurate account which located the Russians at 69°91′S on 18 November 1819, when by either calendar they had not yet departed Rio (Fonvielle 1889: 134).

exploration from Cook to Ross. They were not listed under primary sources above, because they have a complex relationship with the original events three decades earlier. The text of the first, *The South Pole*, probably relies on Bellingshausen's book for the main narrative, with lively additions which may have been based on Novosil'skij's journal, now lost, or else were simply taken from memory. His second pamphlet, *The Sixth Continent*, was a secondary commentary. It went into several editions and is the more significant of the two for present purposes because it compared the achievements of different explorers (Novosil'skij 1853a, b, c; 1854a, b, c).[5,6]

In *The South Pole* Novosil'skij echoed his commander's account of 28 January 1820, changing only one word, so that 'small hills' (*prigorkami*) instead of 'hillocks' (*bugrami*) were scattered across the impassable ice field (see for example Novosil'skij 1853c: 29). In the third, expanded edition of *The Sixth Continent*, however, he also described the scene as an encounter with 'an immobile ice coast/shore' [8] (Novosil'skij 1854c: 8). But that phrase does not confirm Lazarev's '*main ice of extraordinary height*' (Chap. 7), because Novosilskij added that from some way off they could see that countless overlapping ice fragments were scattered across the ice field. Lazarev noted that large icebergs could rise 91–122 m above sea level, adding that one was unusually high at 143 m (Bulkeley 2014a: 168). The extraordinarily high ice coast that he thought was seen on 28 January 1820 should therefore have measured 70 m to 80 m at the very least. But *Mirnyj*'s lookouts, at somewhere between 35 m and 40 m (Luchininov 1973b: 4), would not have been able to survey the surface of such a feature in the way described by Novosil'skij. It follows that the two men reported quite different situations for the same ship and the same date.

Furthermore, although they were still rare, 'ice coast' and similar phrases had been applied to an ice front well before Novosil'skij. For example, though perhaps not used by Cook himself that expression occurs in a multi-authored account of his three Pacific voyages which went

---

[5] The bibliography is complicated because the original magazine articles were signed but the pamphlet versions were not. No Soviet author ever referred to the original versions in the magazine *Panteon*.

[6] Despite the eight months between their publication dates in 1949, Bogatkina (see Chap. 2) seems not to have consulted Ostrovskij's survey of the literature (Ostrovskij 1949b) while preparing her bibliography. As between Novosil'skij's two pamphlets, she listed only *The Sixth Continent*, whereas Ostrovskij discussed only *The South Pole*. Andreev republished the latter, and was the only commentator in 1949 who referred to both (Chap. 6).

through numerous editions in the late eighteenth and early nineteenth centuries (Cook and Others 1814: 323). Bellingshausen had also used the term *bereg* (coast or shore) for the vast main body of ice, from which he surmised that the great tabular ice islands, observed on 17 February 1820, had broken away (Bellingshausen 1831 1: 189). It would be wrong to suppose, therefore, that in his final description of 28 January 1820 Novosil'skij was suggesting that they were in the presence of land. The word 'immobile' tells us just the opposite—they were confronted by ice, but ice which did not behave like most sea ice. (Buffon had virtually predicted just such an encounter.)

Even in the first, magazine version of *The Sixth Continent*, in November 1853, Novosil'skij expressed his conviction that the ensemble of expeditions to date had demonstrated the existence of a south polar continent, and that the Russian discoveries of Peter I I. and Alexander I Coast, in particular, had been a key contribution, because they were the most southerly achieved before the Ross expedition. Both in that version and in the first two pamphlet editions (1854a, b), Novosil'skij gave only a brief account of the Russian expedition with hardly any dates. Doubtless readers were expected to refer to his companion piece, *The South Pole*, but by mentioning the Russian expedition before the explorations of Trinity Land etc. by Bransfield, Palmer, and others in 1820, and by describing it as the first Antarctic expedition since Cook, Novosil'skij created the impression that the Russian discoveries in January 1821 had preceded the Anglo-Saxon discoveries of Trinity Land in 1820, which they did not.[7] Only in the third, deluxe edition of *The Sixth Continent* (1854c), which was about 50% longer and appeared six months after the second, did Novosil'skij provide a detailed account of the Russian voyage with dates for their discoveries, though not for Bransfield's which may not have been easy for him to obtain. But Novosil'skij did at least mention that Bransfield began his voyage in 1819 from Valparaíso, far closer to Antarctica than Kronstadt, and stated that 'he saw a high mountain, covered with snow, at 63°20'S 62°W' [9] (1854c: 16–17). That placed Bransfield's discovery squarely within the conjectured mainland shown on his (Novosil'skij's)

---

[7] It was of course true that Bellingshausen sailed from Kronstadt in July 1819 before Bransfield sailed from Valparaíso in December 1819, but that had no bearing on the order of their discoveries.

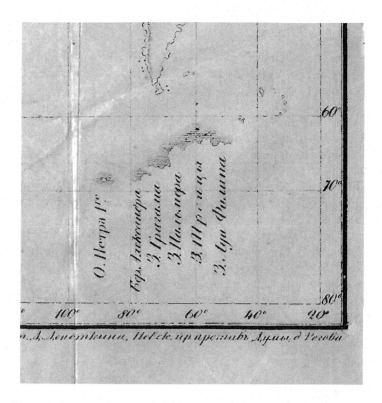

Fig. 4.1 Detail from Novosil'skij's map (drawn by his son Midshipman Andrej Pavlovich Novosil'skij). From left to right, the place names are: Peter I Island; Alexander Coast; Graham Land; Palmer Land; Trinity Land; Louis Philippe Land. (Source: Novosil'skij 1854c)

map (Fig. 4.1).[8] Novosil'skij also surmised in the text of every edition that Alexander Island, as we know it today, was continental and that Trinity, Palmer, and Graham Lands were all connected to it and thus also

---

[8] Assuming this to be a reference to Mount Bransfield, the contrast between the reasonably accurate longitudes found in documents from the voyage (Campbell 2000: 133–134, 185) and Novosil'skij's error of some +5°W may have been due to deficiencies in some of the early published accounts. However some of the early maps which Novosil'skij could have seen also gave accurate locations for the future Mount Bransfield (Campbell 2000: 76–77). (Novosil'skij cited a map published by R. Laurie, but too vaguely for it to be identified.)

continental.[9] By doing so, however, he contradicted his own conclusion that 'The honour of first discovery should be given, in all justice, to the Russian navigator Bellingshausen' [10] (1854c: 55). If Bransfield had sighted part of the mainland, Trinity Land, in 1820, then the Russian expedition could not have been the first to do so in 1821.[10]

The Sixth Continent was undoubtedly the source for Zherebtsov's even bolder version: '... Captain Bellingshausen, assisted by Captain Lazarev ... probed the most heavily glaciated parts of the Antarctic, and fixed the dimensions ... of a large continent, which probably stretches down to the Pole' [11] (Gerebtzoff 1858: 310; see also Russwurm 1870: 17). In his final word on the subject, however, and doubtless influenced by the revival of scepticism after the contretemps between Wilkes and Ross, Novosil'skij echoed the British explorer's reservations: 'Do all these discovered coasts make up one and the same undivided continent, or are they, rather, separate lands? This question will be decided by *future* voyagers in the south polar sea' [12] (Novosil'skij 1855: 30—emphasis added).

Despite his faulty reasoning Novosil'skij's knowledge of early British and American exploration had been exemplary. For the next 50 years no Russian authors could match him, though foreign histories of exploration, such as Jules Verne (general) or the unreliable Fonvielle (Antarctic), were translated. The few notable Russian treatments of the expedition usually described its achievements without making comparisons or using words like 'first' (Ivashintsov" 1872: 54; Lyalina 1898: 438–439; S. Grigor'ev" 1906: 7). Shokal'skij, however, pointed out that Bellingshausen had been 'the first to circumnavigate the Southern Ice continent [*materika*][11] while keeping mainly between 60° and 70°' [13] (Shokal'skij 1898: 493). Other encyclopedia entries kept their language neutral or sometimes fell into

[9] Although he described Dumont d'Urville's 1838–1840 expedition at some length, Novosil'skij seems to have been unaware of the Frenchman's claim to have discovered and named a strait, the 'Canal d'Orléans', which if present would have rendered Trinity Land an island and disqualified Bransfield from any claim to mainland discovery. That line of thinking would have to wait 54 years for another Russian commentator, Rabinovich.

[10] Although Novosil'skij gave no dates for Bransfield's voyage his description of it was quite detailed (1854c: 19), and by placing it before Palmer's voyage to the Peninsula, which he rightly dated to 1820 (the same: 19–20), he showed that he knew very well that Bransfield's survey had preceded the Russian discovery of the Alexander I Coast in January 1821.

[11] The translation follows the usage in the encyclopedia concerned, but should not be backdated to Bellingshausen's reference to a *materik*, or main, of ice. See Chaps. 3 and 8.

well-worn errors such as that Bellingshausen had gone two degrees further south than Cook [14] (Polovtsov" 1900: 682).

RUSAKOV AND RABINOVICH

The two boldest champions of the expedition, between Novosil'skij and the Bolshevik Revolution in 1917, were Rusakov and Rabinovich. In 1903, under the pseudonym Viktor Rusakov, Sigizmund Feliksovich Librovich (1855–1918) published perhaps the strangest account of the expedition ever given. He failed even to mention the discoveries of Peter I I. and the Alexander I Coast, and did not claim that the expedition discovered Antarctica. Nevertheless Rusakov (as he is generally known) devoted eight short pages to elevating Bellingshausen above all other Russian explorers. Placing him at the front of the book, ahead of earlier figures, Rusakov saluted him as 'one of the most distinguished Russian Columbuses' [15] (Rusakov" 1903: 10).[12] Bellingshausen had earned this status, he thought, by discovering 29 islands,[13] which were made to seem

[12] In Russian literature, the Columbus trope was possibly coined by Lomonosov in his 1747 ode to Vitus Bering. Mindful of Bering's Danish origins, Lomonosov marked his status as a subject of the Russian state, acquired by more than 30 years of service in the Russian Navy, by attaching the appropriate adjective *rossijskij* to *Kolumb*. In a later, unfinished poem about Peter the Great, Lomonosov described the native Russians who had explored the Northern Sea Route as 'Russian Columbuses', replacing *rossijskij* with the ethnic epithet *rosskij*. The German geographer August Petermann (1822–1878) was perhaps the first person to liken Bellingshausen to Magellan and Columbus. He did so, however, on the grounds of the Russian discoveries in 1821, not 1820, and for the short-lived rhetorical purpose of spurring German mariners to go forth and do likewise (Petermann 1865–1867: 6). Petermann was then quoted and endorsed by Men'shikov in the entry on Bellingshausen in a biographical dictionary (Men'shikov" 1891), a likely source for Rusakov's deployment of the trope a few years later.

[13] The number was taken from the last paragraph of *Two Seasons*, where Bellingshausen, unlike Rusakov, carefully pointed out that only two were Antarctic, with eight in the southern temperate zone, and the rest tropical. The corresponding numbers from his final table of discoveries (Bellinsgauzen" 1821e) are: Antarctic 2; sub-Antarctic 10; tropical 16; Bellingshausen perhaps included a reef, item O, to reach his total of 29. It would have been difficult for him to find out, before finishing his book in 1824, that he had been preceded by Bransfield at the Elephant and Clarence sub-group of the South Shetland Islands. Of those seven islands, six had already been discovered by Bransfield, but Rozhnov/Gibbs I. had not. Four more islands in the South Pacific, which Bellingshausen found independently, had also been discovered already. The actual number of islands which the expedition were probably the first Europeans to see was therefore 18, plus another 10 which they found independently, a highly creditable achievement (see also Chap. 8). Just like Rusakov, however, authoritative

Antarctic, partly by omitting any reference to the tropical phase of the voyage and partly by a very misleading juxtaposition:

> Bellingshausen crossed the Antarctic Circle three times and discovered in all 29 islands … [16] (Rusakov″ 1903: 8)[14]

In fact only six were new Antarctic or sub-Antarctic discoveries, namely the Marquis de Traversay group, Peter I I., the Alexander I Coast, and Rozhnov/Gibbs I. Rusakov also exaggerated the importance of the Traversay Is. by claiming that Cook had found nothing in the region, whereas in fact they are a northern extension of the South Sandwich Is.—discovered by Cook.[15]

Rusakov unwisely boosted the Antarctic achievements of his hero by claiming the main part of the South Shetland Is. for him, despite the fact that Bellingshausen never said anything of the sort. (That should have added at least eight more islands to the tally of 29 which Rusakov had already accepted on the previous page). But as Novosil'skij had explained in the 1850s the islands were discovered and widely explored by British and American sealers before the Russians reached them (Novosil'skij 1854c: 19–20). Rusakov also increased Bellingshausen's furthest south by more than half a degree, from 69°53′ to 70°30′, instead of checking the known facts for himself.

Indifference to sources is a good indicator for the presence of historical myth, and Rusakov seems barely to have consulted the works of Bellingshausen, Simonov, or Novosil'skij, which might have prevented his more egregious mistakes. His treatment of the expedition has been discussed in detail here because it was reprinted in 1996 and remains in print, so that it probably influences some Russian perceptions of the expedition today.

---

information sources in Russia continue to disseminate the larger number of places as 'discovered' by the expedition without qualification, presumably for no better reason than that 29 is a larger number than 18 (Bulkeley 2014a: 120–123; Gidromettsentr Rossii 2015; Yeltsin Presidential Library 2018).

[14] The expedition actually entered Antarctic waters, by crossing the Circle southwards, six times.

[15] Simonov's diary, which reveals that Bellingshausen was probably told by sealers where to find the Traversay Islands, had been published in the 1820s and so could have been consulted by Rusakov (Bulkeley 2014a: 146). But he was evidently not a research historian.

Before 1917 the single most positive, and influential, Russian account of the expedition came from Ivan Osipovich Rabinovich. Like Novosil'skij he focused on the evidence for an Antarctic continent, including Peter I I. and the Alexander I Coast, a question which had by then been settled for other geographers by the first expeditions of the heroic age (Adams 1904; Penck 1904). Unlike Novosil'skij but like Rusakov, Rabinovich paid no regard to early British and American exploration, around the time of the Bellingshausen expedition. He probably had good reasons for that. He was writing after the Norwegian explorer Carl Anton Larsen claimed to have confirmed Dumont d'Urville's 'Canal d'Orléans'. Larsen's claim was challenged in turn by the Belgian explorer Adrien de Gerlache de Gomery, but the latter then reported a similar strait further to the south (Yelverton 2004: xiii).[16] Any such channel meant that places discovered in 1820 on the south side of the Bransfield Strait, such as Trinity Land or Palmer Land, comprised one or more large islands, irrelevant to the question of who first saw the mainland. As the following passage shows, Rabinovich also believed that the Alexander I Coast was part of the mainland, and together those two perceptions would have justified his conclusion that:

> … in January 1821 our brave navigators discovered an island and the coast of a land which form[s] [indeterminate grammar—RB] part of the Antarctic continent, as confirmed by modern exploration. On that basis, the honour of discovering the continent evidently belongs to Russians … [17] (Rabinovich" 1908: 16)

It should be emphasized that Rabinovich's contention was that the continent had been discovered by Russian explorers in 1821, nothing to do with later claims in respect of 1820.[17]

---

[16] De Gerlache was rebutted in turn by Charcot, but the latter's first account of his 1903–1905 expedition may have appeared too late, in December 1906, for Rabinovich to notice that the status of Trinity Land had switched (temporarily) back from insular to continental (Charcot 1906: 253–254, 338). However Rabinovich did give some details of the Charcot expedition, possibly taken from newspaper accounts.

[17] Rabinovich's argument has to be reconstructed because he himself never explained why he disregarded the Antarctic discoveries of British and American mariners in 1820, which were just beginning to be taken more seriously elsewhere (Nordenskjöld and Anderson 1905: 70).

Early Soviet Assessments

Not surprisingly the Bolshevik revolution had no immediate effect on perceptions of the Bellingshausen expedition. Twelve letters from Lazarev to his friend Shestakov were published during those turbulent times, but the significance of the first of them was not realized until much later (Lazarev 1918). Meanwhile Russian Antarctic literature was largely taken up with the recent achievements of foreigners (Pimenova 1925).

Bellingshausen's admirer Shokal'skij was president of the Geographical Society from 1917 to 1931, during which time he was also tasked with aiding the development first of Black Sea fisheries and then of the Northern Sea Route, no small commitments. His commemorative article for the centenary of the expedition was a thorough analysis, though in a sign of the times it was published several years late. Even after stressing that the Russians had discovered the first land south of the Circle in 1821, his evaluation of their achievement was characteristically internationalist:

> The expeditions referred to [Cook & Bellingshausen] not only resolved the question before them but did so in such a comprehensive fashion as to leave scope thereafter only for investigation into details and singularities of the wider topic [18]. (Shokal'skij 1928: 193)[18]

It would be 33 years before another Soviet author portrayed Cook's Antarctic exploration in such a positive light (V. Lebedev 1961: 8).

In a history of Russian geographical exploration down to 1923, Berg noted that the expedition had been near Enderby Land in February (thus) 1820, and had then gone on to discover Antarctica, in the shape of Alexander I Land, on 17 January 1821 [O.S. & s.d.] (Berg 1929: 47).

Original Texts of Quotations

[1] ... открытіе острововъ *Александра* и *Петра* ...

[2] Открытие новаго материка въ южномъ полушаріи.

[3] Cap. Bellingshausen, der weiter als frühere Seefahrer gegen den Südpol vorgedrungen ist,...

[4] Мы углублялись во многихъ местахъ далѣе сего мореходца, оставались не въ примеръ долѣе его за полярнымъ кругомъ ...

---

[18] Krasheninnikov had made the same point 75 years earlier (1853: 242).

[5] ... dans la seconde [campagne], il poussa jusqu'à 70°,...

[6] ... въ 70° южной широты ...

[7] *Российскіе мореходцы ... первые разрѣшили важный вопросъ, открывъ землю подъ 70 градусомъ южной широты ...*

[8] ... неподвижнымъ ледянымъ берегомъ ...

[9] ... онъ видѣлъ высокую гору, покрытую снегомъ, подъ 63°20′ ю. ш. и 62° з. д....

[10] Честь перваго открытія его [материка], по справедливости, должна принадлежать Русскому мореплавателю Беллинсгаузену.

[11] ... le capitaine Bellingshausen, aidé du capitaine Lazareff ... descendirent dans les régions les plus glaciales du pôle antarctique, et y déterminèrent la configuration ... d'un grand continent, qui avance probablement jusqu'au pôle.

[12] Составляютъ ли свѣ эти выдавшіеся берега одинъ и тотъ же непрерывный материкъ или напротивъ отдѣльныя земли? Вопросъ этотъ разрѣшатъ будущіе въ Южномъ Полярномъ морѣ плаватели.

[13] ... Беллинсгаузенъ первый совершилъ полное плаваніе вокругъ Южнаго Ледовитаго материка, почти все время между ш. 60°–70°, ...

[14] ... черты, до которой достигъ Кукъ (67°15′ ю. ш.); ... доходили до широты 69°48′.

[15] ... одного изъ самыхъ выдающихся "русскихъ Колумбовъ".

[16] Три раза заходилъ Беллинсгаузен за южный полярный кругъ и открылъ всего 29 острововъ ...

[17] ... наши смѣлые мореплаватели открыли въ январѣ 1821 г. островъ и берегъ страны, которая, какъ увѣрены современные изслѣдователи, представляютъ собой часть Антарктическаго материка. Такимъ образомъ, русскимъ принадлежить, повидимому, честь открытія этого материка ...

[18] Цель и задачи обеих экспедиций [Cook and Bellingshausen] одинаковы: необходимо было разрешить вопрос: существует ли Южный материк и до каких широт он простирается в умеренном поясе.

Названные экспедиции и разрешили поставленный вопрос и притом настолько широко и всесторонне, что после них осталось место только для обследования подробностей и частностей этой обширной задачи.

REFERENCES[19]

Adams, C.C. 1904. "Antarctica" The New Continent Larger than Europe or the United States. *The New York Times*, March 13: SM2.

Anon. 1824. Return of the Russian Antarctic Expedition. *Philosophical Magazine* 43: 462.

———. 1830. Südpolarländer [Antarctic Lands]. *Allgemeine deutsche Real–Encyklopädie für die gebildeten Stände*. Vol. 10. Leipzig: Brockhaus.

———. 1841. Geograficheskiya otkrytiya v" pervyya tridtsat' let" tekushhago stolyetiya [Geographical Discoveries in the First 30 Years of the Present Century]. *Zhurnal" Ministerstva Narodnago Prosvyeshheniya* 32 (5): 1–24.

———. 1842. Otkrytie novago materika v Yuzhnom" polusharii [Discovery of a New Mainland in the Southern Hemisphere]. *Syn" Otechestva* 1: 24–32.

———. 1868. Expeditions to the North Pole. *Hours at Home* 7: 138–146.

Bellinsgauzen", F.F. 1821e. Report to the Marquis de Traversay from Kronstadt, August 5 1821. St Petersburg: SARN—F–203, S–1, P–826, fos 1–18v.

Bellinsgauzen", Kapitan". 1831. *Dvukratnyya izyskaniya v" yuzhnom" ledovitom" okeanye i plavanie vokrug" svyeta v" prodolzhenii 1819, 20 i 21 godov" [Two Seasons of Exploration in the Southern Ice Ocean and a Voyage around the World, During the Years 1819, 1820 and 1821]*, ed. L.I. Golenishhev"-Kutuzov". 2 vols. plus *Atlas*. St Petersburg: Glazunovs.

Berg, L.S. 1929. *Ocherk istorii russkoj geograficheskoj nauki (vplot' do 1923 g.) [Towards a History of Russian Geography Down to 1923]*. Leningrad: Academy of Sciences.

Bruce, William S. 1894. The Story of the Antarctic. *Scottish Geographical Magazine* 10: 57–62.

Bulkeley, Rip. 2014a. *Bellingshausen and the Russian Antarctic Expedition, 1819–21*. Basingstoke: Palgrave.

Campbell, R.J. 2000. *The Discovery of the South Shetland Islands*. London: Hakluyt Society.

Charcot, J.-B. 1906. *Le "Français" au Pôle Sud [The Français at the South Pole]*. Paris: Flammarion.

Cook, James, and Others. 1814. *Captain Cook's Three Voyages to the Pacific Ocean*. Vol. 1. New York: Duyckinck.

de Fonvielle, Wilfrid. 1889. *Le Pôle Sud [The South Pole]*. Paris: Hachette.

de Gerebtzoff, Nicholas. 1858. *Essai sur l'histoire de la civilisation en Russie [On the History and Civilization of Russia]*. Vol. 2. Paris: Amyot.

---

[19] Key to archival references: SARN = State Archives of the Russian Navy, F = Fond, S = Series [*Opis'*], P = Piece [*Delo*], fo./fos = folio/s [*list/y*], v = verso [*oborotnoe*]. (The latter is necessary because verso pages were usually unnumbered.) Dates on manuscript documents are given in Universal Time (CE).

Gidromettsentr Rossii. 2015. 195 let nazad 28 yanvarya Russkaya èkspeditsiya otkryla Antarktidu [On January 28 195 Years Ago a Russian Expedition Discovered Antarctica]. https://meteoinfo.ru/news/1-2009-10-01-09-03-06/10474-28012015-195-28-.

Grigor'ev", S.G. 1906. *Vokrug" yuzhnago polyusa [Around the South Pole]*. Moscow: Ryabushinskie.

Ivashintsov", N. 1872. *Russkiya krugosvyetnyya puteshestviya, c" 1803 po 1849 god" [Russian Circumnavigations from 1803 to 1849]*. St Petersburg: Ministry of Marine.

Kemtts", Dr. 1848. Ob" uspyekhakh" zemlevyedyeniya s" pervoj poloviny XVIII stolyetiya [Results of Exploration since the First Half of the 18th Century]. In *Karmannaya knizhka dlya lyubitelej zemlevyedyeniya, izdavaemaya ot" Russkago Geografitcheskago Obshhestva*, ed. Anon, 7–142. St Petersburg: Imperial Chancellery.

Khvostov", D.I. 1825. *Russkie morekhodtsy na ledovitom" okeane [Russian Seafarers in the Ice Ocean]*. St Petersburg: Ministry of Education.

Kornilovich", A. 1825. Izvyestie ob" èkspeditsiyakh" v" Syeverovostochnuyu Sibir' flota Lejtenantov" Barona Vrangelya i Anzhu v" 1821, 1822, i 1823 godakh" [News of the Expeditions of Lieutenants Baron Wrangell and Anjou to North-Eastern Siberia in 1821–23]. *Severnyj Arkhiv* 13 (4): 334–378.

Krasheninnikov", S.P. 1853. Bellinsgauzena Èkspeditsiya v Yuzhnyj Ledovityj okean" [Bellingshausen's Expedition to the Southern Ice Ocean]. In *Voennyj Èntsiklopedicheskij Leksikon"*, vol. 2, 241–242. St Petersburg: Institute of Military Education.

von Krusenstern, Vice-Admiral. 1833. Über die Entdeckung des südlichen Continents [On the Discovery of the Southern Continent]. *Annalen der Erd-, Völker- und Staatenkunde* 8 (4): 95–96.

Laurie, James. 1842. *System of Universal Geography: founded on the works of Malte-Brun and Balbi*. Edinburgh: Black.

Lazarev", M.P. 1918. Letter 1 in: Pis'ma Mikhaila Petrovicha Lazareva k" Aleksyeyu Antonovichu Shestakovu v" g. Krasnyj Smolenskoj gubernii [Letters from Mikhail Petrovich Lazarev to Aleksyej Antonovich Shestakov at Krasnyj in the Smolensk Gubernorate]. *Morskoj sbornik"* 403 (1): 51–66.

Lebedev, V.L. 1961. Geograficheskie nablyudenya v Antarktike èkspeditsij Kuka 1772–1775 gg. i Bellinsgauzena–Lazareva 1819–1821 gg. [Geographical Observations in the Antarctic by the Cook, 1772–1775, and Bellingshausen–Lazarev, 1819–1821, Expeditions]. *Antarktika—doklady kommisii* 1960 (1): 7–24.

Leskov", A.S. 1823. Letter to Admiral Moller, April 2 1823. St Petersburg: SARN—F-116, S-1, P-2596, fo. 3. (Text in Belov, 1963a: 8.)

Lowe, F. 1842. Bellingshausens Reise nach der Südsee und Entdeckungen im südlichen Eismeer [Bellingshausen's Voyage to the Pacific and Discoveries in

62   R. BULKELEY

the Southern Ice Ocean]. *Archiv für wissenschaftliche Kunde von Russland* 2: 125–174.

Luchininov, S.T. 1973b. *Shlyup Mirnyj [Sloop Mirnyj]*. Moscow: DOSAAF.

Lyalin, A.Ya., and Others. 2007. *Morskoj Muzej Rossii [The Russian Naval Museum]*. St Petersburg: Central Naval Museum.

Lyalina, M.A. 1898. *Russkie moreplavateli arkticheskie i krugosvyetnye [Russian Arctic Voyages and Circumnavigations]*. St Petersburg: Devrien.

Men'shikov", M. 1891. Bellingsgauzen" [*thus*], Fhaddej (Fabian") Fhaddeevich" [Bellingshausen, Faddej (Fabian) Faddeevich]. In *Kritiko-biograficheskij slovar' russkikh" pisatelej i uchonykh"*, ed. S.A. Vengerov", vol. 2, 388–392. St Petersburg: Efron.

Nordenskjöld, N.O.G., and G. Anderson. 1905. *Antarctica, or Two Years Amongst the Ice of the South Pole*. London: Hurst and Blackett.

Novosil'skij, P.M. 1853a. Yuzhnyj polyus": iz" zapisok" byvshago morskago ofitsera [The South Pole: From the Memoirs of a Former Naval Officer]. *Panteon"* 11 (9): 31–80; (10): 19–62.

———. 1853b. *Shestoj kontinent" [The Sixth Continent]. Panteon"* 12 (11): 99–116.

———. 1853c. *Yuzhnyj polyus": iz" zapisok" byvshago morskago ofitsera [The South Pole: From the Memoirs of a Former Naval Officer]*. St Petersburg: Vejmar".

———. 1854a. *Shestoj kontinent" [The Sixth Continent]*. 1st ed. St Petersburg: Vejmar".

———. 1854b. *Shestoj kontinent"*. 2nd ed. St Petersburg: Vejmar".

———. 1854c. *Shestoj kontinent"*. 3rd ed. St Petersburg: Imperial Academy of Sciences.

———. 1855. O geograficheskikh" otkrytiyakh" v" Yuzhnom" Polyarnom" morye, v" istekshej polovinye XIX stolyetiya [On Geographical Discoveries in the South Polar Sea in the First Half of the 19th Century]. *Zhurnal" Ministerstva Narodnago Prosvyeshheniya* 87: 16–31.

Ostrovskij, B.G. 1949b. O pozabytykh istochnikakh i uchastnikakh antarkticheskoj èkspeditsii Bellinsgauzena–Lazareva [On Forgotten Sources for and Members of the Bellingshausen–Lazarev Antarctic Expedition]. *Izvestiya vsesoyuznogo geograficheskogo obshhestva* 81 (2): 239–249.

Penck, A. 1904. Antarktika. *Deutsche geographische Blätter* 37: 1–9.

Petermann, A. 1863. Neue Karte der Süd-Polar-Regionen [A New Map of the Antarctic]. *Petermann's Geographische Mittheilungen* 9: 407–428.

———. 1865–1867. Die Erforschung der arktischen Central-Region durch eine Deutsche Nordfahrt [Exploration of the Central Arctic by a German Expedition]. *Petermann's Geographische Mittheilungen* Ergänzungsband IV: 1–14.

Pimenova, È.K. 1925. *Geroi Yuzhnogo Polyusa (Lejtenant Shekl'ton i Kapitan Skott) [Heroes of the South Pole (Lieutenant Shackleton and Captain Scott)]*. Leningrad–Moscow: Kniga.

Polevoj, N.A. 1833. Retsenziya na *Poyezdku k Ledovytomu moryu Fr. Belyanskogo i Poyezdku v Yakutsk"* [Review of *A Journey to the Arctic Ocean by Fr. Belyanskij* and *Journey to Yakutsk*]. *Moskovskij Telegraf"* 52 (14): 216–252.

Polovtsov", A.A., ed. 1900. *Russkij biograficheskij slovar' [Russian Biographical Dictionary]*. Vol. 2. St Petersburg: Imperial Estate.

Rabinovich", I.O. 1908. *Shestaya chast' svyeta [The Sixth Continent]*. St Petersburg: Stepanova.

Rusakov", V. (pseudonym of S.F. Librovich). 1903. *Russkie Kolumby i Robinzony [Russian Columbuses and Crusoes]*. Moscow: Vol'f.

Russwurm, Carl. 1870. *Nachrichten über die adeliche und freiherrliche Familie von Bellingshausen [The Noble and Landed Bellingshausen Family]*. Reval: Lindfors' Erben.

Shhedrovskij, I.S. 1844. *Ego vysokoprevoskhoditel'stvu Admiralu Faddeyu Faddeyevichu Bellinsgauzenu [For His Excellency Admiral Faddej Faddeevich Bellingshausen]*. Kronstadt: Self—not seen.

Shokal'skij, Yu.M. 1898. Polyarnyya strany Yuzhnago polushariya [Polar Countries of the Southern Hemisphere]. In *Èntsiklopedicheskij slovar'*, ed. K.K. Arsen'ev" and F.F. Petrushevskij, vol. 24, 489–495. St Petersburg: Brokgauz and Efron.

———. 1928. Stoletie so vremeni otpravleniya Russkoj antarkticheskoj èkspeditsij pod komandoyu F. Bellinsgauzena i M. Lazareva 4 iyulya 1819 g. iz Kronshtadta [Centenary of the Departure from Kronstadt of the Russian Antarctic Expedition, Commanded by F. Bellingshausen and M. Lazarev, on July 4 1819 [O.S.]]. *Izvestiya gosudarstvennogo russkogo geograficheskogo obshhestva* 60 (2): 176–212.

Simonov", Prof i Kav. 1822b. *Slovo o uspyekhakh" plavaniya shlyupov" Vostoka i Mirnago okolo svyeta i osobenno v" Yuzhnom" Ledovitom" morye, v" 1819, 1820 i 1821 godakh" [An Address about the Results from the Voyage of the Sloops Vostok and Mirnyj around the World and Especially in the Southern Ice Ocean, in 1819, 1820 and 1821]*. Kazan: Kazan University Press.

Suris, B. 1957. *I. Shhedrovskij*. Moscow: Izogiz.

Tammiksaar, E. 2016. The Russian Antarctic Expedition under the Command of Fabian Gottlieb von Bellingshausen and its Reception in Russia and the World. *Polar Record* 52 (5): 578–600.

V.I. 1881. Mikhail" Petrovich" Lazarev". *Russkij arkhiv"* 2 (2): 347–361.

Yeltsin Presidential Library. 2018. Russkoj èkspeditsiej otkryta Antarktida [Antarctica Discovered by a Russian Expedition], January 28. https://www.prlib.ru/history/618985.

Yelverton, David E. 2004. *The Quest for a Phantom Strait*. Guildford: Polar Publishing.

Zubov, V.P. 1956. *Istoriografiya estestvennykh nauk v Rossii (XVIII—pervaya polovina XIX v.) [The Historiography of the Natural Sciences in Russia: 1700–1850]*. Moscow: Academy of Sciences.

CHAPTER 5

Shifting Grounds (1929–1947)

The rebranding of the Bellingshausen expedition in 1949 was preceded by a period of almost two decades during which the accepted headline—discovery of the first Antarctic land in January 1821—began to come under pressure from conceptual, factual, and political developments. Vast and complex events, including the Five Year Plans, the attempted Sovietization of intellectual life and institutions, famine, terror, purges, the Great Patriotic War, and the incipient Cold War, also racked the Soviet Union during those years, affecting even so small a thing as the reputation of the expedition. One result of those upheavals was that the exclusion of Soviet intellectuals from international scientific bodies, a ban which had been imposed on Russia as well as Germany by the victorious allies in 1919, continued for longer than it might otherwise have done. The turbulence of Soviet culture was reflected in the epithets of the Geographical Society: from 'Russian' in 1919 it became 'Russian State' and then 'State Russian' under Shokal'skij, then plain 'State' from 1931 under the geneticist Nikolaj Ivanovich Vavilov (1887–1943), and finally 'All-Union' (a euphemism for 'Soviet') in 1940, immediately after the death of Shokal'skij and shortly before Vavilov was arrested (for opposing Lysenkoism) and replaced by Berg.[1,2]

The most important new cultural factor for the Bellingshausen historiography was the notion that major, quasi-permanent ice features along the coasts of Antarctica should be treated as parts of the mainland. This was not something discussed and decided at a particular international scientific meeting. It simply became part of the way Antarctica was mapped and described. It can be detected as early as 1905 in a map drawn by John George Bartholomew (Mill 1905: enclosure), and appeared regularly on maps during the interwar period (Bird 1930, 1935; Houben 1934). The ease of its acceptance doubtless owed something to the experience of explorers who established their hutted camps on such 'terrain', most often on the Ross Ice Shelf. By that twentieth-century criterion, as popularly applied, catching sight of a major coastal ice feature, such as an ice shelf, ice barrier, or ice tongue, would amount to discovering Antarctica just as much as sighting part of its rocky coastline. That criterion has been accepted in this book, because it is evidently the one applied in Russian accounts of the expedition since 1949.

Two further general points about this transitional period should be mentioned. First, published Soviet assessments of the Bellingshausen expedition went largely unchanged. Whether in 1929 or in 1947, the expedition had discovered the mainland of Antarctica in January 1821, not 1820 (Berg 1929: 47; Gvozdetskij 1947: 89). From 1936, however, commentators such as Vvedenskij and Gvozdetskij had their work cut out to preserve Rabinovich's assessment of the expedition after its foundations, the mainland status of the Alexander I Coast and the island status of Trinity Land, had been transposed and therefore undermined. Shokal'skij had made it clear, in June 1937, that the findings of the British Graham Land Expedition (BGLE) were beyond all doubt: 'they found that the actual situation was completely different' [1] (Shokal'skij 1937: 666), but few of his colleagues ever understood this and, as we shall see in the case of Vvedenskij, Shokal'skij's influence diminished after his death in 1940.

There was one small change. Reviving the usage first tried out by Krusenstern almost 100 years earlier, and then copied elsewhere in Europe

---

[1] The original texts of translated quotations, indicated by numbers in square brackets, are provided at the end of each chapter. For the system of romanization used, the rendering of dates, and a few other matters, please consult the Apparatus at the end of the book.

[2] Vavilov died in prison from tuberculosis.

(Fricker 1898: 41), Soviet authors began to refer to the expedition's second discovery in January 1821 as Alexander I Land, replacing Bellingshausen's 'coast' [*bereg*] with the more substantial and prestigious 'Land' [*Zemlya*] (Berg 1929: 89). That meant ignoring Bellingshausen's categorization of it as an island, at the end of *Two Seasons*, in order to rank it alongside Enderby Land, Victoria Land, Coats Land, etc., place names which had been chosen to suit stretches of mainland coast, in case they should turn out to be so.

Another persistent feature of this period was the neglect by all parties of a potentially important piece of evidence. In 1821, in a letter to Shestakov first published in 1918, Lazarev had described a dramatic encounter with ice in January 1820:

> On 16 January [O.S.] we reached latitude 69°23′S, where we met main ice of extraordinary height. It was a fine evening, and looking out from the crosstrees it stretched just as far as our gaze could reach, but we had not long to enjoy that amazing spectacle, because the murk quickly came over again and the usual snow set in. That was in longitude 2°35′ W of Greenwich. From there we held our course east, pushing south at every opportunity, but we always met an ice floe main before we reached 70° [2]. (Lazarev″ 1821: fo. 2)

If or when someone noticed that source, and if by general consensus such an encounter with main ice amounted to a sighting of the mainland, such a Russian achievement might arguably have occurred two days before the first British sighting of Trinity Land on 30 January 1820 (R. Gould 1925: 222; Campbell 2000: 131).

INCREASING RIVALRY

Turning to publications and other events which can be treated chronologically, during the 1930s some leaders of Soviet science tried hard to overcome their exclusion from the International Research Council. Their efforts, supported by many Western scientists but frustrated by the purges and the war, were not finally successful until the 1950s. Meanwhile the Soviet Union's extensive participation in the Second International Polar Year of 1932–1933 (IPY2) was an important milestone. In February 1929 a party of Norwegian whalers had landed on Peter I I. and claimed it for their country, and two years later Norway asserted its sovereignty over the island by royal proclamation (Anon. 1939; Shokal'skij 1939).[3] With explicit reference to recent Australian, Norwegian, and American exploration, plans were drawn up for a party of Soviet scientists to carry out research for IPY2 in the Antarctic on board the *Aleut* whaling flotilla, normally based at Kamchatka in the Soviet Far East. The venture might have brought about the first official Russian presence in the Southern Ocean since Bellingshausen, but it came to nothing when the South African government declined a request for logistical facilities (Belov 1966a).[4]

In the late 1930s, however, it was not Soviet but American and British commentators who revived the question of priority in Antarctica. British governments had done much to politicize the continent since before World War I (Auburn 1982: 6–7), and in 1928 a survey identified the British Empire and France as the only known territorial claimants, with Britain, Australia, and New Zealand together aspiring to 'nearly all of the Antarctic Continent' (Miller 1928: 248).[5] Meanwhile British and American authors had already mooted the priority of 'their men' in sighting Trinity Land and Palmer Land, still widely taken to be islands (Balch 1909, 1925; R. Gould 1925). That debate was resumed after Colonel Lawrence Martin (1880–1955) presented a paper claiming American priority to the 15th International Geographical Congress, held at Amsterdam in July 1938 (L. Martin 1938, 1940; Hobbs 1939a, b; A.R.H. 1939; R. Gould 1941).[6] No Soviet delegates were present, but a few months later the Geographical Society created a 90-strong section for the history of geography (Ovlashhenko 2013: 80). Two members of its bureau, Berg and Aleksandr Ignat'evich Andreev (1887–1959), were to play important roles in rebranding the Bellingshausen expedition in 1949.

---

[3] The timing of these articles suggests that they were responding, not to the original Norwegian action, but to the mini-crisis triggered by the German expedition in 1939, as too did the Soviet government (below).

[4] Another factor, not considered by Belov, may have been *Aleut*'s lack of the size and strength required to endure several months of service in the Southern Ocean. She was less than half the tonnage of a typical Norwegian pelagic whale factory of the period. Meanwhile the proposed Soviet expedition had been duly noted by the British government (UK National Archives, CO 78/208/3).

[5] At that point Australia and New Zealand had been dominions within the British Empire for less than 30 years, and their Antarctic policies would be heavily influenced by Britain for decades to come.

[6] Martin's intervention may have been provoked by an early sight of Debenham's innovative thoughts about Bellingshausen, a draft of which passed through Washington in 1937 (below).

The historical section of the Soviet Geographical Society was formally constituted on 29 March 1939 in Leningrad, where both the Naval Archives and, until 1963, the archives of the Academy of Sciences were located. The chairman was the vice-president of the Society, Ignatij Yulianovich Krachkovskij (1883–1951) and the secretary was Andreev. (A separate Moscow historical section was established after the war and held its first meeting in October 1947—below.) On 24 March 1940 the section discussed a report on the 1938 Geographical Congress and, presumably, related matters from the cartographer Pavel Petrovich Pomerantsev (1899–1979) (Anon. 1940). No details of that important meeting have survived, but to judge from Vvedenskij's work (below), Pomerantsev may have confused the British discoveries which Shokal'skij had carefully explained in 1937. (Although Shokal'skij was still the honorary president of the Society he must have been absent, since he was ill for several days before dying of a heart attack, aged 83, two days later. The Society's name was changed to 'All-Union', in other words 'Soviet', within days.)

DEBENHAM'S INPUT

Meanwhile by 1936, while annotating his translation of Bellingshausen's book, Debenham had surmised that the expedition might possibly have sighted 'the land ice which, everywhere along this coast, marks the edge of the continent' in January 1820. A few pages later he concluded that they definitely did so about three weeks later (Bellingshausen 1945 1: 117, 128).[7] As mentioned in Chap. 2, Debenham drew the attention of the British historian Mill to Bellingshausen's 'virtual discovery of Rag[n]hild Land as far as seeing is concerned, even tho' not believing' (Debenham 1937). Although he did not tie that particular remark to an expedition date, the reference to Ragnhild Land, which lies to the east of Crown Princess Martha Land and therefore further along Bellingshausen's course, confirms that he already found the sighting of coastal ice on 17 February 1820 more convincing than the approach on 28 January 1820.[8]

---

[7] The fact that Debenham made no reference to Bransfield, in those notes to his translation, suggests that he reached this conclusion about Bellingshausen's work in 1820 before the first report from the BGLE, at the end of 1936, had put Bransfield into contention, and of course well before the debate about priority was revived at the 1938 Geographical Congress.

[8] Debenham was using the original Norwegian nomenclature in which, reading from west to east, Kronprinsesse Märtha Land, Prinsesse Ragnhild Land, and Dronning (Queen) Maud

Two years later, soon enough for his statement to come to the notice of the Soviet Geographical Society's historical section in 1940 (above), he publicly specified the later date (Debenham 1939).

Debenham's translation had been the work of many hands. A version completed by Edward Bullough in 1924 was 'examined and revised by other experts, of whom the chief was a Russian student of the editor, N. Volkov, a descendant of Lazarev, who rendered valuable service'.[9] Meanwhile Debenham kept up a 'most courteous' correspondence with the Russian Admiralty in a frustrated search for accounts by other members of the expedition (Bellingshausen 1945: 1, vii–viii). The publication history also disseminated Debenham's ideas, because after the book was reluctantly rejected by Cambridge University Press in January 1937, as unlikely to repay its costs, it was offered unsuccessfully to the American Geographical Society, whose members were informally hosted for many years in the Maps Division of the Library of Congress by its head, their former president Lawrence Martin.

As we have seen however, Debenham published his conclusion, that Bellingshausen had sighted the Antarctic mainland in February 1820, as early as 1939, six years before the translation appeared. Furthermore his statement was immediately aired at the Pacific Science Congress, which convened in California a few months later. Bjarne Aagaard (1873–1956) did not believe it was possible as yet to determine who discovered Antarctica, but he also accepted from Debenham that, unless Trinity Land and Graham Land were continental (which he considered still unproven), then 'Antarctica was discovered by von Bellingshausen' (Aagaard 1940: 691).

A Soviet Note

Given that several people in or from different countries had seen the relevant part of Debenham's MS before 1937, then even before his views about Bellingshausen appeared in print, there were various routes by which word of them could have reached the Soviet Ministry of Foreign Affairs in

Land were flanking coastal regions of equal status (Isachsen 1931: 346). The last name soon came to be applied to the whole region.

[9] Nicholas Volkov studied Modern and Medieval Languages at Cambridge from 1928 to 1931 and Debenham was probably his moral tutor. The meagre evidence suggests that he was a White Russian emigré who arrived in Britain in the 1920s and settled in London after going down from Cambridge.

time, perhaps, to influence its actions. On 17 December 1938 a German Antarctic expedition sailed from Hamburg. It was bound for Queen Maud Land in East Antarctica, the region whose waters Norwegians had explored and in which Norway had announced a national interest, but without declaring an actual claim. An Antarctic mini-crisis was triggered when news of the German move reached the Norwegian Foreign Ministry, prematurely from the German point of view, a few days later. On 14 January 1939, a day before the German expedition sighted the coast, the Norwegian parliament proclaimed sovereignty over the region (Lüdecke and Summerhayes 2012: 49–51). Then on 27 January 1939 the Soviet Foreign Ministry responded to the Norwegian claim with a Note stating that:

> ...the Soviet government finds it necessary to reserve its position on the question of sovereignty over this territory, which was discovered by the Russian navigator Bellingshausen during his expedition in the years 1819–21. (Bush 1982: 152)[10]

Since Bellingshausen sailed through the longitudes of Queen Maud Land between 20 January and 6 March 1820, his discovery of 'territory', as claimed in the Note, would have to have occurred a year earlier than anyone except for Debenham had previously supposed, and might imply an absolute priority over all other sightings. But although Lazarev's significant description of 28 January 1820 was now in print, no Soviet commentator had yet paid any attention to it. Nor had any Soviet scholar referred to Queen Maud Land in print with reference to Bellingshausen.[11]

The official but apparently still secret Soviet position in January 1939 is therefore a mystery. So far only Debenham, in as yet unpublished footnotes, had suggested that Bellingshausen had sighted an ice coast of Antarctica, perhaps not in January, but certainly in February 1820.[12] The

---

[10] The author has been unable to locate the Russian text of this Note, or even to discover when it was made public in the Soviet Union. No hint of it was given by likely commentators (Anon. 1939; Afonin 1948). Nevertheless Berg was able to refer to it in his 1949 address to the Geographical Society (Chap. 6).

[11] Writing before the Norwegian claim, Shokal'skij came close to doing so when he mentioned that Bellingshausen had been about 150 nm west of Enderby Land in late February 1820, which would place the Russian squadron somewhere north of Queen Maud Land (Shokal'skij 1928: 185). (The point was first made by Fricker [1898: 40].)

[12] We can surely discount the notion that the January 1939 issue of *Polar Record*, in which Debenham first published his idea, was rushed to the Soviet Foreign Ministry in time to influence events.

intriguing possibility, that the Soviet Foreign Ministry had somehow got wind of Debenham's ideas by late 1938, is consistent with the significance later attributed to Debenham by Soviet authors.

VVEDENSKIJ

The work of N.V. Vvedenskij illustrates the challenge to traditional assessments of the Bellingshausen expedition that was posed by the events described in the last few paragraphs.[13] In 1939 he published an article about the expedition in which he twice stated that both Peter I and Alexander I were islands (Vvedenskij 1939: 46–47). A year later, in the first ever book about the expedition by a non-participant, he did so again, citing exploration by Charcot in 1908 and Wilkins in 1928, but not the BGLE; Wilkins, he mentioned, had also established that Graham Land was an island (Vvedenskij 1940: 88, 76).[14] So far his views were partly out-of-date but in line with the best available information before 1936. As we have seen his take on Alexander I. agreed with Bellingshausen and other early sources, but despite considerable input into his book from Shokal'skij he appears to have been unaware, so far, of the discoveries of the BGLE.

A few months later, however, Vvedenskij reversed his position completely: 'On 17 January 1821 [O.S. & s.d.] the Alexander I Coast was discovered ... Bellingshausen and his companions had sighted the true coast of the Southern Continent—Antarctica' [3] (Vvedenskij 1941: 119). Amongst the 'latest research' which he cited in support of this revival of the outdated Rabinovich position was the BGLE, which, he claimed, had made it 'absolutely certain that the Alexander I Coast constitutes a north-pointing peninsula of the continent of Antarctica' [4] (the same). Unfortunately that was the exact opposite of what was thought to have been discovered by a BGLE sledging party, led by Alfred Stephenson (1908–1999), which travelled 370 km south on the sea ice of George VI Sound, between what is now known as Alexander Island and the mainland. Their exploration was reported in detail. However the expedition

---

[13] The author has tentatively identified this person as Nikolaj Vladimirovich Vvedenskij, a Civil War veteran and at this point Associate Professor of the Faculty of Geography, Leningrad State Pedagogical Institute (his *alma mater*), who returned to military service after the German invasion in June 1941.

[14] Names for sections and subsections of the Antarctic Peninsula, the northern tip of which had been named 'Trinity Land' by Bransfield in 1820, remained in flux until after the Antarctic Treaty was signed in 1959.

leader John Rymill (1905–1968) probably misled non-English readers by providing a map of the area in his book which showed Alexander I. without defining its western and southern limits, which had not been visited on foot. To add to the confusion the map used the legend 'Alexander I Land' with a rather convoluted explanation, whereas 'Alexander I Island' was used throughout the text (Rymill 1938a, b: 161–166, 191–194). Vvedenskij never directly quoted Rymill (or Shokal'skij), but he published a crude and unattributed copy of Rymill's map in his article (1941: 120).[15]

One has to wonder what changed for Vvedenskij between his 1940 book and his second article. Perhaps he was influenced in some way by the discussion of Pomerantsev's report, of which we have no details. Perhaps he simply emerged from under the wing of his late mentor Shokal'skij to discover and revive the Rabinovich position. But perhaps also he caught some intimation that his government (not yet at war with Germany) was starting to pay more attention to its historic interests in Antarctica. The notion that those interests might date from 1820 rather than 1821, however, no more occurred to Vvedenskij than to any other Soviet commentator before 1949.

* * *

During the Great Patriotic War the demand for morale-boosting literature did not increase the frequency of articles about the expedition. The numbers of items about or substantially referring to the expedition immediately before the German invasion were: 1939–5; 1940–4; 1941–8. After the invasion they were: 1941–2; 1942–0; 1943–1; 1944–5; 1945–0 (sources: Ovlashhenko 2013 and the author's files). A more significant effect of the war was the dispersal of intellectuals such as the poet and historian Sergej Markov (Chap. 10), invalided out of his regiment, to rear areas like Kazan. They discovered that provincial libraries and archives held a neglected wealth of materials for maritime history (S. Markov 1944).

In January 1947 the Soviet Union began whaling in the Antarctic with an ex-German, ex-British whaling flotilla seized in reparations (Bulkeley 2011). Later that year, while working on the second edition of *Two Seasons*, Admiral Shvede reviewed Debenham's translation in largely positive terms, but without mentioning the latter's views about what might have been

---

[15] The 'peninsula' misreading of Rymill's map of 'Alexander I Land' persisted; even a historian of real quality could repeat it eight years later (Andreev 1949a: 77).

seen in 1820 (Shvede 1947). In November an article in *Zvezda* about American activity in the Arctic and Antarctic said nothing about any Soviet interests in the latter (Golant 1947).

ORIGINAL TEXTS OF QUOTATIONS

[1] … нашла действительное положение в совершенно ином виде…

[2] *16-го генваря* [ст.ст.] *достигли мы широты 69°23′S гдѣ встретили матерой ледъ чрезвычайной высоты и въ прекрасный тогда вечеръ, смотря на саленгу, простирался оный такъ далеко какъ могло только достигать зрѣніе, но удивительнымъ симъ зрѣлищемъ наслаждались мы недолго, ибо вскорѣ опятъ напасмурило и пошелъ по обыкновенію снѣгъ. Это было въ долготѣ 2.°35.′W^{ъй} отъ Гринвича.—отъ Сюда продолжали мы путь свой къ Осту покушаясь при всякой возможности къ Зюйду, но всѣгда встрѣчали льдяной материкъ не доходя 70°.*

[3] 17 января 1821 года [ст.ст.], был открыт Берег Александра I … Беллинсгаузен и его спутники увидели подлинный берег Южного материка—Антарктиды.

[4] …доподлинно известно, что Берег Александра I является выдвинутым к северу полуостровом материка Антарктиды …

REFERENCES[16]

A.R.H. 1939. On Some Misrepresentations of Antarctic History. *Geographical Journal* 94: 309–330.

Aagaard, Bjarne. 1940. *Who Discovered Antarctica?* Proceedings of the Sixth Pacific Science Congress of the Pacific Science Association, held at the University of California, Berkeley, Stanford University, and San Francisco, July 24th–August 12th 1939, vol. 2, 675–707. Berkeley, CA: University of California Press.

Afonin, M. 1948. Bor'ba vokrug Antarktiki [The Dispute About the Antarctic]. *Izvestiya*, March 19: 3.

Andreev, A.I., ed. 1949a. Russkie v Antarktike v 1819–1821 gg. [Russians in the Antarctic 1819–1821]. *Izvestiya Akademii Nauk: seriya istorii i filosofii* 6 (1): 77–78.

---

[16]Key to archival references: SARN = State Archives of the Russian Navy, F = Fond, S = Series [*Opis'*], P = Piece [*Delo*], fo./fos = folio/s [*list/y*], v = verso [*oborotnoe*]. (The latter is necessary because verso pages were usually unnumbered.) Dates on manuscript documents are given in Universal Time (CE).

Anon. 1939. Razdel Antarktiki [Partition of Antarctica]. *Problemy Arktiki* 3 (5): 107.

———. 1940. Otdelenie istorii geograficheskikh znanii [Section for the History of Geographical Knowledge]. *Izvestiya vsesoyuznogo geograficheskogo obshhestva* 72 (2): 845–846.

Auburn, F.M. 1982. *Antarctic Law and Politics.* London: Hurst.

Balch, Edwin Swift. 1909. Stonington Antarctic Explorers. *Bulletin of the American Geographical Society* 41 (8): 473–492.

———. 1925. The First Sighting of Western Antarctica. *Geographical Review* 15 (4): 650–653.

Bellingshausen, Captain. 1945. *The Voyage of Captain Bellingshausen to the Antarctic Seas, 1819–1821*, ed. Frank Debenham, 2 vols. London: Hakluyt Society.

Belov, M.I. 1966a. Proekt pervoj sovetskoj èkspeditsii v Antarktidu [A Proposal for the First Soviet Expedition to Antarctica]. *Byulleten' Sovetskoj Antarkticheskoj Èkspeditsii* 58: 64–67.

Berg, L.S. 1929. *Ocherk istorii russkoj geograficheskoj nauki (vplot' do 1923 g.) [Towards a History of Russian Geography Down to 1923].* Leningrad: Academy of Sciences.

Bird, Richard Evelyn. 1930. *Little America.* New York: Putnam's.

———. 1935. *Discovery.* New York: Putnam's.

Bulkeley, Rip. 2011. Cold War Whaling: Bellingshausen and the *Slava* Flotilla. *Polar Record* 47 (2): 135–155.

Bush, W.M. 1982. *Antarctica and International Law.* Vol. 3. Dobbs Ferry, NY: Oceana.

Campbell, R.J. 2000. *The Discovery of the South Shetland Islands.* London: Hakluyt Society.

Debenham, F. 1937. Letter to H.R. Mill, 13 February 1937. Cambridge: Archives of the Scott Polar Research Institute: MS/100/23/48.

———. 1939. Foreword. *Polar Record* 3 (17): 1–2.

Fricker, Karl. 1898. *Antarktis [The Antarctic].* Berlin: Schaff & Grund.

Golant, V. 1947. Polyarnaya likhoradka v Amerike [Pole Fever in America]. *Zvezda* 11: 173–179.

Gould, R.T. 1925. The First Sighting of the Antarctic Continent. *Geographical Journal* 65: 220–225.

———. 1941. The Charting of the South Shetlands, 1819–28. *The Mariner's Mirror* 27: 206–242.

Gvozdetskij, N.A. 1947. Pervoe morskoe puteshestvie rossiyan vokrug sveta [The First Russian Circumnavigation]. *Priroda* 1: 85–89.

Hobbs, William Herbert. 1939a. The Discoveries of Antarctica within the American Sector, as Revealed by Maps and Documents. *Transactions of the American Philosophical Society* 31 (1): 1–71.

————. 1939b. The Discovery of Antarctica: A Reply to Professor R.N. Rudmose Brown. *Science* 89: 580–582.

Houben, H.H. 1934. *Sturm auf den Südpol [Assault on the South Pole].* Berlin: Ullstein.

Isachsen, Gunnar. 1931. Norske undersøkelser ved Sydpollandet 1929–1931 [Norwegian Explorers in the Antarctic 1929–1931]. *Norsk geografisk tidsskrift* 3 (5–8): 345–351.

Lazarev″, M.P. 1821. Pis'mo Mikhaila Petrovicha Lazareva k″ Aleksyeyu Antonovichu Shestakovu, 24 sentyabrya 1821 [A letter from Mikhail Petrovich Lazarev to Aleksyej Antonovich Shestakov, September 24, 1821 [O.S]]. St Petersburg: SARN—F–315, S–1, P–775, fos 1–6v.

Lüdecke, Cornelia, and Colin Summerhayes. 2012. *The Third Reich in Antarctica.* Norwich: Erskine.

Markov, S.N. 1944. Klady «Kolumbov Rossijskikh»: dokumenty o russkoj morskoj slave [Treasures of the 'Russian Columbuses': Documents on Russian Naval Glory]. *Morskoj sbornik* 8–9: 76–81; 10: 81–88.

Martin, Lawrence. 1938. An American Discovered Antarctica. In *Comptes rendus du Congrés international de géographie, Amsterdam, 1938,* pt 2 (4): 215–218. Leiden: International Geographical Union.

————. 1940. Antarctica Discovered by a Connecticut Yankee, Captain Nathaniel Brown Palmer. *Geographical Review* 30 (4): 529–552.

Mill, Hugh Robert. 1905. *The Siege of the South Pole.* London: Alston Rivers.

Miller, D.H. 1928. Political Rights in Polar Regions. In *Problems of Polar Research,* ed. W.L.G. Joerg. New York: American Geographical Society.

Ovlashhenko, Aleksandr. 2013. *Materik l'da: pervaya russkaya antarkticheskaya èkspeditsiya i eyo otrazhenie v sovetskoj istoriografii (1920-e–1940-e gody) [The Continent of Ice: The First Russian Antarctic Expedition and its Footprint in Soviet Historiography (1920s to 1940s)].* Saarbrücken: Palmarium.

Rymill, J.R. 1938a. British Graham Land Expedition, 1934–37. *Geographical Journal* 91 (4): 297–312; 424–438.

————. 1938b. *Southern Lights: the Official Account of the British Graham Land Expedition, 1934–1937.* London: Chatto and Windus.

Shokal'skij, Yu.M. 1928. Stoletie so vremeni otpravleniya Russkoj antarkticheskoj èkspeditsij pod komandoyu F. Bellingsgauzena i M. Lazareva 4 iyulya 1819 g. iz Kronshtadta [Centenary of the Departure from Kronstadt of the Russian Antarctic Expedition, Commanded by F. Bellingshausen and M. Lazarev, on July 4 1819 [O.S.]]. *Izvestiya gosudarstvennogo russkogo geograficheskogo obshhestva* 60 (2): 176–212.

————. 1937. Novosti ob Antarktide [News about Antarctica]. *Izvestiya gosudarstvennogo geograficheskogo obshhestva* 69 (4): 666–667.

————. 1939. Ostrov Petra I [Peter I Island]. *Izvestiya gosudarstvennogo geograficheskogo obshhestva* 71 (9): 1393–1396.

Shvede, E.E. 1947. Puteshestvie kapitana Bellinsgauzena v antarkticheskie morya 1819–1821 [The Voyage of Captain Bellingshausen to the Antarctic Seas 1819–1821]. *Izvestiya vsesoyuznogo geograficheskogo obshhestva* 79 (3): 357–358.

Vvedenskij, N. 1939. Russkaya krugosvetnaya antarkticheskaya ėkspeditsiya 1819–1821 gg. [The Antarctic Circumnavigation by a Russian Expedition, 1819–1821]. *Geografiya v shkole* 3: 43–47.

———. 1940. *V poiskakh yuzhnogo materika [The Quest for a Southern Continent]*. Leningrad–Moscow: GlavSevMorPut.

———. 1941. K voprosu o russkikh otkrytiyakh v Antarktike v 1819–1821 godakh, v svete novejshikh geograficheskikh issledovanij [On the Question of Russian Discoveries in the Antarctic in 1819–1821, in the Light of Recent Geographical Research]. *Izvestiya vsesoyuznogo geograficheskogo obshhestva* 73 (1): 118–122.

CHAPTER 6

The Fix (1948–1949)

1948

In public, the Soviet view of the Bellingshausen expedition continued unchanged in 1948, with no sign of the backdating of Bellingshausen's achievement from 1821 to 1820 which was hinted at by the unpublished Note to Norway in 1939 (Magidovich 1948: 33; Kasimenko 1948: 18). An official version of what had become almost the standard (Rabinovich) assessment featured in the Soviet Navy's *Draft of Antarctic Sailing Directions*, a manual which happened to coincide with the American condominium proposal in August (below). According to a 'General Outline' contributed by Captain V.I. Vorob'ev, the Bellingshausen expedition discovered Antarctica in January 1821 when they sighted Alexander I Land— 'the northern coast of the continent' [1]. (The phrasing neatly implied that anything further north, such as the Peninsula, was not part of the continent.) As for 1820, they had been prevented from seeing the coast by bad weather (V. Vorob'ev 1948: 3, 7). Apparently the Soviet Hydrographic Department were either unconvinced by Debenham's suggestion of an 1820 sighting or unaware of it. No coastal information about the mainland or in-shore islands, apart from the Antarctic Peninsula, was included, although the Department had consulted foreign sources filled with such

© The Author(s), under exclusive license to Springer Nature Switzerland AG 2021
R. Bulkeley, *The Historiography of the First Russian Antarctic Expedition, 1819–21*,
https://doi.org/10.1007/978-3-030-59546-3_6

information from the discoveries of all nations including Russia.[1] In particular, nothing was provided for Peter I I. or Alexander I Is/Land. The book was probably intended to meet the needs of the *Slava* whaling flotilla which, after employing Norwegian experts for two seasons, was about to head south without them for the first time. It was replaced with a complete *Antarctic Sailing Directions* in 1961.[2]

Meanwhile tension was growing. Stalin had already been incensed by British obstruction over the transfer of an ex-German whaling flotilla as reparations (Bulkeley 2011), and matters came to a head when the US State Department announced in August 1948 that they had approached their allies with a proposal for a Western condominium over Antarctica (US State Department 1948). One purpose of such a regime was to exclude the Soviet Union from the region. In January 1949 the Central Committee of the Soviet Communist Party decided on a two-stage response, recalled a year later, during the preparation of the June 1950 memorandum (Chap. 2), by Foreign Minister Andrej Yanuar'evich Vyshinskij (1883–1954):

> In connection with the Central Committee decision of 29 January 1949, it was considered expedient to indicate, at first unofficially, that the Soviet Union had interests regarding the question of a regime in Antarctica, and then later, depending on how the situation developed, to give such an indication officially as well [2]. (Vyshinskij 1950)

The Central Committee decision reached in January 1949 must have been the culmination of a process of deliberation, partly because such bodies rarely act without seeing a prepared dossier, but also because the nominally unofficial Geographical Society meeting took place only twelve days later, and would have taken time to prepare. There are some scraps of evidence for this. It was later reported that a Soviet delegate to the International Geological Congress, meeting in London between 25 August and 1 September 1948, had raised the question of Bellingshausen's priority (L. Gould 1960: 77). Although it cannot be verified from the

---

[1] The careful acknowledgements of Western sources in the *Draft* show that, so far at least, the ideology of cultural self-sufficiency (below) could safely be ignored in a work not destined for the general public.

[2] The original texts of translated quotations, indicated by numbers in square brackets, are provided at the end of each chapter. For the system of romanization used, the rendering of dates, and a few other matters, please consult the Apparatus at the end of the book.

Proceedings the story may have been true; for one thing three of the nine Soviet delegates present were diplomats from the London Embassy. The head of delegation, geologist and geophysicist Vladimir Vladimirovich Belousov (1907–1990), was also likely to take an interest in Antarctic matters. But without any details about what was actually said, it is impossible to know whether the unidentified Soviet delegate was suggesting an 1820 sighting or merely repeating the Rabinovich position, that Russians had discovered Antarctica in 1821. Another document which is consistent with preparations being made in 1948 is the painting '*Vostok* and *Mirnyj* off the coast of Antarctica' [3], completed that year by the artist M.M. Semyonov from sail-plan drawings which he prepared for Shvede's second edition of *Two Seasons* (below).[3] Despite his cosmetic replacement of the actual weather and ice conditions with hazy sunshine and a rocky shore, Semyonov may only have intended to represent the ships off Peter I I. or the Alexander I Coast in 1821.[4] But the date of the work is at least suggestive.[5] Meanwhile in Britain the second edition of the *Antarctic Pilot* accepted Debenham's finding, first announced in 1939, that Bellingshausen had sighted Antarctica in February 1820, about three weeks after Bransfield (Hydrographic Department 1948: 294).

As of January 1949, Soviet historians of geography such as Berg and Shvede were giving nothing away at a mid-January symposium on the history of Russian science. The only hint of what was about to happen came from Dmitrij Mikhajlovich Lebedev (1892–1978):

> The role of the geographer-historian comprises … For example, we need to give concrete examples …, 5) of how the self-sufficiency and primacy of Russian geographical knowledge have been manifested [4]. (D. Lebedev 1949: 868)

[3] The painting is now in the Naval Museum at St Petersburg which, in a sign of modern times, recently decided to expand the title by adding the phrase 'which was discovered by them'.

[4] A strikingly unrealistic element in the picture is its depiction of both ships with all sail set. That was very unlikely, because *Vostok* usually had to shorten sail to allow the much slower *Mirnyj* to keep up. Furthermore full sail would never have been carried so close to an unknown ice front in waters which, as the artist shows, were fraught with ice hazards.

[5] There is a fine Soviet-era pen-and-ink drawing, depicting the ships in front of a large tabular iceberg and now in the archives of the Scott Polar Research Institute at Cambridge, which may also date from this important conjuncture. Unfortunately however it is unsigned, no accession details have been kept, and the back has been covered over (Bulkeley 2014a: Fig. 6).

Berg

On the morning of the Geographical Society meeting in February (Fig. 6.1) the only solid evidence for the new version of Bellingshausen's discovery was Lazarev's letter (Chap. 7), published three decades earlier in a naval journal. Soon after the press reports of the following day at least six more versions or digests of the lecture by the Society's president, Lev

Fig. 6.1 Members of the Geographical Society arriving for their meeting in Demidov (now Grivtsov) Lane, Leningrad, on 10 February 1949 From (Kublitskij 1949b)—artist unknown

Berg, appeared in popular or academic periodicals, while other literature remained sparse for several months. Berg began and ended by discussing current political issues, in particular the US condominium proposal and the economic potential of Antarctica, from which the Soviet Union would be excluded by such a regime. He described the Bellingshausen expedition, listed its discoveries in the Antarctic and sub-Antarctic, and strongly urged that the Soviet Union be included in international negotiations. He said very little, however, about an absolute Russian priority in discovering the continent. He did pay special attention to 28 January 1820, when they had been 'near the Antarctic mainland' [5] (Anon. 1949), but like Debenham he made any achievement a matter of hindsight: '*Nowadays* it is clear that the hill[ock]y ice described by Bellingshausen as extending from east to west in fact comprised the margin of the Antarctic continent' [6] (Berg 1949a: 43—emphasis added; the sentence was repeated in a book for children—Berg 1950: 142). His wording closely reflected that of Debenham; for example his 'in fact' corresponded to Debenham's 'indeed' (Bellingshausen 1945 1: 117). He mentioned Lazarev's letter, but ignored its description of 28 January 1820. His overall conclusion was a cautious periphrasis, reminiscent of Shokal'skij: 'they were the first ... to discover *that* [the] *Antarctic*[a] *existed*' [7] (Berg 1949a: 40—emphasis added).[6]

Berg's paraphrase of Bellingshausen's description of 28 January 1820 departed from the original in two important respects. First, Bellingshausen had described the field as 'continuous ice [which] extended from East through South to West' [8] (Bellinsgauzen'' 1831 1: 172). Ignoring the 'South', Berg amended this with language which suggests to landsmen an ice front roughly aligned from east to west, whereas the original wording would normally be taken to mean three bearings taken from the ship, unless otherwise stated, rather than from one point on the ice to another.[7]

---

[6] One aspect of Berg's reserve towards the new assessment of the expedition was his tendency to use the name of the region, *Antarktika*, which only rarely denotes the continent, rather than the usual name for the continent, *Antarktida*. The most he ever conceded was 'Antarctic continent', but never with 'discovered' in front of it. There were always periphrases such as 'existed' here or 'a number of lands' (below).

[7] Debenham took the same, landsman's view before Berg, and Jones did so afterwards (Jones 1982: 89). But even if we stretch the language into a description of an ice front running first east, then south, and then west, that would have been either a three-sided box that threatened embayment (with a different orientation), or else, on a highly implausible reading, a large ice mass meeting the ocean on three sides. Neither reading suggests a simple linear ice front.

Bellingshausen was explaining that the expedition was at risk of becoming embayed, thus giving rise to his next order, which veered the ships to NWbW.[8] Second, Berg also adjusted the text by condensing the original 'continuous ice ... an ice field scattered with hillocks' [9] to 'hill[ock]y ice' [10] (Berg 1949a: 43).[9] The problem was that by 1949 the adjective *bugristyj* had come to mean simply 'hilly' to most Russians, whereas in referring to scattered hillocks as an additional feature of the ice field (besides its continuity), Bellingshausen had not gone so far as to call it hilly (for which in his day he might have used *kholmistoj*). With a few deft strokes what seems to have been an embayment of continuous (pack) ice scattered with either pressure domes or large fragments of icebergs, or both, became for his twentieth-century Russian audience a rectilinear, barrier-like expanse of hilly (even mountainous) ice. Thus from the outset the new assessment of the expedition was supported by tweaking the evidence. Berg also slightly misled his audience and readers of his much reprinted text by stating that the position reached on 28 January 1820 was Bellingshausen's furthest south of the whole voyage, another way to suggest that something important might fittingly have been seen that day. In fact it was only his furthest south so far; he went a little further five days later, and his absolute furthest south came during their second Antarctic season in 1821.

Members of the Society duly hailed the expedition as the first discoverers, the 'Columbuses', of Antarctica. But one incident highlighted the sharpness of the historiographic 'turn' that was being demanded of intellectuals. Stanislav Vikent'evich Kalesnik (1901–1977), a professor of geography and pro-rector of Leningrad University, took a firm grasp of the wrong end of the stick when he supported his 'Columbuses' remark by citing the expedition's discoveries in January 1821, the traditional assessment (Anon. 1949). By paying no attention to positions possibly reached by the expedition a year earlier, Kalesnik was undermining the

---

[8] The situation, and Bellingshausen's response to it, resembled an encounter with impassable pack ice at 60°25'20"S on 16 January 1820, twelve days earlier, which has never been hailed as the discovery of Antarctica (Bellinsgauzen" 1831: 1, 157). In 1853 Novosil'skij confirmed that embayment was indeed the issue on 28 January 1820: 'it was necessary to make our way out of that bay' [...надо было выйти изъ этой бухты...] (1853c: 29).

[9] The reader is referred to the discussion of Bellingshausen's word for 'hillocks', *bugry*, in Chap. 3.

whole purpose of the meeting, which was to promote the work of the expedition *in January 1820* as the first ever discovery of Antarctica and a glorious achievement by Russian seamen. But his divergence from the new line was not accidental; he repeated it in print a few months later (Kalesnik 1949a: 81).

CAUTION AND CONFUSION

Berg's caution was reflected in the resolution passed by the meeting, which said merely that 'they were the first to approach its shores' [11] (Anon. 1949). Nevertheless the lecture had combined the two requisites for absolute priority: an interpretation of main ice as the edge of the continent, and a date prior to Bransfield's sighting of Trinity Land.[10]

For at least a year after the Geographical Society meeting leading scholars, such as Andreev or Shvede, worded their endorsements of the new discovery claim with evident reserve. Explicitly or implicitly, they continued to rely on Debenham's view of 28 January 1820, rather than Lazarev's, and met the political demands of the Soviet state by simply turning Debenham's 'might have been' into a fact. Andreev, for example, supported the claim that the expedition 'reached the very edge of the immovable ice' by citing Debenham for their proximity to Princess Martha Land. After that he needed only to strengthen his language a little further to 'they were first to reach *solid land*' [12], a feat which, he explained, had occurred a year earlier than previously supposed (Andreev 1949b: 7–8— emphasis added; see also A. Grigor'ev and D. Lebedev 1949). Andreev gave no explanation or source for revising Debenham's miss into a hit. It was almost as if he wanted intelligent readers to draw their own inferences from such an arbitrary and unscholarly procedure by a reputable historian. Andreev republished Lazarev's letter from the 1918 text, but described it as confirming Bellingshausen's account rather than adding something to it (Andreev 1949c: 168).

[10] Three accounts of Bransfield's voyage had appeared in the first half of the twentieth century, namely (R. Gould 1925, 1941; and Anon. 1946b), of which the first two included contemporary charts and reconstructed charts of the voyage. All three gave the date of Bransfield's sighting and naming of Trinity Land, but the second only did so in its chart of the voyage. The third simply republished the account given in 1820 by the *Literary Gazette*.

Examples of the confusion that followed in 1949 are not far to seek. As its title underlines, an article in the *USSR Information Bulletin*, purporting to present the new Soviet case to the world, achieved the direct opposite by stating merely that 'they encountered masses of ice on 16 January [1820—O.S.]', whereas in January 1821 'it was ascertained ... that the coast had been sighted' (Orlov 1949: 297). In February, despite a recent political reprimand (below), *Zvezda* published an article by the Arctic historian Boris Genrikhovich Ostrovskij (1890–19??) which also took the officially outmoded 'discovery in 1821' position (Ostrovskij 1949a), perhaps because it focused on Novosil'skij's account, next to a version of Berg's article which was at least trying to make some sort of case for 1820. In the April issue of *Around the World* Georgij Ivanovich Kublitskij (1911–1989), a former geodesist and *Pravda* journalist who became a popular author of books on travel, geography, and general science for older children, also took the outdated line (Kublitskij 1949a: 9). In the same month, so did an article in an authoritative naval magazine, despite referring to press reports and quoting the resolution of the Geographical Society meeting (Chernous'ko 1949: 96–97).[11] And on 11 June, just two days before Andreev's anthology (above) was sent to press, a note by him upholding the old 1821 position was also sent to press in a journal of the Academy of Sciences: 'On 17 January 1821 [O.S. & s.d.] ... Russians were the first Europeans to reach the southern continent and discover its coast' [13] (Andreev 1949a: 77).

The August issue of *Priroda* (*Nature*) contained two articles which essentially took the 1821 position, although Kalesnik made what he could of the expedition's encounters with ice fields in 1820 by using the term 'fast ice' (*pripaj*), which usually implies the presence of land (Aleksandrov 1949; Kalesnik 1949a: 81). As for Berg, in the second edition of his history of Russian geographical discoveries, which went to press at the end of November, he blurred his version of the discovery still further. The sighting on 28 January 1820 was downgraded to 'very probably' continental, and Debenham's language was now applied only to the post-Bransfield probe of 17 February 1820, retrospectively discovered to have been 'just by the coast of the Antarctic continent'. Despite which the expedition had somehow gained for Russia 'the right of priority in discovering a number

---

[11] The magazine remedied its mistake in September (Morozov and Nikul'chenkov 1949: 56–57).

of lands in the Antarctic' [14] (Berg 1949b: 181, 187).[12] In short the cautious wording in his Leningrad lecture had been no accident. He died at the end of 1950, leaving his virtual adoption of Debenham's position as his final statement on the subject, and the non sequitur of his obfuscation about priority for readers to make of what they would.

SHVEDE

In his long-awaited and much needed second edition of *Two Seasons*, which appeared in November 1949, Admiral Shvede, a former student and lifelong admirer of Shokal'skij, consigned the obligatory political remarks, along the lines laid down at the Geographical Society meeting, to a separate foreword. He acknowledged the assistance of staff at the Naval Archives, who were busy preparing the first volume of the Lazarev *Dokumenty* (Chap. 8), with its extensive documentation of the expedition. Surviving registers at the Archives, however, suggest that he did not himself work with the originals of Bellingshausen's reports or Lazarev's letter (personal communication from E. Tammiksaar, 18 December 2015), which raises the possibility that he may have been provided only with copies of the edited transcripts that were being prepared for the Naval Press, and in which significant cuts were made.

When it came to the discovery question, Shvede made no use of Lazarev's letter to Shestakov despite quoting it elsewhere in the book. By comparison with the other two prominent historical treatments in 1949, by Berg and Andreev, he presented the most careful and for that reason the least consistent case for an 1820 discovery. His statement that 'Had it not been for poor conditions of visibility, then Bellingshausen and Lazarev might have been able to provide entirely accurate information about the geography of the Antarctic continent by as early as 16 January [O.S.]' [15] (Shvede 1949: 26) was a close paraphrase of Debenham's assessment (Bellingshausen 1945 1: 117), and just like Debenham it implied not that they discovered Antarctica that day, but that they were prevented from doing so. Without citing Lazarev's letter, which might have helped him considerably, Shvede struggled through to the obligatory conclusion that 'Russian seamen … were the first to discover Antarctica' [16] (Shvede 1949: 30), partly by distracting the reader's attention with later and more

---

[12] Almost certainly nothing more than a reference to the discoveries made in January 1821.

plausible sightings, irrelevant to the priority question, and partly by avoiding all mention of Edward Bransfield.

Shvede's 1949 edition of *Two Seasons* cut short passages from Bellingshausen's descriptions of New South Wales and Brazil, in both cases with explanatory footnotes (Bellinsgauzen 1949: 251, 332). Unlike other cuts to primary sources in this period, neither omission bears any sort of political interpretation.

THE CONDEMNATION OF ANDREEV

The revised evaluation of the Bellingshausen expedition was launched during a period known to historians as the second Stalinist cultural revolution, sometimes referred to as the *Zhdanovshhina* after its most notorious protagonist, when the Soviet state renewed its efforts to bring all aspects of thought in the Soviet Union in line with the values and aims of the Communist Party. People working in every cultural field were required to show their commitment to socialism, as currently defined by the Party, whether or not they were party members. In the late Stalinist period Soviet patriotism was seen as essential for such a commitment, because the socialism in question was under attack in its embodiment, the Soviet Union, supposedly the first fully socialist society on the planet. A crude but effective test for the necessary patriotism was that cultural work should be based on and uphold the cultural self-sufficiency not merely of the Soviet Union but also of pre-revolutionary Russia, one sign of that self-sufficiency being the priority of Russian cultural achievements over those of other societies. During the post-war years young Soviet citizens were presented with lists of Russian firsts in culture and technology, some (though by no means all) of which were historically implausible (V. Bolkhovitinov and Others 1950), and within which the Bellingshausen expedition duly took its place (V. Yakovlev 1953: 25).[13]

---

[13] This essay necessarily refers without much explanation to Soviet incarnations of abiding major or minor components of Russian culture, such as the cult of priority, components which predated the Bolshevik revolution, and which have survived the restoration of capitalism in 1991. One important feature, so much so that it has long pervaded most aspects of national life with the possible exception of religion, has been its exotropic character, its need to define, indeed to become, itself by struggling towards and against a cultural other, typically labelled with a geographical nickname such as the West, the East, or even the South. One has only to compare the history of Russia with that of China to see how Russian this exotropy, sometimes coloured with resentment or pessimism, has over centuries become.

In practice, the second cultural revolution proceeded mainly by denouncing and penalizing individuals and institutions whose work was deemed to be deficient in Soviet patriotism, and it opened in 1946 with just such an indictment of the literary magazines *Zvezda* and *Leningrad* from Politburo member Andrej Aleksandrovich Zhdanov (1896–1948) (Anon. 1946a; Zhdanov 1946). The notion of kowtowing to foreign thinkers (*rabolepie* or *nizkopoklontsvo*) was frequently deployed in such indictments (Zhdanov 1952: 19). The corresponding positive term was *partijnost'*, meaning commitment to the ideas and goals of the Communist Party in its role of vanguard in the global class struggle.

In 1947 the cultural revolution afflicted the distinguished historian of science, exploration, and the colonization of Siberia Aleksandr Andreev, who at that point was professor of ancillary historical disciplines at the Moscow Institute of Historical Archives.[14] Andreev had already clashed with the Soviet state in 1929, when he was indicted as a minor figure in the 'Academy case' and sent into internal exile, but was rehabilitated in 1935 (Tikhonov 2012). The first volume of a prestigious series of studies of the reign of Emperor Peter I, edited by Andreev, appeared in June 1947 (Andreev 1947a). But danger loomed when *Pravda* published an article on Soviet patriotism on 11 August 1947. In it Dmitrij Trofimovich Shepilov (1905–1995), a senior official in the Central Committee's agit-prop department, expounded the superior, because class-based, nature of Soviet patriotism, which meant that:

> The Soviet Union stands as a bastion of world civilization and progress. There can be no question, nowadays, of any civilization without the Russian language, without the learning and culture of the Soviet peoples. *Priority belongs to us.* Nowadays the light of renovation and renaissance shines out to the West and to the East from the land of socialism. The dialectic of history has transposed the roles of 'teachers' and 'students' [17]. (Shepilov 1947: 3—emphasis added; see also Baltijskij 1945)

And much more of the same. Significantly, Shepilov extended the privileged status of Russian culture retrospectively, citing a string of eighteenth- and nineteenth-century geniuses who would have made a greater

---

Another enduring theme, of some relevance to the subject of this essay, has been a tendency to regard belief as superior to knowledge.

[14] For a summary of Andreev's career, including his fall from grace in 1947, see Afanas'ev and Others (1996).

contribution to world civilization had it not been for the shackles of tsarism. History was not mentioned, but the implications of the call to root out every vestige of servility to Western culture from Soviet thought, every trace of such subversive values as objectivity, humanism, or universality, were clear.

In September the magazine *Military Thought* duly published a critical review of Andreev's Peter I volume by Aleksandr Aleksandrovich Strokov (1907–1987), which declared that:

> The chief deficiency in the contents of the collection consists in the fact that several articles make an open attempt to revive the long since condemned 'theory' which reduces all Peter I's creative activity to the transplantation of foreign models onto Russian soil [18]. (Strokov 1947—text from Tikhonov 2012: 121)

Strokov went on to condemn Andreev's article about Peter I's visit to England in 1698 (Andreev 1947b) in the ominous catchwords of the new orthodoxy: '...the article by the editor propagandizes kowtowing and fawning by Russian figures before the outside world. It offers nothing but harm to anyone reading the collection' [19] (the same: 122). Andreev and his colleagues managed to resist their critics for several months, but their academic fates were sealed by a second, even more hostile, attack on Andreev and another contributor, Sof'ia Aronovna Fejgina (1897–1983), from Grigorij Nikolaevich Anpilogov (1902–1987) (Anpilogov 1948).

On 6 October 1948 the editor of *Studies in History* and Stalin prize-winner Anna Mikhajlovna Pankratova (1897–1957) renewed the condemnation of Andreev and Fejgina at a senior level in the Academy of Sciences (Tikhonov 2013: 203). A few days later the academic council of the Institute of History held a three-day meeting to hear and discuss a report from its director, Boris Dmitrievich Grekov (1882–1953), on the political shortcomings of several historians (Anon. 1948a). The meeting was attended by 28 members of the Institute and six other comrades, unidentified. Andreev's chapter on Peter I in England was the second item on Grekov's list. Andreev had not only committed the grievous error of 'kowtowing to the West' [20]; he had also been reluctant to understand and acknowledge his mistake. Furthermore, he had given space in the collection to another scholar (Fejgina) who retailed the libellous opinions of foreign authors, including fascists. The Institute found fault with another of his recent articles (Andreev 1947c), and his work on the fifth volume of a *History of Russia* was also condemned; that book appears never to have been published.

The castigation of Andreev and others continued at a second meeting and in a major editorial published at the end of the year. The accusations were much the same for all those arraigned and were explicitly brought to the attention of all Soviet historians: lack of *partijnost'*, subservience to foreign sources and historians, lack of Soviet patriotism, and failure to uphold the self-sufficiency and superiority of Russian culture both before and after the October Revolution:

> In practice, the Institute has responded feebly to the Party's call to organize the struggle against kowtowing to the West, and has done nothing to unmask the mendacious claim that Russian culture and knowledge are not self-sufficient [21]. (Anon. 1948b: 5)

It was doubtless this 1948 campaign against objectivism and for Soviet patriotism in cultural work that spurred D. Lebedev, in January 1949, to summon his colleagues in the history of geography to demonstrate the 'the self-sufficiency and primacy of Russian geographical knowledge' (above).

Andreev was spared further public rebukes in 1949 because the ideological coordination of history moved on to a new and more intense phase, the anti-Semitic struggle against 'cosmopolitanism'. Meanwhile the outcomes for Andreev were serious, though milder than the penalties which might have been inflicted ten years earlier. The series of studies of Peter I was cancelled; his health deteriorated; and he lost his position at the Institute of Historical Archives and was obliged to return to Leningrad and find other work. He also lost both his editorship of the flagship title in Russian history, *Letters and Papers of the Emperor Peter the Great* (after seeing two volumes through the press—the first to appear since the formation of the Soviet Union in 1922), and the editorship of a textbook series published by the Archives Institute.[15]

Andreev's rate of publication fell noticeably after 1948. No political charges were ever levied against his 1949 Bellingshausen anthology (Andreev 1949c), but he was not retained as editor for the second edition,

---

[15] The editorship of the *Letters and Papers* passed to a colleague, B.B. Kafengauz, who had also been criticized, but less severely. Posthumously reinstated as a co-editor in the 1960s, today Andreev is sometimes credited with having edited the entire series, which began in the year of his birth and continues in the present century.

enlarged and furnished with a title better suited to the patriotic requirements of the day (Sementovskij 1951). The selection of his successor,
Vladimir Nikolaevich Sementovskij (1882–1969), at that point still a
somewhat undistinguished geographer from Kazan (Sementovskij and
N. Vorob'ev 1940), may have been prompted by a recent article in which
the latter had demonstrated his ideological soundness (Sementovskij 1950).

All in all it is hardly surprising that in 1948 and early 1949, as Andreev
was preparing the Bellingshausen anthology for press with a storm of
political criticism breaking over his head, he failed to rectify the inconsistency between his introduction and his note in the Academy's *Proceedings*
(above), as to whether Antarctica had been discovered in 1820 or 1821.
As he put it in a letter to one of his students, Russian exploration in the
late eighteenth and early nineteenth centuries was a 'more complicated'
[22] historical challenge than that of earlier times (Andreev 2012: 13).

To round out this sketch of the cultural circumstances in which the
reassessment of the Bellingshausen expedition was launched, in April 1948
the president of the Soviet Union of Writers, Aleksandr Aleksandrovich
Fadeev, ordered a new edition of Boris Pasternak's poems (25,000 copies)
to be pulped shortly before it was due to go on sale, and at the beginning
of March 1949, just three weeks after the Geographical Society meeting,
Bonifatij Mikhajlovich Kedrov (1903–1985), editor of the Soviet Union's
ideological flagship journal *Studies in Philosophy*, was dismissed on political
grounds. Soviet intellectuals could not lightly ignore the wishes of their
rulers in those times.

APPLYING THE *ZHDANOVSHHINA* TO BELLINGSHAUSEN

As we have seen, Soviet authorities sometimes objected to the conclusions
of historians after they were published. The much stronger claim made by
Konstantin Shteppa (1896–1958), that the Soviet historian was 'told in
advance what final results he must obtain in his research', has been more
often repeated than substantiated with example (Shteppa 1962: xi–xii).
But the sequence of the Central Committee's deliberations leading to its
decision in January 1949, the Geographical Society meeting two weeks
later, and then the use made of the Society's findings in the Soviet Note of
June 1950 (all above), does seem to be a case of people being 'told in
advance'.

For historians to fulfil their socialist duty, however, it would not be
enough to bring in a bald empirical finding that Russians happened to be

the first people to discover Antarctica. What was needed was a view of the circumstances suggesting that it was the right and proper thing to have happened. Such an approach was mandatory according to the principles of socialist realism, a world view established in the Soviet Union in the 1930s and revived in the second cultural revolution. As John McCannon has remarked:

> At the heart of socialist realism was a synergistic relationship between fact and fantasy. … What this symbiosis fed on best was heroism; bold deeds and epic achievements became the cornerstones of Soviet culture during the 1930s. … Simply stated, socialist realism, while purporting to depict reality, was really about portraying what *should be* in the language of what *actually was*. (McCannon 1998: 81–82—emphases original)

McCannon's telling study of the Soviet cult of Arctic development was mainly focused on the creative arts and the overall media blitz through which that cult was sustained, and his only examples of misrepresentation of fact concerned the omission of the negative realities of the *gulag*, rather than outright inaccuracy. The present study, by contrast, focuses on the factual narrative of the Bellingshausen expedition as it was presented in Russian texts. From 1949 the creative arts certainly played a supporting role, but that will only be touched on occasionally.

To return to our narrative, Andreev and his colleagues were working in a complex cultural and political nexus, the elements of which (for historians) fell under three main headings. First (a) there was the long-standing question for Soviet Marxism of how the numerically weak and inexperienced Russian working class, rather than their more likely-seeming German or British comrades, had made a successful proletarian revolution, established a socialist state in the teeth of violent domestic and foreign opposition, and constituted themselves and their leadership, the Soviet Communist Party, as the agents of world history for the foreseeable future. Whatever their private opinions, Soviet historians of the post-war period would have been familiarized with (a) during their professional formation; Shvede, for instance, was a senior figure in the educational hierarchy of the Soviet Navy. The Stalinist answer, that Russian workers, peasants, and lower-middle-class elements had been innately more progressive and humane than people elsewhere for centuries before the revolution, only to find their moral intentions repeatedly frustrated by the imperial autarchy and aristocracy, was at least rhetorically satisfactory. Thus it informed the practice of dating the history of the Soviet Union 'from the most ancient times', rather than from 1922 (Grekov and Others 1947). It was a sadly parochial device, given the opportunity for revolutionary internationalism provided by the real history of early communist ideas and practices, namely that they evolved within exploited groups such as slaves and peasants across many different societies, and were feared and rumoured by their oppressors even more widely than they actually occurred. But such considerations would have abandoned the Stalinist dogma of the Russianness of progressive values in favour of a more general enquiry into the parallel experiences of exploited groups, and any investigation along those lines would have left the original question, namely why the revolution happened in Russia, without an answer.

During the Cold War Soviet historians and other authors treating pre-revolutionary topics also needed (b) to adapt to and contribute to the late Stalinist re-shaping of (a) for purposes of state economic and security policy (Vujacic 2007; Brandenberger 2010), and in particular (c) to avoid falling foul of Pankratova, Grekov, Shepilov, and other Zhdanovite enforcers of *partijnost'* and patriotism in Soviet historiography. In the case of the Bellingshausen expedition they used several devices to achieve this. They portrayed not just the leaders but all members of the expedition as enlightened agents of humanity; they provided them with bourgeois, imperialist opponents; they continued the practice, initiated by Rabinovich and then taken up by Shokal'skij, of Russifying the expedition by treating the ethnic Russian Lazarev as a co-commander alongside the Baltic German Bellingshausen, which Lazarev was not;[16] and they exaggerated the political significance of a putative early sighting by ignoring other, sounder bases for Soviet rights in Antarctica.

Berg, Andreev, and Shvede all opened their 1949 treatments of the expedition with a summary of the current political situation in respect of Antarctica, and all included the highly political Geographical Society resolution, though in Berg's case that only happened in some versions of his lecture, such as that in *Zvezda*, which was doubtless trying hard after its recent reprimand (Berg 1949c). If Christopher Lloyd (1906–1986) had done anything of the sort with his selections from Cook's journals that

---

[16] As the captain of *Mirnyj*, Lazarev's real counterpart in the expedition was Bellingshausen's senior aide on *Vostok*, Captain Lieutenant Ivan Ivanovich Zavodovskij (1780–1837). Zavodovskij and Lazarev ran the ships; Bellingshausen directed the expedition. *Vostok's* Master, Yakov Poryadin, was probably the fourth member of this senior management team.

year (Lloyd 1949), his readers and reviewers would have expressed their irritation at his irrelevancy in no uncertain terms. But the holistic Marxist approach to historical explanation takes a different view of the relationship between past and present. From that standpoint, bourgeois capitalist rejection of Soviet rights in Antarctica and non-acceptance of Bellingshausen's priority in discovery were two sides of the same coin; *both* made his priority at least likely to have been the case (Kalesnik 1949b).

Even Andreev supported this approach by claiming that the West no longer believed in the existence of a southern continent after Cook (Andreev 1949b: 3). There was of course some truth in that. Alexander Dalrymple, the leading British advocate of voyages to discover the southern continent, fell silent after Cook's second voyage, and there was no question of sending the French explorer Lapérouse to the far south in 1785, even though his mission was to complete the work of Cook. But Andreev was presumably unaware that not only Cook's widely read biographer Andrew Kippis but, for example, the highly esteemed historian William Robertson had been convinced of the existence of a southern continent *by reading* Cook (Robertson 1780: 467). He also overlooked the British and American sealers who reached the South Sandwich Is. (from South Georgia) soon after the Battle of Waterloo, and the South Shetlands, Antarctic Peninsula, and South Orkneys between 1819 and 1821. Such men had not waited for a lead from Russia before taking up the quest for land (and prey) in the far south.[17]

In 1949, the first to come to the aid of the evidently lukewarm Berg and flounderers like Kalesnik were the geographers D. Lebedev and Academician Andrej Aleksandrovich Grigor'ev (1883–1968), the first director of the Institute of Geography.[18] In their article, based on a lecture given about three weeks after the Geographical Society meeting, they gave several quotations from the first edition of *Two Seasons* and projected a deceptive air of historical competence. Deceptive, because, for example, they described Lazarev as receiving orders from the Emperor which (a) have not survived to show their addressee,[19] and (b) would in any case

---

[17] After 1960, however, Belov pointed out that sealers had continued to push south after Cook (1963a: 7), that belief in the possibility of a southern continent was not extinguished by Cook's voyage (Belov 1970a: 201), and that Cook had been misrepresented in Russia (1969: 37–38).

[18] If Lebedev could have foreseen the ideological reprimand that awaited Grigor'ev in 1950 he might have chosen another collaborator (K. Markov 1950).

[19] Nothing of the sort appears in the Lazarev *Dokumenty*.

have been addressed to the highly regarded minister, de Traversay, rather than a lowly lieutenant; next, they enhanced the events of 28 January 1820 by throwing in a sighting of petrels, which most readers would not know that Bellingshausen had been recording for two months by that point; and third, they concocted a spurious precedence for Palmer over Smith at the South Shetlands by dating Palmer's arrival to 1818, before Smith discovered the islands, citing no evidence whatsoever, after which they dismissed Palmer's claim to have sighted the mainland (in November 1820) on the outdated grounds that Graham Land was an island (A. Grigor'ev and D. Lebedev 1949: 186, 187, 190). But their main aim was not historical or geographical accuracy so much as bolstering the case for priority. They achieved this by amalgamating Bellingshausen's inconclusive description of 28 January 1820 with his accounts of later southings, notably that of 17 February 1820, and then treating the latter as if it were (a) relevant to priority and (b) contested by Western historians, neither of which was the case. (The device has been known since ancient times as the fallacy of *ignoratio elenchi*, aka the red herring, diversion, or distraction.) Like everyone else in 1949 they overlooked Lazarev's more positive description of 28 January 1820. In their article they had the excuse that Lazarev's letter had not yet been republished in Andreev's anthology or listed in Bogatkina's bibliography; but they also ignored Lazarev's letter in the pamphlet version a year later, which did at least drop the nonsense about Lazarev's personal correspondence with the Emperor (A. Grigor'ev and D. Lebedev 1950). By publishing such a strongly worded article directly after the Geographical Society meeting, Grigor'ev and Lebedev became quite influential (see Chap. 8). Shvede, for one, appears to have played safe by borrowing several of their points.

One well-written early treatment was a booklet by Kublitskij. He set his account of the expedition within a lively description of the Geographical Society meeting, and with a print run of 45,000 and a price of 5 rub. it may well have attracted adults in search of something more readable than Andreev's anthology and more substantial than Berg's lecture (Kublitskij 1949b). Kublitskij was perhaps the first author to use fictional passages to paint his Antarctic heroes in the requisite colours. A few years after the voyage one of *Vostok*'s lieutenants, Konstantin Petrovich Torson (1793–1851), had been involved in the Decembrist uprising against the succession of Emperor Nicholas I, an endeavour retrospectively commended by the Soviet authorities. So Kublitskij selected Torson for the part when he wanted to establish the moral superiority of the Russian expedition in comparison with Cook, by depicting Bellingshausen as unwilling to frighten the islanders with gunfire (the same: 95–96, 100, 125). In reality, (a) some expedition members secured profitable specimens, such as shrunken heads, by 'unequal exchange' with their owners; (b) Bellingshausen did use gunfire to intimidate the islanders as or more often than he abstained from doing so, going so far as to inflict a flesh wound on one occasion; and (c) the Russians were only able to get away with *showing* weapons to Maoris at Queen Charlotte Sound because Cook and Furneaux had previously been obliged to *use* them. One might also point to the invaluable support provided to the expedition (and fully acknowledged by Bellingshausen) by European settlements which had been established in South America and Australia by force of arms. But such 'objectivist' considerations were evidently beside the point.

Kublitskij was very well informed. He seems to have had access to some of Bellingshausen's reports, to the still unpublished MS of Simonov's last account of the voyage (Simonov 1951, 1990), and to other documents being prepared for the Lazarev *Dokumenty* (Chap. 8). But his willingness to deceive his young readers is evident from his Cold War version of Bellingshausen's words (supposedly to his men) on the subject of modern weapons: 'Remember, we do not use gunpowder to harm the islanders. Let us postpone their better acquaintance with the seamen of Western Europe' [23] (the same: 100). The commander's actual words, in a report but not repeated in his book, had been: 'At the sight of so much stubbornness, born of ignorance, I postponed their better acquaintance with Europeans' [24]—no 'Western' about it (Bulkeley 2014a: 90). The sentence came at the end of a passage which recorded the firing of muskets and cannon over the heads of recalcitrant islanders, the very tactic which Kublitskij airbrushed out of the story.

Lastly, in this miscellany of examples of *partijnost'* from this period, Aleksandr Ivanovich Solov'ev (1907–1983) replaced the real but evidently inconvenient Bellingshausen, who told his Emperor on the eve of the voyage that he doubted whether there was anything left in the world to discover (Novosil'skij 1853a: 34), with conveniently unidentified 'Russian scholars' who 'were convinced of the existence of a southern continent' [25] (Solov'ev 1951: 3).

THE DENIGRATION OF COOK

Soviet authors repeatedly cast the Bellingshausen expedition in a patriotic and progressive light by contrasting it favourably with its only precedent, the second circumnavigation by James Cook, RN, carried out between 1772 and 1776. Bellingshausen had certainly surpassed Cook by discovering land south of the Circle, and by sailing through many more longitudes in or close to the ice fields (at or below 65°S), a feat which earned him the soubriquet 'intrepid' (first conferred by J.C. Ross) with foreign commentators, who were also the first to express their admiration numerically, providing calculations of Bellingshausen's southerliness that were later refined by Shokal'skij (Petermann 1865–1867: 5–6; Fricker 1898: 41). But the transition from Cook to Bellingshausen was the normal progress of European maritime exploration, in which the later expedition benefitted from intellectual and material developments, such as modern port facilities in the region. Bellingshausen himself was a firm admirer of 'the great navigator', as he sometimes called him (Bulkeley 2014b).

Cook's belief, that the heavy ice which he encountered in the Antarctic was a sure sign of the existence of land further south than he could reach, was known in Russia in the nineteenth century (Anon. 1842: 24–25). In the twentieth century, however, and despite the moderating influence of Shokal'skij, Russian commentators began to claim, as some foreign writers had done before them, that Cook's lack of Antarctic discoveries, together with his ill-judged remark that his voyage had put 'a final end … to the searching after a southern continent' (Cook 1777 2: 192), had indeed imposed a complete suspension of Antarctic exploration (Rabinovich'' 1908: 12–13; S. Grigor'ev 1937: 13). But although Buffon, in particular, welcomed Cook's views about a largely oceanic southern hemisphere—'I was very satisfied to have my conjectures confirmed by the facts' [26] (Buffon 1778: 604 and map)—no one inside or outside Russia has ever identified a government or an explorer that was deterred from seeking Antarctica by Cook. Indeed they were unlikely to do so, first because foolish remarks are seldom as influential as was claimed for this one, and second because the fact that France and Britain were at war for 33 of the 40 years between 1775 and 1815 offers a more likely explanation for the slackening, though not entire cessation, of maritime exploration, then largely in their hands. German and Russian authors, whose armies had fought more intermittently in the Napoleonic Wars, may not have appreciated this fully. Furthermore when Soviet authors began to deploy the ideology of late Stalinism by painting Bellingshausen's bold challenge to Cook as a sign of the superiority of Russian civilization, they laid themselves open to much the same question, namely why in that case Russia had needed to wait until 1819, by which time British and American sealing skippers were already pushing south.[20]

Vvedenskij formulated the view that Cook had somehow banned further Antarctic exploration in fairly measured terms and with a long quotation. He cut Cook off, however, just before the sentence: 'That there may be a continent, or large tract of land, near the Pole, I will not deny; on the contrary I am of opinion there is; and it is probable that we have seen a part of it' (Cook 1777 2: 239), thus permitting himself the falsehood that Cook 'could only conclude that the ocean extended further to the South Pole' [27] (Vvedenskij 1940: 13). M. Rajkhenberg simplified this a year later to 'Cook reached the conclusion that the southern continent did not exist' [28]. But instead of (mis)quoting Cook he quoted Petermann (Rajkhenberg 1941: 62, 68), doubtless unaware that the latter had contrived the tale of Cook's negative influence and Bellingshausen's doughty challenge to it—as if Bellingshausen had anything to do with mounting the project—for the short-lived rhetorical purpose, in the mid-1860s, of inciting Germans to emulate the polar voyages of seamen from other parts of Europe. In his introduction to the 1948 translation of Cook's second voyage, however, Iosif Petrovich Magidovich (1892–1976) conceded that Cook had believed that a continent existed (Magidovich 1948: 33–34). He could hardly have avoided doing so, since although he made some cuts he included the passage in which Cook not only said that, but also surmised that the continent lay more towards the Atlantic and Indian Oceans, rather than the Pacific (Kuk 1948: 440–443), a brilliant conjecture. Instead Magidovich reserved his scorn for Cook's second ill-judged remark in the same chapter: 'I can be bold enough to say that no man will ever venture farther than I have done; and that the lands which may lie to the south will never be explored' (Cook 1777 2: 206). Magidovich portrayed Cook as wilfully indifferent to science, instead of reflecting that the British navigator never received the sort of education that he himself had enjoyed (Magidovich 1948: 34).

---

[20] In the early 1960s it became possible for Mikhail Belov to point out that Russian exploration had been hampered by the wars with Napoleon. That he did not extend this explanation to other countries was perhaps because at that point he was not trying to compare them with Russia (Belov 1963a: 5).

In a lecture about European conceptions of Antarctica before the Bellingshausen expedition, given two weeks after the Geographical Society meeting and so probably prepared before it, the historian of cartography A.Z. Alejner barely mentioned Cook except to say that his voyages had disproved the previous notion that there was a vast and undiscovered southern continent which extended as far north as the Tropic of Capricorn, or even the Equator (Alejner 1949: 496).

Thereafter, however, Cook's Soviet detractors focused mainly on his less sensible remarks and often misrepresented him, with or without misquotations. Authors were especially given to repeating Berg's phrase 'Cook's mistake' [29] from the Geographical Society meeting (Adamov 1949: 26). A. Grigor'ev even went as far as 'Captain Cook was mistaken about everything...' [30] (A. Grigor'ev 1949: 10). But whether 'Cook's mistake' was his (alleged) disbelief in the existence of a continent, or his failure of imagination with respect to future exploration, was seldom made clear.

Of the leading authors in this period, Berg quite properly quoted Cook's proviso 'unless near the Pole' in the original lecture (Anon. 1949), but in some later versions Cook was first softened up, so to speak, by quoting the 1948 Russian translation (which leaves out several things at this point)—'I was now well satisfied no continent was to be found in this ocean,...' [31] (Kuk 1948: 209), and then omitting what followed: 'but what must lie so far to the south, as to be wholly inaccessible on account of ice; and that if one should be found in the southern Atlantic Ocean, it would be necessary to have the whole summer before us to explore it' (Cook 1777 1: 269–270; Berg 1949c: 92, 1949d: 40). In the most condensed version of Berg, Cook's ideas were simply dropped (Berg 1949e). As mentioned above, Andreev quoted Cook's unfortunate remark about putting a 'final end' to Antarctic exploration, and drew attention to the relative northerliness of much of Cook's track, but he did not discuss Cook's ideas about the existence of a continent (1949b: 5–6). Careful as ever, Shvede allowed Cook his 'unless near the Pole' and accurately rendered him as denying 'the accessibility of the southern continent', [32] rather than its existence. However he added an adjective which also became popular: 'Cook's fateful mistake' [33] (Shvede 1949: 8). Grigor'ev and Lebedev chose to provide their own translation of the 'unless near the Pole' passage, but rather spoiled the effect by having Cook say there was no continent anywhere in the southern hemisphere (A. Grigor'ev and

D. Lebedev 1949: 185). In the 1950 pamphlet version they stopped trying to summarize Cook's ideas, perhaps wisely.

No one cared (or dared) to understand that it was precisely from his in general more northerly track (which at other times they relished explaining) that Cook had suggested there was almost certainly land to the south. And like Vvedenskij, most Soviet authors avoided Cook's clearest affirmative passage: 'That there may be a continent, or large tract of land, near the Pole, I will not deny; on the contrary I am of opinion there is; and it is probable that we have seen a part of it' (Cook 1777 2: 239). They had no alternative, because Cook's actual words would have made a nonsense of the simplistic melodrama they were presenting.

Soviet authors sometimes scoffed at one particular event in Cook's voyage, his discovery of the South Sandwich Is., because he had provisionally named them, after an imperfect sighting, as 'Sandwich Land'. But Cook had been careful to specify that it was 'either a group of islands, or else a point of the continent', in that order (the same: 230). When he reached the area in December 1819 Bellingshausen charted the islands with great skill, thus settling the matter in favour of the first of Cook's two possibilities. The mischievous tale of Bellingshausen giving the lie to Cook was launched by Simonov's 1822 lecture (Simonov" 1822b: 16), but in such a case the historian should always examine the original words of the explorers concerned. That is what Shvede and others failed to do in this case (Shvede 1949: 25). It also helps to stop and think. If Russia alone contains no fewer than three archipelagos each called a 'Land' (Novaya, Severnaya, and Franz Josef), what was so very wrong with Cook's 'Sandwich Land'? After all, no Russian author has ever found fault with Krusenstern for applying the word 'Land' to a single small island, Peter I, in 1833.

The Abuse of Sources

The treatment meted out to Cook was extended to Simonov and Bellingshausen. Andreev's 1949 anthology was completed just two months after the Geographical Society meeting, so that most of the preparation of texts, including his excellent transcription of the seaman's sometimes barely legible diary, must have been done before it. That explains why two passages in Simonov's 1822 lecture which did not support the new 1820 discovery line were retained. The first was: 'The [South] Sandwich Is. were the last land seen by us during that passage up until we reached New Holland' [34]. And later came Simonov's guarded suggestion, repeated here, that the continent might have been sighted, not in 1820 but in 1821: '...unless the Alexander Coast is the edge of such a land, we find ourselves obliged to confirm his [Cook's] findings by stating that we saw no signs of the supposed Southern Continent...' [35] (Simonov" 1822b: 18, 55).

But as we have seen, Andreev would have been painfully aware in 1948 of the ban on kowtowing to foreigners, and that is evident from the cuts he made in Simonov's lecture, some of them substantial and none of them signalled in any way. Minor cuts included Simonov's opening and closing tributes to the Emperor's benevolent patronage of learning and personal interest in the expedition (the same: 3–4, 58–59). More radically, Andreev withheld Simonov's enthusiastically positive description of New South Wales and part of his acknowledgement of the help given to the expedition by Governor Lachlan Macquarie (1762–1824) and other colonists, as well as positive accounts of the work of British missionaries in Tahiti and the friendly reception extended to the Russians, as fellow-Christians, by the Tahitians (the same: 22–26, 36–39). Altogether, Andreev's cuts amounted to some ten short pages, or 17.5% of Simonov's text.

A subtle political guidance can perhaps also be detected in the choice of primary sources, beyond the reports and *Two Seasons*, for republication. It is no criticism of the anthology, for which Andreev quite reasonably selected four different primary sources including Simonov's most complete early narrative, to point out that while the lively but historically speaking unimportant seaman's diary was published, Simonov's shipboard diary for the first season, one of the very few immediate accounts of those weeks that we have, albeit from a landsman, was never republished. Just possibly, that text was unattractive because Simonov recorded the day on which Antarctica had allegedly been discovered as a fairly routine encounter with an impassable ice field, nothing like land or a vast permanent ice cap, the very thing which, as he had carefully explained a few days earlier, he was actively hoping to see Simonov" (1822a, pt 4: 181, 178–180).

THE LIMITS OF EARLY SCHOLARSHIP

One intriguing aspect of the 1949 revision of Bellingshausen's achievement was actually a non-event, namely the absence of an ur-study, a new game-changing analysis of all the surviving sources and the relationship of the expedition to the world in which it took place, from which fresh lines of investigation and alternative historical perspectives would naturally develop. The example of Paul Forman's seminal paper on Weimar physics may be setting the bar rather high (Forman 1971), but it reminds us that new perspectives in the history of science need to be based on a solid foundation of scholarship. Shvede's edition of *Two Seasons* and Andreev's anthology were substantial achievements. But despite having very little in the way of evidence for priority before Lazarev's letter was brought into play, the first exponents of the 1820 discovery thesis were notably reluctant to get to grips with their subject.

For one thing, despite the involvement of naval officers, a thorough nautical history of the expedition was not attempted. The identity and date of the expedition's first fatality, for example, remained unknown because he happened to die on *Mirnyj*, Bellingshausen's reports and book were focused on *Vostok*, and no one ever looked at the crew status report submitted at the end of the voyage. On the scientific side, no attempt was made to clarify the embryonic European glaciology of the day, a vital tool for the interpretation of sometimes cryptic sources. As for scientific personnel, almost every commentator seized the opportunity to put the expedition in a slightly more scientific light (and Germans in a worse one), by declaring that *two* German naturalists had failed to join Bellingshausen at Copenhagen, whereas the orders cut for Bellingshausen's southern and Vasil'ev's simultaneous northern squadron, both published in this period, show clearly that they were assigned *one* naturalist each. The archival record confirms this. A single, rather humble voice, that of the children's author Kublitskij, got that part of the story right, and explained the absence of the young man in question, Karl Heinrich Mertens (1796–1830), as due not to cowardice as Shvede alleged (1949: 13) but rather to parental opposition (Kublitskij 1949b: 103).[21] As for descriptions and discussion of the expedition's scientific results and publications, they were cursory at best.

Mention should also be made of an early pamphlet by Dmitrij Golubev. It was a competent and well-illustrated account of the whole voyage, indeed perhaps the best available when it appeared a few weeks before Shvede's second edition of Bellingshausen (Golubev 1949). At 3 rub. it

---

[21] Kublitskij's exceptional insight suggests that along with Bellingshausen's reports he had seen archival documents which gave the true story about Mertens. But he was ignored by other commentators.

cost little more than Andreev's anthology and the 50,000 print run was more than those of Andreev and Shvede combined.

* * *

In 1949 Magidovich began his multi-volume history of geographical discoveries with this definition of his subject:

> Geographical discovery means: a visit by representatives of one or other civilized people to a new portion of the earth's surface, previously unknown to the world of learning, or the establishment of [new] routes between previously known portions of the earth's surface [36]. (Magidovich 1949: 3)

The phrase 'earth's surface' was somewhat opaque, given that Magidovich offered no examples of the discovery of oceans; the last part of the sentence was presumably meant to cover endeavours like the Northern Sea Route. The definition combined two elements, first, the suggestion, if nothing more, that discoveries had to be made officially on behalf of this or that nation, hence no sealers, trappers, etc., and second, the requirement that explorers should actually visit or set foot in the place. The wording strongly suggests that merely to look on it from afar, as Moses viewed the Promised Land and Bellingshausen Antarctica, would not suffice. In decidedly un-Marxist fashion, Magidovich said nothing about the socialization, after the event, of new geographical knowledge. Be that as it may, no Soviet commentators on Bellingshausen ever took any notice of Magidovich's definition, despite his considerable reputation.

* * *

The single most striking feature of the 1949 revision of the expedition's achievements was the failure of those initially tasked with its formulation, Berg, Shvede, and Andreev, to show much enthusiasm for it. Berg was inclined to prefer Debenham's assessment, that 28 January 1820 had been a near miss, and Shvede openly declared as much (Shvede 1949: 26). Andreev's position was rather like the ancient Zen koan about the egg and the bottle, in which the student is asked how a fresh hen's egg can emerge from a narrow-necked bottle without breaking. There are many possible answers, such as 'The way in is also the way out', but in one famous solution the master simply claps his hands and says, 'There, it's out!'. The

Soviet government had indeed clapped its hands, but in 1949 and 1950 the egg of Antarctic priority was still having difficulty in emerging unscathed from the Bellingshausen bottle. What was needed was stronger evidence, rather than wishful thinking about proximity, and that would soon come into play.

ORIGINAL TEXTS OF QUOTATIONS

[1] … северный берег материка…

[2] В связи с этим решением ЦК ВКП(б) от 29 января 1949 г. было признано целесообразным заявить, сначала в неофициальной форме, о заинтересованности Советского Союза в вопросе о режиме Антарктики, в дальнейшем, в зависимости от последующих обстоятельств, сделать такое заявление и в официальной форме.

[3] «Восток» и «Мирный» у берегов Антарктиды.

[4] Задача географика-историка заключается в том … В частности, надо конкретно показать … 5) в чем проявились самостоятельность и приоритет русской географической науки…

[5] … вблизи Антарктического материка …

[6] *Теперь ясно*, что описываемые Беллинсгаузеном бугристые льды, прострашиеся с востока на запад, представляли собою именно окраину антарктического материка … [разрядка наша].

[7] …впервые … открыли *существование Антарктики*… [разрядка наша].

[8] … сплошные льды простираются отъ Востока чрезъ Югъ на Западъ…

[9] …сплошные льды … льдяное поле, усѣянное буграми.

[10] … бугристые льды …

[11] …впервые подошли к его берегам…

[12] … подошли к самой кромке неподвижного льда … подошли впервые *к твердой земле* … [разрядка наша].

[13] 17 января 1821 г. … русские первыми из европейцев достигли южного материка и открыли берег его…

[14] Весьма вероятно … вплотную у берега антарктического материка. … право приоритета открытия ряда земель Антарктики.

[15] Если бы не плохие условия видимости, то уже 16 января [ст. ст.] Беллинсгаузен и Лазарев смогли бы дать совершенно точные сведения о землях антарктического материка…

[16] Русские мореплаватели … первыми открыли Антарктиду…

[17] Советский Союз стал оплотом мировой цивилизации и прогресса. Теперь не может итти речь ни о какой цивилизации без русского языка, без науки и культуры народов советской страны. *За нами приоритет*. От страны социализма на Запад и Восток иддут теперь лучи обновления и возрождения. Диалектика истории изменила роли «учителей» и «учеников». [разрядка наша].

[18] Крупнейшим же недостатком содержания сборника является то, что в некоторых статьях делается открытая попытка воскресить давно осужденную "теорию", сводившую всю созидательную деятельность Петра I к перенесению на русскую почву иностранных образцов.

[19] ...статья редактора сборника пропагандирует низкопоклонство и раболепие русских деятелей перед иностранщиной и что читателю сборника, кроме вреда, она ничего дать не может.

[20] низкопоклонство перед Западом.

[21] Институт практически слабо откликнулся на призыв партии организовать борьбу с низкопоклонство перед Западом и ничего не сделал для разоблачения лживой версии несамостоятельности русской культуры и науки.

[22] ...Сложнее...

[23] Запомните, что мы не будем употреблять пороха во вред островитянам. Предоставим времени познакомить их с мореплавателями Запданой Европы.

[24] *Видя таковую опорность отъ незнанія, предоставилъ времени знакомства ихъ со Европейцами.*

[25] Русские учены были уверены в существовании Южного материка...

[26] ... j'ai vu avec la plus grand satisfaction mes conjectures confirmées par les faits.

[27] ...убеждался лишь в том, что океан и далее простирается к Южному полюсу.

[28] Кук пришел к убеждению, что Южный материк не существует.

[29] Ошибка Кука.

[30] Капитан Кук ошибался во всем...

[31] Теперь я твердо убежден, что на юге Тихого океана материка нет, ...

[32] ...наличие южного материка ...

[33] ...роковая ошибка Кука ...

[34] Сандвичевы острова были послѣднею землею, выдѣнною нами въ продолженіи сего плаванія, до самой новой Голландіи.

[35] …если берегъ Александра не есть оконечность земли сей, то принужденными найдемся подвердить слова его [Кука], должны будемъ сказать, что мы не видѣли никакихъ признаковъ предполагаемаго южнаго материка…

[36] Географическое открытие означает: посещение представителями какого-либо цивилизованного народа новой, ранее неизвестной культурному человечеству части земной поверхности или установление пространственной связи между известными уже частями земной поверхности.

REFERENCES[22]

Adamov, Arkadij. 1949. Antarktida—russkoe otkrytie [Antarctica—A Russian Discovery]. *Znanie–sila* 7: 26–29.

Afanas'ev, Yu.N., and Others. 1996. *Sovetskaya istoriografiya [Soviet Historiography]*. Moscow: Russian State University for Humanities.

Alejner, A.Z. 1949. Geograficheskie predstavleniya ob Antarktike s drevnejshikh vremen do pervoj russkoj antarkticheskoj èkspeditsii i ikh otrazhenie na kartakh [Geographical Representations of the Antarctic from Ancient Times to the First Russian Antarctic Expedition, and their Reflection in Maps]. *Izvestiya vsesoyuznogo geograficheskogo obshhestva* 81 (5): 484–496.

Aleksandrov, A.I. 1949. Antarktika [The Antarctic]. *Priroda* 8: 25–31.

Andreev, A.I., ed. 1947a. *Pyotr Velikij—sbornik statej [Peter the Great—An Anthology]*. Moscow–Leningrad: Academy of Sciences.

———. 1947b. Pyotr I v Anglii v 1698 g. [Peter I in England in 1698]. In *Pyotr Velikij—sbornik statej*, ed. A.I. Andreev, 63–103. Moscow–Leningrad: Academy of Sciences.

———. 1947c. Rabota S.M. Solov'eva nad «Istoriej Rossij» [S.M. Solov'ev's Work on *The History of Russia*]. *Trudy Istoriko-arkhivnogo instituta* 3: 3–16.

———. 1949a. Russkie v Antarktike v 1819–1821 gg. [Russians in the Antarctic 1819–1821]. *Izvestiya Akademii Nauk: seriya istorii i filosofii* 6 (1): 77–78.

———. 1949b. Èkspeditsiya F.F. Bellinsgauzena—M.P. Lazareva v Yuzhnyj Ledovityj okean v 1819–1821 gg. i otkrytie russkimi moryakami Antarktidy [The Expedition of F.F. Bellingshausen and M.P. Lazarev to the Southern Ice Ocean in 1819–1821, and the Discovery of Antarctica by Russian Seamen]. In

---

[22] Key to archival references: SARN = State Archives of the Russian Navy, F = Fond, S = Series [*Opis'*], P = Piece [*Delo*], fo./fos = folio/s [*list/y*], v = verso [*oborotnoe*]. (The latter is necessary because verso pages were usually unnumbered.) Dates on manuscript documents are given in Universal Time (CE).

108   R. BULKELEY

Plavanie shlyupov «Vostok» i «Mirnyj» v Antarktiku v 1819, 1820 i 1821 godakh, ed. A.I. Andreev, 5–16. Moscow: Geografgiz.

———. 1949c. *Plavanie shlyupov «Vostok» i «Mirnyj» v Antarktiku v 1819, 1820 i 1821 godakh [The Antarctic Voyage of the Sloops Vostok and Mirnyj in 1819, 1820 and 1821]*. Moscow: Geografgiz.

———. 2012. Pis'mo A.I. Andreeva O.M. Medushevske, iyunya 1949 [A Letter from A.I. Andreev to O.M. Medushevskaya, June 1949]. In Pis'ma O.M. Medushevskoj A.I. Andreevu, ed. V.G. Anan'ev. *Vestnik Rossijskogo Gosudarstvennogo Gumanitarnogo Universiteta* 21: 11–23.

Anon. 1842. Otkrytie novago materika v Yuzhnom" polusharii [Discovery of a New Mainland in the Southern Hemisphere]. *Syn" Otechestva* 1: 24–32.

———. 1946a. O zhurnalakh «Zvezda» i «Leningrad» [On the Magazines *Zvezda* and *Leningrad*]. *Pravda*, August 21: 1.

———. 1946b. Edward Bransfield's Antarctic Voyage, 1819–1820, and the Discovery of the Antarctic Continent. *Polar Record* 4: 385–393.

———. 1948a. O rabote Instituta istorii Akademii nauk SSSR [The Work of the Institute of History, Soviet Academy of Sciences]. *Voprosy istorii* 11: 144–149.

———. 1948b. Protiv ob"ektivizma v istoricheskoj nauki [Against Objectivism in Historiography]. *Voprosy istorii* 12: 3–12.

———. 1949. Russkie otkrytiya v Antarktike [Russian Discoveries in the Antarctic]. *Izvestiya*, February 11: 3.

Anpilogov, G. 1948. «*Pyotr Velikij»—sbornik statej* pod redaktsiej d-ra istoricheskikh nauk A.I. Andreeva [*Peter the Great—An Anthology*, ed. Dr of History A.I. Andreev]. *Voprosy istorii* 4: 120–123.

Baltijskij, N. (pseudonym of O.V. Kuusinen). 1945. O patriotizme [On Patriotism]. *Novoe vremya*, 1: 3–10.

Bellingshausen, Captain. 1945. *The Voyage of Captain Bellingshausen to the Antarctic Seas, 1819–1821*, ed. Frank Debenham, 2 vols. London: Hakluyt Society.

Bellinsgauzen, F.F. 1949. *Dvukratnye izyskaniya v yuzhnom ledovitom Okeene i plavanie vokrug sveta v prodolzhenie 1819, 1820, i 1821 godov*, ed. E.E. Shvede. Moscow: Geografgiz.

Bellinsgauzen", Kapitan". 1831. *Dvukratnyya izyskaniya v" yuzhnom" ledovitom" okeanye i plavanie vokrug" svyeta v" prodolzhenii 1819, 20 i 21 godov" [Two Seasons of Exploration in the Southern Ice Ocean and a Voyage Around the World, During the Years 1819, 1820 and 1821]*. ed. L.I. Golenishhev"-Kutuzov", 2 vols. plus *Atlas*. St Petersburg: Glazunovs.

Belov, M.I. 1963a. O kartakh pervoj russkoj antarkticheskoj ekspeditsii 1819–1821 gg. [The Maps of the First Russian Antarctic Expedition 1819–1821]. In *Pervaya russkaya antarkticheskaya ekspeditsiya 1819–1821 gg. i eyo otchyotnaya navigatsionnaya karta*, ed. M.I. Belov, 5–56. Leningrad: Morskoj Transport.

————. 1969. Istoriya otkrytiya i issledovaniya Antarktiki [History of the Discovery and Exploration of Antarctica]. In *Atlas Antarktiki [Atlas of the Antarctic]*, ed. E.I. Tolstikov, vol. 2, 35–97. Moscow–Leningrad: Geodesic and Cartographic Department.

————. 1970a. Otkrytia ledyanogo kontinenta [The Discovery of the Icy Continent]. *Izvestiya vsesoyuznogo geograficheskogo obshhestva* 102 (3): 201–208.

————. 1970b. The First Maritime Magnetic Survey around Antarctica. In *Problems of Polar Geography*, 244–252. (Trans. of *Trudy Arkticheskogo i Antarkticheskogo nauchnogo issledovatel'skogo instituta*, 185). Jerusalem: Israel Program for Scientific Translations.

Berg, L.S. 1949a. Russkie otkrytiya v Antarktike i sovremennyj interes k nej [Russian Discoveries in the Antarctic and Modern Interest in the Region]. *Vestnik Akademii Nauk: seriya geograficheskaya i geofizicheskaya* 3: 39–46.

————. 1949b. *Ocherki po istorii russkikh geograficheskikh otkrytii [Towards a History of Russian Geographical Discoveries]*. 2nd ed. of (Berg 1929). Moscow: Academy of Sciences.

————. 1949c. Sovremennyj interes k Antarktike [Modern Interest in the Antarctic]. *Zvezda* 2: 91–96.

————. 1949d. Russkie otkrytiya v Antarktike i sovremennyj interes k nej [Russian Discoveries in the Antarctic and Modern Interest in the Region]. *Vestnik Akademii Nauk* 3: 39–46.

————. 1949e. Russkie otkrytiya v Antarktike [Russian Discoveries in the Antarctic]. *Nauka i zhizn'* 3: 22–25.

————. 1950. *Velikie russkie puteshestvenniki [Great Russian Travellers]*. Moscow–Leningrad: Detgiz.

Bolkhovitinov, V., and Others. 1950. *Rasskazy o russkom pervenstve [Tales of Russian Priority]*. Moscow: Young Guard.

Brandenberger, David. 2010. Stalin's Populism and the Accidental Creation of Russian National Identity. *Nationalities Affairs* 38 (5): 723–739.

de Buffon, M. le Comte. 1778. *Histoire naturelle générale et particulière. Supplément*. Vol. 5. Paris: Imprimerie Royale.

Bulkeley, Rip. 2011. Cold War Whaling: Bellingshausen and the *Slava* Flotilla. *Polar Record* 47 (2): 135–155.

————. 2014a. *Bellingshausen and the Russian Antarctic Expedition, 1819–21*. Basingstoke: Palgrave Macmillan.

————. 2014b. Bellingshausen on Cook—'Glorious' or What? *Cook's Log* 37 (1): 20–23.

Chernous'ko, L.D. 1949. Russkie—pervootkryvateli Antarktidy [Russians—The First Discoverers of Antarctica]. *Morskoj sbornik* 4: 95–99.

Cook, James. 1777. *A Voyage Towards the South Pole, and Round the World; Performed in His Majesty's Ships the Resolution and Adventure, in the Years 1772, 3, 4, and 5*. 2 vols. London: Strahan and Cadell.

Forman, Paul. 1971. Weimar Culture, Causality, and Quantum Theory, 1918–1927: Adaptation by German Physicists and Mathematicians to a Hostile Intellectual Environment. *Historical Studies in the Physical Sciences* 3: 1–115.

Fricker, Karl. 1898. *Antarktis [The Antarctic]*. Berlin: Schaff & Grund.

Golubev, D. 1949. *Russkie v Antarktike [Russians in the Antarctic]*. Moscow: Goskult'prosvetizdat.

Gould, Laurence M. 1960. Statement of Dr Laurence M. Gould. *Hearings Before the Committee on Foreign Relations: United States Senate, Eighty-Sixth Congress Second Session*, June 14: 74–77.

Gould, R.T. 1925. The First Sighting of the Antarctic Continent. *Geographical Journal* 65: 220–225.

———. 1941. The Charting of the South Shetlands, 1819–28. *The Mariner's Mirror* 27: 206–242.

Grekov, B.D., and Others. 1947. *Istoriya SSSR. Vol. 1. S drevnejshikh vremen do kontsa XVIII veka [History of the USSR. Vol. 1. From Ancient Times to the End of the 18th Century]*. Moscow: Politizdat.

Grigor'ev, A. 1949. Russkie otkrytiya v Antarktike [Russian Discoveries in the Antarctic]. *Smena* 6: 10–11.

Grigor'ev, A.A., and D.M. Lebedev. 1949. Otkrytie antarkticheskogo materika russkoj èkspeditsiej Bellinsgauzena–Lazareva 1819–1821 gg. [The Discovery of the Antarctic Continent by the Russian Expedition of Bellingshausen and Lazarev 1819–1821]. *Izvestiya Akademii Nauk SSSR: seriya geograficheskaya i geofizicheskaya* 13 (3): 185–193.

———. 1950. *Prioritet russkikh otkrytii v Antarktike [The Priority of Russian Discoveries in the Antarctic]*. Moscow: Pravda.

Grigor'ev, S.G. 1937. *Vokrug yuzhnogo polyusa*. 3rd ed. Moscow: Textbook Press.

Hydrographic Department. 1948. *The Antarctic Pilot*. London: HMSO.

Jones, A.G.E. 1982. *Antarctica Observed*. Whitby: Caedmon.

Kalesnik, S.V. 1949a. K 130-letiyu russkoj èkspeditsii v Antarktidu [The 130th Anniversary of the Russian Antarctic Expedition]. *Priroda* 8: 80–82.

———. 1949b. Russkie otkrytie v Antarktike [Russian Discoveries in the Antarctic]. *Slavyane* 4: 19–22.

Kasimenko, V.A. 1948. *Kak lyudi otkryvali zemlyu [How People Discovered the Earth]*. Moscow: Lenin Library.

Kublitskij, Georgij. 1949a. Kolumby Antarktiki [Columbuses of the Antarctic]. *Vokrug sveta* 4: 4–9.

———. 1949b. *Otkryvateli Antarktidy [The Discoverers of Antarctica]*. Moscow–Leningrad: Children's Press, Ministry of Education.

Kuk, Dzhems. 1948. *Puteshestvie k yuzhnomu polyusu i vokrug sveta [A Journey Towards the South Pole and Around the World]*, ed. I. Magidovich. Moscow: Geografgiz.

Lebedev, D.M. 1949. Obsuzhdenie dokladov [A Comment on the Papers]. In *Voprosy istorii otechestvennoj nauki*, ed. S.I. Vavilov, 819–831. Moscow: Academy of Sciences.

Lloyd, C., ed. 1949. *The Voyages of Captain James Cook Round the World*. London: Chanticleer.

Magidovich, I. 1948. Dzhems Kuk, ego dejstvitel'nye i mnimye otkrytiya [James Cook, his Real and So-called Discoveries]. In *Puteshestvie k yuzhnomu polyusu i vokrug sveta*, ed. I. Magidovich, 3–34. Moscow: Geografgiz.

Magidovich, I.P. 1949. *Ocherki po istorii geograficheskikh otkrytii [Towards a History of Geographical Discoveries]*. Vol. 1. Moscow: Textbook Press.

Markov, K.K. 1950. Oshibki Akademika A.A. Grigor'eva [The Errors of Academician A.A. Grigor'ev]. *Izvestiya vsesoyuznogo geograficheskogo obshhestva* 82 (5): 453–471.

McCannon, John. 1998. *Red Arctic*. Oxford: Oxford University Press.

Morozov, P.F., and K.I. Nikul'chenkov. 1949. Èkspeditsiya F.F. Bellinsgauzena—M.P. Lazareva v Yuzhnyj Ledovityj okean i otkrytie Antarktidy [The Expedition of F.F. Bellingshausen and M.P. Lazarev to the Southern Ice Ocean and the Discovery of Antarctica]. *Morskoj sbornik* 9: 51–62.

Novosil'skij, P.M. 1853a. Yuzhnyj polyus": iz" zapisok" byvshago morskago ofitsera [The South Pole: From the Memoirs of a Former Naval Officer]. *Panteon"* 11 (9): 31–80; (10): 19–62.

———. 1853c. *Yuzhnyj polyus": iz" zapisok" byvshago morskago ofitsera [The South Pole: From the Memoirs of a Former Naval Officer]*. St Petersburg: Vejmar".

Orlov, B. 1949. Russian Antarctic Discoveries of 1821 are Basis of Soviet Claim. *USSR Information Bulletin* 9: 296–297.

Ostrovskij, B.G. 1949a. Novoe ob istoricheskom pokhode Bellinsgauzena–Lazareva v Antarktiku [A New Source for Bellingshausen and Lazarev's Historic Voyage to the Antarctic]. *Zvezda* 2: 96–99.

Petermann, A. 1865–1867. Die Erforschung der arktischen Central-Region durch eine Deutsche Nordfahrt [Exploration of the Central Arctic by a German Expedition]. *Petermann's Geographische Mittheilungen Ergänzungsband IV*: 1–14.

Rabinovich", I.O. 1908. *Shestaya chast' svyeta [The Sixth Continent]*. St Petersburg: Stepanova.

Rajkhenberg, M. 1941. Otkrytie pervoj zemli v Antarktike [The Discovery of the First Land in the Antarctic]. *Sovetskaya Arktika* 2: 61–69.

Robertson, William. 1780. *The History of America*. Vol. 2. 3rd ed. London: Strahan and Cadell.

Sementovskij, V.N. 1950. Ideo-politicheskaya napravlennost' kursa fizicheskoj geografii SSSR [The Politico-ideological Orientation of a Course in the Physical Geography of the USSR]. *Voprosy geografii* 18: 21–33.

———., ed. 1951. *Russkie otkrytiya v Antarktike [Russian Discoveries in the Antarctic]*. Moscow: Geografgiz.

Sementovskij, V.N., and N.N. Vorob'ev. 1940. *Fiziko-geograficheskie èkskursii v okrestnostyakh g*. In *Kazani [Excursions in Physical Geography around Kazan]*. Kazan: Tatgosizdat.

Shepilov, D. 1947. Sovetskij patriotizm [Soviet Patriotism]. *Pravda*, August 11: 2–3.

Shteppa, Konstantin F. 1962. *Russian Historians and the Soviet State*. New Brunswick, NJ: Rutgers University Press.

Shvede, E.E. 1949. Pervaya russkaya antarkticheskaya èkspeditsiya 1819–1821 gg. [The First Russian Antarctic Expedition 1819–1821]. In *Dvukratnye izyskaniya v yuzhnom ledovitom okeane i plavanie vokrug sveta v prodolzhenie 1819, 20 i 21 godov*, ed. E.E. Shvede, 7–30. Moscow: Geografgiz.

Simonov, I.M. 1951. Shlyupy «Vostok» i «Mirnyj» ili plavanie rossiyan v Yuzhnom Ledovitom okeane i okolo sveta [Sloops *Vostok* and *Mirnyj*, or the Voyage by Russians in the Southern Ice Ocean and around the World]. In *Russkie otkrytiya v Antarktike*, ed. V.N. Sementovskij, 51–175. Moscow: Geografgiz.

———. 1990. «Vostok» i «Mirnyj» [*Vostok* and *Mirnyj*]. In *Dva plavaniya vokrug Antarktidy*, ed. T.Ya. Sharipova, 46–248. Kazan: Kazan University Press.

Simonov″, Prof. i Kav. 1822a. Plavanie shlyupa *Vostoka* v″ Yuzhnom″ Ledovitom″ Morye [The Voyage of the Sloop *Vostok* in the Southern Ice Ocean]. *Kazanskij vyestnik″* 4 (3): 156–165, 4 (4): 211–216, 5 (5): 38–42, 5 (7): 174–181, 6 (10): 107–116, 6 (12): 226–232.

———. 1822b. *Slovo o uspyekhakh″ plavaniya shlyupov″ Vostoka i Mirnago okolo svyeta i osobenno v″ Yuzhnom″ Ledovitom″ morye, v″ 1819, 1820 i 1821 godakh″ [An Address about the Results from the Voyage of the Sloops Vostok and Mirnyj around the World and Especially in the Southern Ice Ocean, in 1819, 1820 and 1821]*. Kazan: Kazan University Press.

Solov'ev, A.I. 1951. Predislovie [Foreword]. In *Russkie otkrytiya v Antarktike*, ed. V.N. Sementovskij, 3–6. Moscow: Geografgiz.

Strokov, A. 1947. O sbornike statej «Pyotr Velikij» [On the Anthology 'Peter the Great']. *Voennaya mysl'* 10: 84–87.

Tikhonov, V.V. 2012. Iz istorii ideologicheskikh kampanij v sovetskoj istoricheskoj nauke: sbornik «Pyotr Velikij» i sud'ba ego avtorov [From the History of Ideological Campaigns in Soviet Historiography: The Anthology *Peter the Great* and the Fate of its Authors]. *Istoriya i istoriki* 2009–2010: 118–133.

———. 2013. «Khudshij obraznik prekloneniya pered inostranshhinoj»: ideologicheskie kampanii «pozdnego stalinizma» i sud'ba istorika S.A. Fejginoj ['The Worst Example of Subservience before the Outside World': The Ideological Campaigns of 'late Stalinism' and the Fate of the Historian S.A. Fejgina]. *Novejshaya istoriya Rossii* 1: 199–207.

US State Department. 1948. Press Release, 28 August 1948. *Department of State Bulletin* 19 (479): 301.

Vorob'ev, V.I. 1948. Obshhij ocherk [Overview]. In *Materialy po lotsii Antarktiki*, ed. I.F. Novoselov, 1–38. Moscow: Hydrographic Department, Soviet Navy.

Vujacic, Veljko. 2007. Stalinism and Russian Nationalism: A Reconceptualization. *Post-Soviet Affairs* 23 (2): 156–183.

Vvedenskij, N. 1940. *V poiskakh yuzhnogo materika [The Quest for a Southern Continent]*. Leningrad–Moscow: GlavSevMorPut.

Vyshinskij, A.Ya. 1950. Dokladnaya zapiska ministra inostrannykh del SSSR A.Ya. Vyshinskogo I.V. Stalinu po voprosu o memorandume ryadu pravitel'stvo rezhime Antarktiki, February 20 1950 [An Official Memo from Soviet Minister of Foreign Affairs A.Ya. Vyshinskij to I.V. Stalin on the question of the Memorandum to Several Governments about an Antarctic Regime]. Arkhiv Aleksandra N. Yakovleva, Sovetsko–Amerikanskie otnosheniya, 1949–1952. Document No. 49. http://www.alexanderyakovlev.org/fond/issues-doc/71711.

Yakovlev, V.G. 1953. O sisteme pionerskikh sborov [On the Training System for Pioneers]. *Sovetskaya pedagogika* 1: 15–30.

Zhdanov, A.A. 1946. Doklad t. Zhdanova o zhurnalakh «Zvezda» i «Leningrad» [Comrade Zhdanov's Report on the Magazines *Zvezda* and *Leningrad*]. *Pravda*, September 21: 2–3.

———. 1952. *Vystuplenie na diskussii po knige G.F. Aleksandrova «Istoriya zapadnoevropejskoj filosofii», 24 iyunya 1947 [Intervention in the Discussion of the Book History of Western Philosophy by G.F. Aleksandrov, 24 June 1947]*. Moscow: Politizdat.

CHAPTER 7

A Forgotten Letter (1821)

On 6 October 1821 (24 September [O.S.]), two months after the expedition returned to Russia, Junior Captain Lazarev, as he now was, wrote a letter about it to his friend Aleksej Shestakov. That letter became a key document for Soviet treatments of the expedition from the 1950s onwards because, alone of all the primary sources, it recorded a possible sighting of an ice coast of the Antarctic mainland on 28 January 1820, thus prior to Bransfield's sighting of the mountainous northern tip of the Antarctic Peninsula two days later (Campbell 2000: 131–133). Instead of scattering various remarks about the letter through the book, or referring readers to the author's Russian article on the subject (Balkli 2013), it will be analysed fully here.[1]

AUTHENTICITY

Because it spent many years in the Shestakov family, it is worth comparing the letter with others held at the Naval Archives which were certainly written by Lazarev, having remained in Navy files. The author has now made that comparison (the first historian to do so), and can report that, at least

---

[1] The original texts of translated quotations, indicated by numbers in square brackets, are provided at the end of each chapter. For the system of romanization used, the rendering of dates, and a few other matters, please consult the Apparatus at the end of the book.

© The Author(s), under exclusive license to Springer Nature Switzerland AG 2021
R. Bulkeley, *The Historiography of the First Russian Antarctic Expedition, 1819–21*,
https://doi.org/10.1007/978-3-030-59546-3_7

to an inexpert eye, the handwriting appears identical. Another reason for checking is that, when the letters to Shestakov were first published in 1918 their editor, Lt. Konstantin Georgievich Zhitkov (1893–about 1920), mentioned that 'The letters are printed from copies, in accordance with modern orthography' [1] (Lazarev 1918: 52). The handwriting check suggests that those copies were made by some mechanical process, such as photostats,[2] rather than by hand.

The date of the letter is another puzzle. After studying the original MS (*podlinnik*) in about 1948, while preparing his anthology of documents from the expedition, Andreev declared that it 'has no date' [2]. He was so positive that such was the case that he responded to the date of 24 September 1824 (O.S.), which had been offered in 1918,[3] by arguing from internal evidence that the letter was probably written shortly after 28 September 1821 (O.S.) (Andreev 1949c: 169–170). Such clear testimony from an experienced archival historian is hard to ignore. However, the handwriting in the date at the top of the first page of the MS, in the copy now shown to researchers at the Naval Archives, looks just like the rest of the document. Andreev's statement about the absence of a date was repeated in the second edition of the anthology (Sementovskij 1951: 286). The disparity between such comments and the document as we now have it is very strange, and seems impossible to resolve.[4]

RELIABILITY

The reliability of the letter should be estimated from its general style and by checking its factual accuracy where possible. The style is relaxed and informal, as one might expect, with exaggerations which are strictly inaccurate in themselves, but should be read in the spirit in which they were

---

[2] This new technology was in use at senior levels of the Bolshevik government during the Brest-Litovsk treaty negotiations at the end of 1917.

[3] Zhitkov seems to have been misled by the ligature between the last two digits of '1821' into reading that as '1824'.

[4] Perhaps, hard pressed as he was by the *Zhdanovshhina* at the time (Chap. 6), Andreev simply made a mistake. For example, it might have been the covering letter from one of Lazarev's descendants, enclosing his ancestor's letters to Shestakov, that had gone undated in the turmoil of revolution. If that or something similar was the case, then Sementovskij probably repeated Andreev's mistake simply because he failed to check it as he should have done. It must be stressed, however, that such thoughts can never be more than conjecture so long after the events.

written. For example Lazarev claimed that the expedition '*blundered about like the spirits of the dead for an entire month*' [3] at the South Sandwich Is., whereas they were actually there for just three weeks (Lazarev" 1821: fo. 1v). When describing the major and urgent repair to *Mirnyj*'s bow which was carried out at Sydney, Lazarev said it was done with only 45 men (fo. 3). That may have been so, but he had over 60 men at his disposal and a few were also sent over from *Vostok*. When describing the last part of the voyage, he wrote that after making their first European landfall at Lisbon '*we set off directly to Russia without calling in anywhere*' [4] (fo. 5). Bellingshausen, however, mentioned that he called at Copenhagen (Bellinsgauzen" 1831 2: 326). Perhaps *Mirnyj* was ordered to sail on, as much the slower ship, while Bellingshausen touched briefly so that despatches could be sent overland by express courier. Another undoubted exaggeration came in a later letter, when Lazarev complained that the *Atlas* volume in Bellingshausen's book had not been lithographed (Lazarev 1918: 34). That was a thoroughly misleading statement, because 45 of the 64 plates in the *Atlas* were lithographed; only the 19 map plates were not. But it was typical of Lazarev to write freely and boldly, leaving it to his old friend to fathom what he really meant.

Besides exaggerations there were several factual errors in the letter, ranging in importance from trivial to significant. In the trivial category, Lazarev listed an officer in the parallel, northern expedition as 'P. Petrovich Stogov' when his first name was in fact Ivan. A slightly more serious mistake concerned navigational errors made by the Russian explorer Otto Kotzebue, who circumnavigated between 1815 and 1818. Lazarev said Kotzebue's errors of longitude were to the east (fo. 3), whereas Bellingshausen, and Kotzebue himself, reported that they were to the west (Kotzebue and Eschscholtz 1830 1: 217). The modern coordinates confirm this.

Another mistake resembles the earlier statement about how long the expedition spent at the South Sandwich Is. After describing the discovery of Peter I I. in January 1821, Lazarev added that the Alexander I Coast was discovered '*six days later*' [5] (fo. 4). But according to Bellingshausen and other sources, including the track chart, the first discovery took place on 10 January and the second on 17 January 1821 (both O.S. and s.d.), seven days later (Bellinsgauzen" 1821e: fo. 13).[5]

---

[5] Lazarev used a phrase meaning 'after [lit. across] five days', where a modern Russian would probably write 'after seven days'. In the older style of the phrase that Lazarev seems to have used, he should have written 'after six days' to be accurate.

Lazarev's most important and provable errors of fact concerned the three seamen who died during the voyage. He concluded his account of the expedition's first Antarctic season by stating that *Mirnyj* reached Sydney '*on 7th April* [O.S.] *after a passage of 138 days during which we not only did not lose a single man, but also had no one ill...*' [6] (fo. 2v). But the crew status report, giving details of any promotions, deaths, or desertions, which was prepared on *Vostok* and then signed off on the day they returned to Kronstadt by Admiral Moller, the port admiral, tells a completely different story. According to that document Able Seaman Fyodor Istomin died of typhoid fever on Monday, 21 February 1820 (Moller 1821a: fo. 26). His illness and death took place on Lazarev's ship, *Mirnyj*, at the end of the first Antarctic season, which Lazarev claimed had been free not merely of fatalities, but of casualties of any sort. Two pages later, he noted that the expedition returned to Rio de Janeiro on 13 March 1821, after their second Antarctic season, and continued: '*That time we were not so fortunate with the people, as the first. One of my seamen died of typhoid fever, and on Vostok one man fell,*[6] *and another went overboard at night ...*' [7] (fo. 4v). But according to the crew status report the second and third fatalities occurred during the second phase of the voyage, which was an exploration of the South Pacific, thus not in the third phase, their second Antarctic season, where Lazarev placed them.[7] Lazarev had distinguished clearly between those two phases (tropical and second Antarctic), pointing out that the Russians spent 50 days at Sydney between them. If he had recalled events correctly, therefore, he should have assigned the second and third deaths to the tropical phase, instead of saying that all three deaths occurred in the third phase, during which no one actually died (Bulkeley 2014a: 169).

The expedition's chaplain was on board *Mirnyj* in February 1820. There was high wind and heavy snow on the day Istomin died, but the weather improved next day and was fine on Wednesday the 23rd, when Bellingshausen noted that *Mirnyj* had fallen behind. We can infer that

[6] Literally 'killed himself', but the expression was regularly applied to people who died from a fall which was in some sense their own fault, whether or not it was intentional. Moller's report reads 'fell'.

[7] Bellingshausen's versions of the three deaths, in his reports and subsequent book, were in line with the crew status report, but he gave only an approximate date and no name for the first one, that of Istomin. Despite the salience of the expedition for Russian naval history, the present author was the first person ever to identify him, in 2014.

Istomin was buried at sea with due ceremony, including the presence of his captain, Lazarev, probably on 23 February. The death and burial must have been entered in the ship's log. Despite that, Lazarev misdated Istomin's death by more than eight months, to somewhere between 2 November 1820 and 11 March 1821. The other two deaths took place on *Vostok* on 23 May and 11 September 1820, but were also misdated by Lazarev to the later period. *Mirnyj* may have been told about the second and third deaths, if only so that the chaplain could offer the appropriate prayers,[8] but they may not have been entered in her log. But Istomin's death alone is sufficient to disprove the claim that Lazarev checked his facts in *Mirnyj*'s log while writing the letter, as was once imagined by Belov (1961b: 11). It also seems unlikely that Lazarev omitted the first, and for all he knew at that point perhaps the only death on the voyage, from his personal journal, which would normally have amplified *Mirnyj*'s log rather than condensing it.

Lazarev's problem with the date of Istomin's death may not have been solely due to the passage of time and the unavailability of ship-board records when he wrote the letter in October 1821. On 17 March 1820 Bellingshausen gave orders for the ships to separate and proceed to Sydney independently. Lazarev's report on *Mirnyj*'s independent passage, which lasted until 19 April, was entered in *Two Seasons*. In the entry for 29 March Lazarev recorded that fine weather had provided a welcome opportunity to air the ship thoroughly, and that all on board were healthy. He went on to praise his surgeon, Nikolaj Galkin (1793–1859), for his untiring labours and 'extreme care for the preservation of the health of all on board' (Bellingshausen 1945 1: 175). Those words were not strictly inaccurate, because Istomin had died over three weeks before the period covered by the report. But they suggested that the health of the entire crew had been preserved during the passage from Rio, not just that of those present, minus Istomin, on 17 March 1820. In short Lazarev may already have hinted, at the time, at the cosmetic revision of *Mirnyj*'s bill of health for 1820 which he later presented to Shestakov.

---

[8] Although Bellingshausen was a Lutheran, and possibly something of a Deist along Enlightenment lines, he mentioned in *Two Seasons* that he was careful to respect the men's beliefs wherever possible.

Main Ice

As was noted in Chap. 5, on the third page of his letter Lazarev wrote as follows:

> *On 16 January* [O.S.] *we reached latitude 69°23′S, where we met main ice of extraordinary height. It was a fine evening, and looking out from the crosstrees it stretched just as far as our gaze could reach, but we had not long to enjoy that amazing spectacle, because the murk quickly came over again and the usual snow set in. That was in longitude 2°35′W of Greenwich. From there we held our course east, pushing south at every opportunity, but we always met an ice floe main before we reached 70°* [8]. (Lazarev″ 1821: fo. 2)

We have to start with Lazarev's language, because he applied a surprising adjective to 'main' in the last sentence. The word *l'dinnyj* had only one meaning, 'pertaining to ice floes' (Dal′ 1865: 846). In Bellingshausen's usage, however, and so probably also in Lazarev's, a 'floe' could be of any size, up to and including ice islands scores of metres high and several kilometres long.[9] If Lazarev meant what he said about an 'ice floe main', and his actual words are all we have to go by, then he was probably in agreement with Bellingshausen, who thought that the 'hilly, solid standing ice' sighted on 17 February 1820 had been formed by ice islands fusing together, and showed as much on his track chart by representing large continuous ice fields with an agglomeration of the dark blue patches with which he represented ice islands (Fig. 3.1). Furthermore the commander once reported an ice floe as looking like a shore in poor visibility, described more than one as 'enormous' or 'high' [9], and himself referred to the fused together 'hilly, solid standing ice' as a 'main coast' (Bellinsgauzen″ 1831 1: 135, 153, 169, 188–189). Soviet and Russian commentators, however, doubtless content with the reference to lofty main ice, paid no attention to Lazarev's actual wording two sentences later, and treated *l'dinnyj* as a slip of the pen for *l'dyanoj* ('made of ice') without discussion.[10]

When he wrote the letter Lazarev was on leave after an extremely strenuous voyage and 'baching' with his brother Andrej at Kronstadt. He described it himself as a 'brief account' [10] rather than a definitive one,

---

[9] In March 1820 Lazarev himself measured something which Bellingshausen called both an ice island and a floe as rising about 122 m above sea level (Bellinsgauzen″ 1831: 1, 223).

[10] The present author made the same mistake himself a few years ago (Bulkeley 2014a: 168).

and it contains some mistakes and exaggerations. The demonstrable inaccuracies suggest that while writing the letter Lazarev did not consult records kept on the voyage. Like other commanders of exploring expeditions, Bellingshausen had been instructed to collect all his officers' journals at the end of the voyage.[11] The squadron's papers were then assembled at the Ministry in St Petersburg. The exact timing of that process is not known, but it can be established from surviving records that Bellingshausen stayed with *Vostok* until 28 August 1821, the day after both ships were taken into the naval harbour at Kronstadt to be decommissioned (Moller 1821b; Anon. 1821/1822, fo. 53v). The equivalent record for *Mirnyj* has not survived, but assuming that Lazarev did the same, then he wrote to Shestakov almost six weeks after parting from the squadron and, almost certainly, all the squadron's records including any created by himself.

Doubtless Lazarev could have made a few notes before handing in his journal, but if he later mistook what one of them referred to, he would not have been the first or last person to do such a thing. That is speculation of course. But however his slips of memory came about, they offer the possibility that the divergence between Lazarev's coast-like description of 28 January 1820 and the less impressive record of impenetrable pack ice that day, left by Bellingshausen and Simonov (and much later also Novosil'skij) for the same day, may have been due to another minor lapse of memory on Lazarev's part. It is also relevant that he was writing informally and spontaneously, whereas Bellingshausen was duty bound to consult the day-by-day records of the expedition while writing his reports and, later, his book.

A persuasive argument against that interpretation, however, is that in his letter Lazarev gave positions for the ships on 28 January 1820 which were close to those given for that day in *Two Seasons*, and nowhere near those given for 18 February in Bellingshausen's reports and book.[12] To be precise, Lazarev's latitude for 28 January was close to Bellingshausen's, but his longitude was closer to his commander's for 30 January. However on the latter date they were becalmed and out of sight of ice. By 2 February

---

[11] Naturally Bellingshausen often referred to Lazarev in *Two Seasons*, and many of those details, such as down which side of the Goodwin Sands Lazarev took *Mirnyj* on 10 August 1819, are likely to have been taken from *Mirnyj*'s log or Lazarev's journal rather than anything noted down on *Vostok*.

[12] There is no position for 17 February in *Two Seasons*, but the dead reckoning position for noon on that day in the navigational chart was quite close to the observed position for the following day, with a southerly probe into the ice fields between them.

1820, with light winds and clear weather, they had returned to a similar latitude to 28 January but about 42 km further East—69°25′S 1°11′W. For that later date Bellingshausen described the ice field as enclosing up to 50 gigantic ice floes and extending further than they could see (Bellinsgauzen″ 1831 1: 177). He added that the ice observed on 2 February was probably a continuation of what they had seen in poor visibility on 28 January (the same).

The brief spell of fine weather recalled by Lazarev on the evening of 28 January was not confirmed by Bellingshausen, who described that afternoon as one of reduced visibility due to snow which continued into a murky and snowy night. There had been occasional glimpses of the sun in the morning, positional coordinates were measured at 9 a.m. and again at noon, and the threatening ice fields were glimpsed in the afternoon (Bellinsgauzen″ 1831 1: 171–172). To sum up these comparisons, Bellingshausen's 2 February resembled Lazarev's 28 January in latitude, visibility, and (roughly) the height of ice formations; their accounts of those two days differed in longitude, and over whether the ice was continuous (Lazarev) or divided into separate, very large icebergs (Bellingshausen—see Table 7.1). It is reasonable to assume that Lazarev agreed with his commander's view that the same ice field was encountered

Table 7.1 Comparison of Lazarev on 28 January with Bellingshausen on 2 February 1820

28 January in Lazarev's letter	2 February in Bellingshausen's book
(a) On 16th January [O.S.] we reached latitude 69°23′ S, where we met main ice of extraordinary height. It was a fine evening, and looking out from the cross-trees it stretched just as far as our gaze could reach, but we had not long to enjoy that amazing spectacle, because the murk quickly came over again and the usual snow set in [11]. (Lazarev″ 1821: fo. 2)	(b) We sailed further South with a gentle breeze from SEbE and clear weather. …; it grew brighter in the South from hour to hour. … further South there were up to 50 ice giants of various shapes, enclosed within an ice field. Surveying that ice field … we were unable to see any limits to it; doubtless it was a continuation of the one we saw in gloomy weather on 16th (28) January, but were unable to survey properly because of the murk and snow … We did not enjoy fine weather for long; at 6 o'clock [p.m.] the sky clouded over, and at 8 gloomy weather, with snow and hail, began … [12] (Bellinsgauzen″ 1831 1: 177–178)

on both days. Given the poor visibility on 28 January, therefore, Lazarev could perhaps have condensed the double probe, which took six days in all, into a single event for the purposes of an informal letter, and have somehow recalled a continuous ice field enclosing massive bergs as if it had been a solid continuous body of land-like ice. Alternatively, and given the scale of his dating errors over the seamen's deaths, he may perhaps have misdated the encounter with main ice by three weeks. Setting speculation to one side, his factual description of their seven weeks of ice navigation in 1820, between leaving the South Sandwich Is. and turning away north to head for Australia, was given in just seven lines of a letter more than 300 lines long, and 28 January was the only date provided in that very brief summary.

The difficulty for those claiming priority of discovery is that *neither* part of the late January to early February probe, as recorded by Bellingshausen, resembles Lazarev's '*main ice of extraordinary height*' all that closely. The only passages from Bellingshausen which do so are the descriptions, in his reports and book, of what the expedition encountered three weeks later, but those differ greatly as to the date and position at which the phenomenon was encountered. It is all very well to observe that memory is notoriously fallible, and may have been the more so for Lazarev after an interval of 20 months which had included much further stressful ice navigation. But rather than resolving the differences between the two officers about this historically significant phase of the voyage, that merely leaves them vexingly unsettled.

Another significant aspect of Lazarev's version of 28 January is that whereas Bellingshausen described only one encounter with main ice, on 17–18 February 1820, Lazarev claimed that more than one such event occurred. Perhaps the two men meant different things by the rarely used expression, but they probably both saw main ice as an extreme case of continuous ice (*sploshnyj lyod*), a continuity of massive ice islands which in Bellingshausen's phrase had 'fused together'. When Lazarev repeated the point about being unable to reach 70° later on, he explained that '*impassable ice fields*' [13], rather than an '*ice floe main*', were responsible (fo. 3v).

Lazarev was fluent in English, after serving in the Royal Navy for several years, but 'main ice', an expression used by Arctic whalers to refer to a very large expanse or 'float' of compact ice, was deployed rather freely in English. Recent accounts of Cook's third voyage to the Pacific made it clear that 'main ice' might apply to ice which only 'reached thirty feet or more, under the water' (Anon. 1793: 221). Even if we suppose that Cook had underestimated the submerged portion of the ice, which he had studied from a boat, its height above water would have been less than three metres. Cook used the expressions 'main ice', 'main field', and 'field' interchangeably for what his abridging editor later simplified as 'main ice', and stated that in his opinion, far from being 'fixed', all the ice he observed at that point was 'moveable' (Cook 1784: 460–464). In short Cook's 'main ice' in the Arctic resembled Bellingshausen's 'continuous ice' (*sploshnyj lyod*) in the Antarctic, whereas Bellingshausen and Lazarev reserved 'main' for high ice, particularly when it extended out of sight.

Whatever Lazarev meant by an 'ice floe main', he did not mean a mainland or continent of the usual sort. Had he done so, he would hardly have refrained from pointing out to his friend, when he (Lazarev) mocked the British for (as he thought) wrongly claiming such a discovery (fo. 4v), that the Russians, by contrast, had actually achieved it. But he made no such comparison or claim. Furthermore, it is literally incredible that a man of Lazarev's proactive temperament should have kept silent about such an achievement, if he believed in it, not merely at the time but for the remaining 30 years of his life.

* * *

The reliability of Lazarev's claim to have sighted main ice on 28 January 1820 cannot be confirmed from the available evidence. The partial resemblance to the expedition's situation on 2 February, together with his mention of multiple encounters with an 'ice floe main' which might include the one recorded in some detail for 17–18 February 1820 by Bellingshausen, suggests that Lazarev's brief summary of events was accurate enough for a private letter to a friend despite some dating errors. The lack of corroboration from Bellingshausen or Simonov and the presence of errors must, however, be counted against the statement about 28 January 1820. As we shall see, some Soviet authors tried to explain the inconsistency between Bellingshausen's and Lazarev's account of 28 January 1820 with the hypothesis that Bellingshausen had also recorded 'main ice' for that day in the draft of his book, only for his editor, or in one version the Minister of Marine, to rewrite the passage before it was published.[13] But no one

---

[13] One document, overlooked by Russian commentators, allows us to see an editorial hand at work on a text by Bellingshausen, namely the autograph covering letter which he sent to the Minister of Marine together with his report on the first Antarctic season (Bellinsgauzen″

putting forward that argument has ever explained either why in that case Bellingshausen's account of 17 February, which resembles Lazarev on 28 January, was *not* removed, or why Bellingshausen made exactly the same distinction—main ice in February but not in January—in his original unedited reports. An alternative reconciliation between the two narratives, according to which Lazarev's memory may have conflated sightings from two probes, each of which extended over several days and one of which, in February, he did not refer to separately, has been proposed on the basis of actual rather than imaginary texts.

It comes down to a balance of probabilities. It is more likely that Lazarev made a minor dating error, much smaller than his dating errors for the deaths, than that Bellingshausen gave such a detailed and emphatic description of land-like ice cliffs seen on 17 February 1820 but stayed silent about a similar phenomenon encountered on 28 January, and did so in three separate texts, one of them written less than three months later and with all the relevant ship's documents to hand. In the judgement of the present author, the evidence against Lazarev's sighting date of 28 January is too strong for it to be accepted as historical fact.

Apart from Andreev's curious remark about the date of Lazarev's letter, no Russian commentators have ever discussed its evidential qualities. As we shall see, however, two of them, Belov and Koryakin, went to some pains either to provide independent arguments for the priority of Russian discovery, or to 'improve' the text of the letter itself. Both seem to have felt that the letter as it stood was insufficient to demonstrate priority. If so, the present author agrees with them.

ORIGINAL TEXTS OF QUOTATIONS

[1] Письма печатаются съ копіи, съ соблюденіемъ современной орѳографіи; ...

[2] ... не имеет даты;

1820b). Small changes were marked up on the original letter, either by the Minister or by the editor of the magazine, before an excerpt was published (Bellinsgauzen″ 1821c). However while the alterations may have improved on Bellingshausen's style, nothing was done to alter his meaning. Changes were also made to the extract from his report which preceded the extract from his letter in the same publication (Bellinsgauzen″ 1821b), for example by transposing it from the first to the third person, but whoever made them, those changes were not marked up on the original document.

[3] *... блуждали какъ тѣни цѣлый мѣсяцъ ...*

[4] *... отставились прямо въ Россию и никуда уже не заходили.*

[5] *... чрезъ пять дней ...*

[6] *... 7-ого апрѣля послѣ 138-ого дневнаго плаванія, въ продолженіи коего не только не лишились мы не одного человѣка, но не имѣли больныхъ ...*

[7] *Въ сей разъ небыли мы однакожъ такъ щастливы, какъ въ первый нащётъ людей у меня умеръ одинъ матрозъ Нервною Горячкою, да на Востокѣ одинъ убился, а другой упалъ въ морѣ ночью ...*

[8] *16-го генваря [ст.ст.] достигли мы широты 69°23′S гдѣ встретили матерой ледъ чрезвычайной высоты и въ прекрасный тогда вечеръ, смотря на саленгу, простирался оный такъ далеко какъ могло только достигать зрѣніе, но удивительнымъ симъ зрѣлищемъ наслаждались мы недолго, ибо вскорѣ опятъ напасмурило и пошелъ по обыкновенію снѣгъ. Это было въ долготѣ 2.°35.′Wbій отъ Гринвича.—отъ Сюда продолжали мы путь свой къ Осту покушаясь при всякой возможности къ Зюйду, но всегда встрѣчали льдяной материкъ не доходя 70°.*

[9] *... огромную ... ... высокихъ ...*

[10] *... скажу кое что вкраткиѣ.*

[11] 16-го генваря [ст.ст.] достигли мы широты 69°23′S гдѣ встретили матерой ледъ чрезвычайной высоты и въ прекрасный тогда вечеръ, смотря на саленгу, простирался оный такъ далеко какъ могло только достигать зрѣніе, но удивительнымъ симъ зрѣлищемъ наслаждались мы недолго, ибо вскорѣ опятъ напасмурило и пошелъ по обыкновенію снѣгъ.

[12] Мы продолжали путь на Югъ, при тихомъ вѣтрѣ отъ SOTO и ясной погодѣ. ... на Югѣ становилось часъ отъ часу свѣтлѣе. ... далѣе къ Югу представилось до пятидесяти льдяныхъ разнообразныхъ громадъ, заключающихся въ срединѣ льдянаго поля. Обозрѣвая пространство сего поля ..., мы не могли видѣть предѣловъ онаго; конечно было продлженіемъ того, которое видѣли въ пасмурную погоду 16го Генваря, но по причинѣ мрачности и снѣга хорошенько разсмотрѣть не могли. ... ... Недолго наслаждались мы ясною погодою, въ 6 часовъ небо покрылось облоками, а въ 8 нашла пасмурность съ снѣгомъ и градомъ, ...

[13] *... неподвижные льды ...*

REFERENCES[14]

Andreev, A.I. 1949a. Russkie v Antarktike v 1819–1821 gg. [Russians in the Antarctic 1819–1821]. *Izvestiya Akademii Nauk: seriya istorii i filosofii* 6 (1): 77–78.

———. 1949b. Èkspeditsiya F.F. Bellinsgauzena—M.P. Lazareva v Yuzhnyj Ledovityj okean v 1819–1821 gg. i otkrytie russkimi moryakami Antarktidy [The Expedition of F.F. Bellingshausen and M.P. Lazarev to the Southern Ice Ocean in 1819–1821, and the Discovery of Antarctica by Russian Seamen]. In *Plavanie shlyupov «Vostok» i «Mirnyj» v Antarktiku v 1819, 1820 i 1821 godakh*, ed. A.I. Andreev, 5–16. Moscow: Geografgiz.

———. 1949c. *Plavanie shlyupov «Vostok» i «Mirnyj» v Antarktiku v 1819, 1820 i 1821 godakh [The Antarctic Voyage of the Sloops Vostok and Mirnyj in 1819, 1820 and 1821]*. Moscow: Geografgiz.

Anon, ed. 1793. *Captain Cook's Third and Last Voyage to the Pacific Ocean*. London: Fielding and Stockdale.

———. 1821/1822. Schedule of Rations Consumed on *Vostok*. n.d. but after August 27. SARN—F-132, S-1, P-813, fos 45–53v.

Balkli, Rip. 2013. Pervye nablyudeniya materikovoj chasti Antarktidy: popytka kriticheskogo analiza [First Sightings of the Mainland of Antarctica: Towards a Critical Analysis]. *Voprosy istorii estestvoznaniya i tekhniki* 4: 41–56.

Bellingshausen, Captain. 1945. *The Voyage of Captain Bellingshausen to the Antarctic Seas, 1819–1821*, ed. Frank Debenham, 2 vols. London: Hakluyt Society.

Bellinsgauzen", F.F. 1820b. Personal letter to the Marquis de Traversay from Sydney, April 20 1820. St Petersburg: SARN—F-166, S-1, P–660b, fos 246–249v.

———. 1820c. Report to the Marquis de Traversay from Sydney, November 2 1820. St Petersburg: SARN—F-166, S-1, P–660b, fos 354–359.

Bellinsgauzen", Kapitan". 1821b. Vypiska iz" doneseniya Kapitana 2 ranga Bellinsgauzena k" Morskomu Ministru ot" 8 Aprelya 1820 goda [O.S.] iz" Porta Zhaksona [Extract from the Report of Junior Captain Bellingshausen from Port Jackson to the Minister of Marine, April 8 1820 [O.S.]]. *Syn" Otechestva*, April 23 1821 (O.S.), 69: 133–135.

———. 1821e. Report to the Marquis de Traversay from Kronstadt, August 5 1821. St Petersburg: SARN—F-203, S-1, P–826, fos 1–18v.

---

[14] Key to archival references: SARN = State Archives of the Russian Navy, F = Fond, S = Series [*Opis'*], P = Piece [*Delo*], fo./fos = folio/s [*list/y*], v = verso [*oborotnoe*]. (The latter is necessary because verso pages were usually unnumbered.) Dates on manuscript documents are given in Universal Time (CE).

128   R. BULKELEY

————. 1831. *Dvukratnyya izyskaniya v" yuzhnom" ledovitom" okeanye i plavanie vokrug" svyeta v" prodolzhenii 1819, 20 i 21 godov"* [*Two Seasons of Exploration in the Southern Ice Ocean and a Voyage around the World, During the Years 1819, 1820 and 1821*], ed. L.I. Golenishhev"-Kutuzov", 2 vols. plus *Atlas*. St Petersburg: Glazunovs.

Belov, M.I. 1961b. Otchyotnaya karta pervoj russkoj antarkticheskoj èkspeditsii [The Official Chart of the First Russian Antarctic Expedition]. *Byulleten' Sovetskoj Antarkticheskoj Èkspeditsii* 31: 5–14.

Bulkeley, Rip. 2014a. *Bellingshausen and the Russian Antarctic Expedition, 1819–21*. Basingstoke: Palgrave Macmillan.

Campbell, R.J. 2000. *The Discovery of the South Shetland Islands*. London: Hakluyt Society.

Cook, James. 1784. *A Voyage to the Pacific Ocean: Undertaken, by the Command of His Majesty, for Making Discoveries in the Northern Hemisphere: ... in the Years 1776, 1777, 1778, 1779, and 1780*. Vol. 2. Dublin: Chamberlain and Others.

Dal', V.I. 1865. *Tolkovyj slovar' zhivago velikorusskago yazyka* [*Reference Dictionary of the Greater Russian Language*]. Vol. 2. St Petersburg–Moscow: Ris'.

von Kotzebue, O., and J.F. Eschscholtz. 1830. *A New Voyage Round the World in the Years 1823, 24, 25, and 26*. London: Colburn and Bentley.

Lazarev", M.P. 1821. Pis'mo Mikhaila Petrovicha Lazareva k" Aleksyeyu Antonovichu Shestakovu, 24 sentyabrya 1821 [A letter from Mikhail Petrovich Lazarev to Aleksyej Antonovich Shestakov, September 24, 1821 [O.S]]. St Petersburg: SARN—F–315, S–1, P–775, fos 1–6v.

————. 1918. Letter 1 in: Pis'ma Mikhaila Petrovicha Lazareva k" Aleksyeyu Antonovichu Shestakovu v" g. Krasnyj Smolenskoj gubernii [Letters from Mikhail Petrovich Lazarev to Aleksyej Antonovich Shestakov at Krasnyj in the Smolensk Gubernorate]. *Morskoj sbornik"* 403 (1): 51–66.

Moller, A.V. 1821a. Raport [Report]. August 5. SARN—F–203, S–1, P–826, fos 23–26.

————. 1821b. Raport [Report]. August 27. SARN—F–203, S–1, P–826, fo. 33.

Sementovskij, V.N., ed. 1951. *Russkie otkrytiya v Antarktike* [*Russian Discoveries in the Antarctic*]. Moscow: Geografgiz.

CHAPTER 8

Digging In, 1: Minor Players (1950–1971)

So far this essay has surveyed, first, the accounts of the Bellingshausen expedition that were given by its commander and the astronomer Simonov; then a long period in which those results were broadly accepted but reinterpreted in 1908 by Rabinovich, on grounds that were valid for almost three decades, as the discovery of the continent in January 1821; third, the combination of new geographical information and category shifts, in the context of which Frank Debenham announced a reassessment of its achievements; and fourth, a dramatic reshaping of Soviet ideas about the expedition so that it could now be portrayed as the discovery of Antarctica. That reshaping was quickly followed by re-publication of some of the primary sources, as well as by popular accounts, some of which had considerable merit. But for about two years after the Geographical Society meeting the new Soviet version of the expedition was poorly supplied with evidence, and even its designated spokespeople made their dissatisfaction with the situation evident.[1]

---

[1] The original texts of translated quotations, indicated by numbers in square brackets, are provided at the end of each chapter. For the system of romanization used, the rendering of dates, and a few other matters, please consult the Apparatus at the end of the book.

© The Author(s), under exclusive license to Springer Nature Switzerland AG 2021
R. Bulkeley, *The Historiography of the First Russian Antarctic Expedition, 1819–21*,
https://doi.org/10.1007/978-3-030-59546-3_8

THE EARLY 1950S

With much still to do, studies of the expedition flourished in the 1950s and 1960s. The possible significance of Lazarev's letter was soon understood (Adamov 1951: 41),[2] although as mentioned in the previous chapter no Russian historian has submitted it to critical analysis from that day to this. Meanwhile *Around the World* continued to underplay the story. In January 1952 a brief notice marked the centenary of Bellingshausen's death, which fell in January like the anniversary of the supposed discovery. He was called a 'hero' and 'discoverer' of Antarctica, rather than 'first discoverer' (*pervootkryvatel*) and no date, whether January or February, 1820 or 1821, was provided for that event (S. Markov 1952).[3]

Between 1949 and 1954 the priority claim, as championed by A. Grigor'ev and D. Lebedev, was taken up by others, usually tertiary authors accustomed to relaying other people's ideas (E.A. 1949; Osipov 1950; Kotukhov 1951—not seen). In the second edition of his pamphlet Kotukhov went far beyond his evidence when he appropriated Novosil'skij's description of the heavy pack encountered on 26 January 1820 (thus) to be 'the Southern Continent' [1] (Kotukhov 1955: 63). (His witness, Novosil'skij, made no such claim himself.) Shvede's revised version of his historical essay in the 1949 edition of Bellingshausen, for a pamphlet, was the most significant of the period. In it he finally quoted Lazarev's description of 28 January 1820, which he had not done in 1949 (Shvede 1952: 18). Even the most Lazarev-centred work from the early 1950s (apart from the *Dokumenty*) had failed to do so (Sokolov and Kushnarev 1951).

---

[2] The author could not discover which Soviet commentator was the first to draw the attention of others to Lazarev's statement about 28 January 1820. It may even have been Adamov, but without realizing what an important step he was taking.

[3] Ovlashhenko's diligent pursuit of dissenters (Chap. 11) has revealed that *Around the World* resisted the revised version of Bellingshausen's achievements throughout the 1950s (Ovlashhenko 2016: 106, 137, 312–313, 338, 394). With a print run which rose from 80,000 to 165,000 copies a month in that decade, that was a significant act of defiance. How a publication belonging to the All-Union Leninist Young Communist League (*Komsomol*) can have got away with such dissent, especially with the *Zhdanovshhina* still at its height between the 1949 Geographical Society meeting and Stalin's death four years later, remains something of a mystery. Given that the magazine succeeded in bridging the interval between the traditional view of the expedition, which was still officially accepted in 1948, and the slightly less restrictive years of the Khrushhev Thaw, its position on Bellingshausen for the rest of the Soviet period might be worth tracing, a project that is probably beyond the resources of the present author.

After Shvede's important second edition of *Two Seasons*, the second most valuable scholarly work on the expedition in this period was the first of three volumes of archival documents setting out the brilliant career of Mikhail Lazarev (Samarov 1952).[4] Besides Lazarev's letter, the *Dokumenty* included seven of Bellingshausen's reports from the voyage, only one of which had been published in the nineteenth century. But all was not what it seemed. Just as Andreev had done with Simonov, Colonel Aleksandr Alekseevich Samarov, the director of the Naval Archives, removed Bellingshausen's acknowledgement of the 'essential requirements' [2] that were provided and for the most part fitted for no charge by the Royal Navy at Sydney, and toned down some of the professions of loyalty to the Emperor. A salute to Cook as '*the glorious navigator*' [3] was revised to 'leading' (Bellinsgauzen" 1820c: fos 358, 356; Samarov 1952: 171, 170). Furthermore, despite the fact that the full text of Lazarev's letter to Shestakov had just been published in both the Andreev and the Sementovskij editions of the anthology, Samarov cut passages in which Lazarev went into detail about the repairs in Sydney Harbour or waxed too fulsome in his gratitude for British support.

Samarov's most serious interference with Bellingshausen's text concerned the discovery question. In his report from Sydney dated 8 April 1820 (O.S.), Bellingshausen summarized that season's work with the remark that: '*I met with no signs anywhere of a great southern land, even though I held most of my course beyond the polar circle or near it...*' [4] (Bellinsgauzen" 1820a: fo. 244v). He repeated that assessment (hereafter 'the Remark') in an accompanying letter to de Traversay and again in the final report which he submitted on returning to Kronstadt in 1821, neither of which was included in the *Dokumenty*. The Remark was then published in the summary and again in the full text of his first report from Sydney (Bellinsgauzen" 1821b, 1823). Under the modern interpretation, first adopted by Debenham, which holds that if at any point Bellingshausen sighted one of the massive and (then) almost permanent coastal ice structures of Antarctica, such as an ice shelf, ice barrier or ice tongue, he should be deemed to have sighted the mainland, the Remark has little or no

[4]After the expedition, Lazarev went on to distinguish himself at the Battle of Navarino before becoming chief of staff of the Black Sea Fleet, in which role he oversaw its modernization into the age of steam and surrounded himself with a group of talented young officers who would lead the Imperial Navy in the Crimean War and beyond, after their mentor's death in 1851. For an overview of his career, see Daly (1991: 44–48).

significance. Despite its innocuousness, however, it was evidently too far off message to be laid before the Soviet public in 1952, and the whole paragraph was suppressed, albeit with an ellipsis (Samarov 1952: 148).

This period also saw the first Soviet map to show Alexander I Land—the name was retained—as an island (Isakov 1950: 6, 7). And in 1951 the second, Sementovskij, edition of the anthology added extensive extracts from Simonov's unfinished narrative of the expedition, preserved in the archives of the University of Kazan (Simonov 1951).

CONTINUED DISSENT

Throughout the early 1950s Soviet accounts of the Bellingshausen expedition were prone to factual error. Unusually, one commentator gave examples, making a mistake himself in the process (Guretskij 1954). More seriously, some Soviet commentators continued to ignore the official 1949 backdating of the alleged discovery of Antarctica from 1821 to 1820. At first that may have been a continuation of the initial confusion described above (Uzin 1950: 8). But with the passage of time that explanation becomes less plausible. In 1951 the authors of a popular book about the natural history of the Antarctic repeated the doubts of Debenham and Shvede about an 1820 sighting: 'the continuous ice and poor visibility prevented the Russian sailors from sighting the hills of the mainland' [5]. They also pointed out that the Alexander I Coast (preferring Bellingshausen's *bereg* over the modern *Zemlya*) was not part of the mainland but an island. They nevertheless concluded, rather lamely, that Russian seamen had somehow disproved the previous notion that there was no continent, because they had discovered 'the main links of Antarctica' [6] (Arsen'ev and Zemskij 1951: 4, 6). Three years later the passage was revised in a second edition. The negative assessment of 28 January 1820 was retained, but Alexander I Coast regained its modern status as a Land (*Zemlya*). The overall conclusion was partially clarified; the Russians had 'discovered the main island links of Antarctica' [7]. Still an obscure phrase, but evidently no continent (Arsen'ev and Zemskij 1954: 10, 11). After the end of the Soviet Union one of the authors, Vyacheslav Alekseevich Zemskij (1919–2012), was responsible for releasing accurate whaling statistics, falsified in the Soviet period (Yablokov and Zemsky 2000). That suggests that his earlier refusal to conform over the discovery question was also deliberate. Another example of dissent came from Mikhail Belov, later a champion of the 1820 priority. In 1954 he too declared that the expedition discovered Antarctica in 1821 (Belov 1954: 10). This was a telling example of the limited foothold that had been gained by the new assessment of the Bellingshausen expedition, even after six years.

The prestigious anthology *Russian Navigators*, published by the Academy of Sciences, encapsulated the divergence of opinion after 1949. In his introduction the oceanographer Nikolaj Nikolaevich Zubov (1885–1960) affirmed the newly established discovery date of 28 January 1820 (N. Zubov 1953: xviii). Elsewhere, in thumbnail biographies of Bellingshausen and Lazarev, Magidovich accepted that the expedition discovered Antarctica at some unspecified point, but did not include a coastal sighting in his list of their 1820 discoveries (Magidovich 1953). Lastly, Nikolaj Yakovlevich Bolotnikov (1897–19??) took a Debenham-like position, asserting that the expedition had seen an ice barrier on 28 January 1820 without understanding what it was, but had nevertheless somehow discovered Antarctica. Bolotnikov provided neither quotations nor citations to support his opinion, and his reference to the situation as an ice field scattered with hillocks, which is what Bellingshausen called it, was not consistent with its having also been a massive ice barrier, the grounds on which he based the discovery claim (Bolotnikov 1953: 195–196). He was perhaps not so much a dissenter as someone too muddled to make an effective contribution.

TRACES OF THE KHRUSHHEV THAW

Much was achieved, especially in publishing or republishing primary sources, in the first few years after the Soviet state imposed a re-assessment of the Bellingshausen expedition. Nevertheless, by focusing on narrative and simple assertions of Russian priority, to the detriment of close analysis and articulated interpretation of the historical issues, the Soviet commentators of those early years may have been partly responsible for the continued reluctance of some of their colleagues to accept that the expedition had discovered Antarctica in January 1820. From about the mid-1950s, however, the manner in which some of this work was presented began to change, in response to such developments as the death of Stalin in March 1953, the accession of the Soviet Academy of Sciences to the IGY in 1954 and to the International Council of Scientific Unions in 1955, and the establishment of the Antarctic Treaty, with the Soviet Union as a founding signatory, in December 1959. That cultural shift was perhaps most evident in the work of Vladimir Lebedev and Mikhail Belov (especially Lebedev) in the 1960s, which was so extensive that it will be treated in the following chapter. But even before that it can be detected in the work of minor authors.

The notion of a political thaw has often been applied to the Soviet Union between 1954 and 1964. Because the term has been used with different periodizations about different things, either Soviet foreign relations or relations between Soviet citizens and the state, it is hard to read the concept directly into the Bellingshausen historiography. Broadly speaking however, treatments of the expedition were more open-minded during that phase of Soviet history than before or after it.

From September 1952 onwards the International Council of Scientific Unions sent the Soviet Academy of Sciences repeated invitations to take part in the IGY. After delaying for more than a year the Academy's president, Aleksandr Nikolaevich Nesmeyanov (1899–1980), replied in the affirmative. A Soviet IGY committee was formed in August 1954 and a delegation attended a preparatory conference at Rome the following month. More significantly in the present context, in July 1955 the Soviet IGY committee finessed its way into the first IGY Antarctic conference at Paris despite an attempt by the American vice-president of the IGY, Lloyd Berkner (1905–1967), to exclude them (Bulkeley 2008). Because the uninvited Soviet representative, Belousov, reached Paris too late to compete with the American delegation for the South Pole Station, the Soviet Antarctic Expedition (SAE) selected a coastal site near the medial meridian of the Soviet Union for its support base, *Mirnyj*, and went on to set up a major inland station called *Vostok* in East Antarctica. The names and roles of Bellingshausen's ships were thus commemorated.

Once the Soviet Union set about transferring much of its extensive polar science capability, both human and material, from the Arctic to the Antarctic in late 1955, commentary on the Bellingshausen expedition began to lose some of its xenophobic edge.[5] In the late 1950s scientific reports from the SAE began to appear and the Antarctic Treaty was signed by the Soviet Union as a founding member. Simultaneously Soviet achievements in space, some of them under the nominal aegis of the IGY, such as the launch of Sputnik 1 on 4 October 1957, greatly boosted the national self-confidence. The choice of the name *Vostok* (*East*) for Yuri Gagarin's

---

[5] The process mirrored the redirection of Arctic historians to the Antarctic a decade earlier (Chap. 9).

orbital space capsule in April 1961 may also have referenced the expedition, but that is conjecture.[6]

From the outset, the literature on Soviet Antarctic research, while properly enthusiastic, showed few of the exaggerations and distortions of the Arctic literature of the 1930s (McCannon 1998), for several reasons: Stalin was dead; Antarctica was not a national Soviet backyard, inhabited by native and conscript citizens; Soviet Antarctic activity was not a net contributor to the national economy; and Antarctic research was an internationally coordinated programme under which the various national teams would share each other's results and, after 1959, were entitled to inspect each other's installations. The book *Mysterious Continent* by Lev Borisovich Khvat, a senior *Pravda* journalist, reflected that cultural shift (Khvat 1956). After opening with a typical semi-fictional treatment of the Bellingshausen expedition, Khvat surveyed the mid-nineteenth-century expeditions, the heroic age, and recent expeditions, including the US Navy's *Operation Highjump*, in an objective spirit that would have been anathema a few years earlier. That was possible because the book could end with a section on the first SAE, complete with photographs of Soviet explorers on the ice in Antarctica and mention of a friendly visit by Australian IGY scientists in January 1956.

A scientific report on the SAE was included in *Mysterious Continent* as an appendix (D.I. Shherbakov 1956). Its author, the geologist Dmitrij Ivanovich Shherbakov (1893–1966), was chairman of the Soviet Antarctic IGY committee but did not himself go south. In November and December 1955 his interviews and articles in the popular press were the first written accounts of the SAE to reach the general public. In another sign of the times he said nothing about Bellingshausen either in the appendix or in an earlier article (D.I. Shherbakov 1955), though a companion piece by a journalist had paid the usual tribute (Ryabchikov 1955).

A minor primary source, Simonov's autobiography, was published during this period of readjustment. The MS (since lost) had been transcribed by the daughter of an archivist in 1891, and its only historiographic

---

[6] As with other historical claims in other countries, much non-rational, because essentially tautological, validation has been created for the January 1820 discovery claim through retrospective namings and other memorials, such as medals, paintings, and postage stamps, both in the Soviet period and subsequently. Such memorials and images of the expedition will usually be mentioned here only if they documented historiographic developments. Though fascinating, this socio-psychological dimension of the subject cannot be discussed at length.

interest is that Simonov repeated his earlier claim that Bellingshausen reached 70°S (Simonov 1955: 271).

In 1957 the first historical novel about the expedition chose to represent the 1820 discovery, not as immediate but rather as a retrospective realization in the light of the discoveries of January 1821 (Vadetskij 1957: 171). Much space was devoted to portraying Bellingshausen as reserved and cautious, but Lazarev as bold and daring.

Two other 1957 publications added nothing new. Kublitskij was more polite to Cook this time, almost to the point of sarcasm, but still misinterpreted 'Sandwich Land' and even cut two degrees off Cook's furthest south, thus pulling him back to Bellingshausen's limit. The latter's encounter with solid pack ice on 28 January 1820 was upgraded still further, to 'a low-lying icebound coast' [8] (Kublitskij 1957: 27). In the same year the oceanologist Vladimir L'vovich Lebedev (1930–20??), about whom more later, published a competent survey of all aspects of the Antarctic with nothing remarkable in its historical section (V. Lebedev 1957).

Between 1959 and 1965 Ajzik Vol'fovich Nudel'man (1917–2013), a member of the interdepartmental commission on Antarctica (Chap. 9), published four volumes summarizing the work of the SAE which did much to make it better known (Nudel'man 1959, 1960, 1962, 1965), and which also did so outside Russia after they were translated into English in Israel. The series was discontinued when Nudel'man left the commission to take up the editorship of the journal *Meteorology and Hydrology*.

The third edition of *Two Seasons* appeared in June 1960. Shvede's introduction reflected the new political climate. The indefinitely dated title, 'The discovery of Antarctica by Russian seafarers in 1819–1821' [9], was a lukewarm response to the reappraisal of the expedition in 1949, but no other reference was made to current political affairs, such as the ongoing transition to the Antarctic Treaty System (ATS). The previous cuts in the text were restored, and while still representing Cook's name for the South Sandwich Is. as somehow wrong (regardless of Franz Josef Land etc.), Shvede conceded that Cook had sighted them 'in mist and fog' [10] (Shvede 1960: 34). Furthermore he repeated none of the stock criticisms of Cook's alleged views about a southern continent. The Russian claim to first discovery was of course endorsed, and Simonov's failure to publish most of his scientific results was once again glossed over. But this was for the most part a thorough, scholarly account of the expedition and of Bellingshausen's book, and as such a sound basis for further research. Bogatkina's bibliography was not republished. That may have been due to considerations of cost, but with only 3000 copies as compared with 20,000 in 1949 the third edition was hardly intended for mass distribution. Whatever the reason, the withdrawal of the bibliography made it slightly harder for readers to find either the nineteenth-century sources for Bellingshausen's Remark, or pre-1949 Russian commentators who had accepted the 1821 discovery position.

ALEJNER

Shvede was at his most political in the third edition when he found fault with the British Admiralty for not accepting Bellingshausen's names for the South Shetland Is. (the same: 35). This apparently minor matter is of some importance because after the continental priority issue it is probably the only other aspect of the expedition on which Russian and foreign historians have disagreed, and because the claim that Bellingshausen was the first to name the islands of the South Shetland group is still maintained today, for example, in textbooks and other materials for Russian schoolchildren (Moroz 2001: 39). Shvede's 1960 introduction first appeared two years earlier in an influential collection of studies of the expedition which included three papers by Alejner, from whom Shvede took this claim about the South Shetlands (Alejner 1958a, b). Alejner's argument for Russian priority was that Bellingshausen had mapped them in February 1821, whereas the British names were not published, by George Powell (about 1796–1823), until 1822 (Powell 1822; Alejner 1958b: 411, 413, 435). The obvious flaw in this was the comparison of two different stages in the development of geographical knowledge, field observation for the Russians as against publication for the British. When like is compared with like, as it should be, Alejner's claim is groundless.

With later names in brackets, William Smith began (on this topic) in 1819 by naming Nelson Island, Lloyds (Rugged) Island, Smith's Cape (Island), and the main island or 'mainland' of the group as New South Britain (New South Shetland), and those names were published before Bellingshausen arrived off Smith's Island on 4 February 1821 (Miers 1820). Several more names were added or found to be already in use by sealers during the voyage of exploration commanded by Edward Bransfield, which took the *Williams* to the islands from 16 January to 18 March 1820, and the sealing voyage of the *John*, commander John Walker, a year later. Maps were drawn and in some cases published by British seamen between 1820 and 1822 (R. Gould 1941), and further names from the two voyages just mentioned were published rather haphazardly in 1821 or 1822. They included Livingston, Roberts, Bridgeman's, Desolation, Falcon, Penguin, and Ridley Islands, plus George's Bay at the northeastern end of the main group, which gave rise to the name King George's Island when the latter was found to be separate.[7] Bransfield also discovered and named Tower and Hope Islands and Trinity Land (at the tip of the Antarctic Peninsula) to the south-east of the South Shetlands, and to the north-east the O'Brien Is. and Seal, Clarence, and Cornwallis Islands. Nearby Elephant Island was at first named Belsham (the same; Anon. 1821b; Baird 1821; Purdy 1822: map at 38–39). Cape Aspland was treated at first as a feature of the O'Briens, and later denoted a separate island. All the names listed so far were published between late 1820 and 22 July 1822, a week before Simonov gave his lecture, let alone publishing it.[8] On 1 November 1822 the Powell map referred to by Alejner, which was based on a voyage through the islands between 8 November 1821 and 26 February 1822, added Low, Greenwich, and Deception Islands and the new names for King George's and Elephant Islands.

On the Russian side, the expedition observed the islands between 4 and 9 February 1821, but Bellingshausen's names for them were not published until 1831. He did not name them on the track chart which he drew during the return voyage in 1821, on which he left their north-western coasts blank (Bellinsgauzen″ 1821d; Belov 1963b: Sheet 14)), nor did he do so in his final report (Bellinsgauzen″ 1821e). Unaware that Bransfield and Smith had discovered the Elephant and Clarence group in February 1820, he did claim to have discovered that outlying part of the South Shetlands (Bulkeley 2014a: 116), although only Rozhnov/Gibbs Island had not already been found by the British.[9] But in 1821 Bellingshausen did not name that group either, nor did Simonov a year later. The Russian

[7] The spread of the term 'island' in the sealers' names between 1819 and 1822 demonstrates that contrary to Soviet and modern Russian versions the sealers needed no help from Russian publications, which began (on this topic) with Simonov's Address in the autumn of 1822, to understand that the group was not part of a continental mainland.

[8] See for example (Anon. 1821c), which was published in August 1821, by which point Bellingshausen had barely returned to Russia.

[9] Bellingshausen's claim to have discovered the Elephant and Clarence group was only made explicitly in his final report, which no Russian historians apart from Kuznetsova appear to have consulted. He included those islands without names in the appended table of discoveries and showed them, still unnamed, on the track chart which he prepared at about the same time (Bulkeley 2014a: 116, 122; Belov 1963b: Sheet 14). Although he described his brief exploration and naming of them in some detail in *Two Seasons* he did not repeat the

names were first published, therefore, in Bellingshausen's narrative and accompanying *Atlas* in 1831, and the point at which he started to use them privately, somewhere between submitting his final report in August 1821 and the draft book in October 1824, is therefore impossible to tell.[10]

In 1958, however, Alejner was at several disadvantages. Although excerpts from Bellingshausen's 1821 track chart had begun to appear it would not be published in full until 1963, and V.V. Kuznetsova's discovery of the final report and its table of claimed discoveries came even later. On the British side R. Gould's first, 1925 article was probably in the library of the Geographical Society; at least V. Lebedev managed to consult it by around 1960. (Gould's second, 1941 article is less likely to have reached Soviet libraries.) But Alejner seems largely to have relied on the 1946 reprint of the limited account in the *Literary Gazette* (Anon. 1946b). He also overlooked the map in Purdy's *Memoir*, which had appeared three months before the Powell map which he did know. Had he listened to common sense he might have realized that, with sealers swarming over the islands from the 1819–1820 season onwards, naming would have proceeded apace in order to meet their practical needs for orders and reports, well before the Russians passed briefly through. But given his limited knowledge of both sides of the story it is not surprising that he resorted to the fallacy of false comparison in order to contrive a Russian priority, the only acceptable outcome at the time. Nor is it surprising that Shvede fell into line with his more expert colleague. The persistent repetition of this ill-founded claim down to the present day, despite demonstrations by modern scholars (Hattersley-Smith 1991; Campbell 2000) that with one exception, Rozhnov/Gibbs I., Bellingshausen was not the first person to name the South Shetland Islands, is another matter. But see also Chap. 12.

---

discovery claim there (Bellinsgauzen" 1831 2: 272–277). However it was implicit in his tally of 29 islands at the end of the book.

[10] With no evidence for the Russian names from documents that should have contained them if they had been chosen on the spot in 1821, and because Russian and English conceive and describe the past in slightly different ways, it is impossible to decide whether Bellingshausen's seemingly straightforward 'named' (Bellinsgauzen" 1831 2: 261) should be translated as 'I "named" this X' when I saw it, or as 'I "have named" this X', that is, while subsequently preparing the narrative for publication.

HISTORY OF GEOGRAPHY

At this point it is worth diverging from the sequence of published treatments of the expedition to consider the roles of the Leningrad and Moscow historical sections of the Geographical Society (LHS and MHS). The LHS was a continuation of the pre-war historical section; the need for a separate Moscow group arose when the Society's Moscow Centre was created in 1946. Between 1950 and 1970 the secretary of the MHS, A.V. Sokolov, published six reports on its work from 1947 to 1967 (Sokolov 1950a, b, 1953, 1960, 1969, 1970). During that time the section, chaired by the distinguished historical geographer V.K. Yatsunskij (1893–1966), heard almost 250 presentations, of which about two-thirds were on the history of geographical knowledge and the remainder on historical geography, sources, and new publications. Within the first group several papers discussed the geographical achievements of such eighteenth- and nineteenth-century figures as Stepan Andreev, Chichagov, Dezhnev, Krusenstern, Popov, Rusanov, Sarychev, Wrangell, and Zagoskin. Regular attenders and contributors included N. Zubov, A. Grigor'ev, D. Lebedev, and Solov'ev, all of whom published tertiary items on the Bellingshausen expedition, based on other people's research, during the 1950s. It was not until 1961, however, that two papers on the Bellingshausen expedition were presented at the MHS, both of them on the newly discovered track chart that was about to be published by Belov (Chap. 9).

More may have been heard about the expedition, and sooner, at the LHS, since A.I. Andreev was its chairman throughout the 1950s and Shvede is known to have attended some sessions, but detailed reports on its work were not published. It may seem natural that research into the expedition centred on Leningrad, the home of the Naval Archives, but that did not prevent Moscow scholars from working on other naval explorers, and many primary sources for the Bellingshausen expedition were now in print. However when the Leningrader Belov reported to the MHS in the early 1950s on his discovery of previously unknown Arctic maps from the eighteenth century, from which he concluded that Stepan Andreev had fabricated his claim to have discovered an unidentified island in the 1760s, his argument was vigorously opposed by N. Zubov, D. Lebedev, and A.V. Efimov (Sokolov 1953: 279). Perhaps their response owed something to the ancient rivalry between Russia's two greatest cities, a minor aspect of the Bellingshausen historiography which became more evident in the 1960s. What is certain is that up until 1961 original research into the expedition was produced by only a handful of Soviet scholars, of whom Shokal'skij, Shvede, and Andreev were natives of St Petersburg/Leningrad and Berg had made his career in the city.

* * *

Between 1949 and 1960 expressions such as 'near', 'reach', and 'just by' reflect the critical role of a single factor, proximity. The problem was, however, that there was no way of knowing where the fluctuating ice coast of Antarctica, in the relevant meridian, actually was in January 1820. Soviet authors were at risk of arguing in a circle: the ice coast was at such-and-such a position, because the explorers saw it, and they must have seen it, because it was at such-and-such a position. It is significant that few Russian commentators in this period were familiar with the navigational problems that can sometimes hinder the making of a good landfall even at a well-known and charted destination, let alone when achieving a brand-new discovery. In 1774, for example, the captain of the second ship in Cook's expedition, Tobias Furneaux, sailed through the gap between South Georgia and the main group of the South Sandwich Is. without sighting the Traversay Islands. The leading commentator with navigational experience as a naval officer, Shvede, was extremely doubtful about the January 1820 discovery claim.

THE EARLY 1960S

In 1963 the polar explorer Aleksej Fyodorovich Tryoshnikov (1914–1991), who was also director of the Arctic and Antarctic Research Institute (*AANII*) from 1960 to 1981, published a general history of Antarctic research. Tryoshnikov gave a conventional account of the discovery of Antarctica on 28 January 1820 by relying on Lazarev to supply the lack of evidence from Bellingshausen (Tryoshnikov 1963: 29), but he was well read in the latest work by Belov and Lebedev (Chap. 9) and work by foreign authors. As regards the 1820s, he included a circumstantial account of Bransfield's voyage but failed, uncharacteristically, to list in his bibliography any sources from which he might have learned that 'the map compiled by Bransfield has proved extremely inaccurate' [11] (Tryoshnikov 1963: 29). Taken overall, the book was a competent and well-written survey of 140 years of Antarctic exploration down to the Sixth SAE in 1960–1962, and thanks to its widespread availability in libraries it is still consulted today.

Mention should also be made of one of the first popular Soviet books about Antarctica to make no mention whatsoever of the Bellingshausen expedition, published in the same year (Cherevichnyj 1963). Most of the work on the Bellingshausen expedition by Mikhail Belov, and all of that by Vladimir Lebedev, was also published in the early 1960s (Chap. 9).

Bibliography and Jurisprudence

In 1949 the Soviet Academy of Sciences published the first volume of its *Bibliography of the History of Science* series, covering the period from 1917 to 1947 (Starosel'skaya-Nikitina and Others 1949). Despite important omissions, such as Lazarev's letters (1918) and the third edition of S. Grigor'ev (1937), the proportion of 12 out of 2914 entries (including duplicates) was a fair measure of the scanty Soviet attention to the expedition before 1949. Other bibliographies covered the 1950s (Starosel'skaya-Nikitina and Others 1955; Bogoyavlenskij 1955; Kaminer and Others 1963). Both the 1955 bibliographies gave special prominence to the expedition, and both listed contemporary reading lists, intended to help teachers and students as they adjusted to Bellingshausen's sudden promotion from the background to the foreground of the Russian historical Pantheon. Indeed one of them was itself such a list. The geographical bibliography for secondary schools published by the Ministry of Education reflected the transformation of the political climate from that of a decade earlier (Bogoyavlenskij 1963). It listed works on or by several foreign explorers, including Amundsen, Borchgrevinck, Columbus, Da Gama, Darwin, Franklin, Livingstone, and Magellan, some of whom were allocated two or more titles. However no separate works were listed for the Bellingshausen expedition, which together with others was covered merely by listing anthologies such as Lupach's *Russian Navigators* (above).

During this period Soviet jurisprudents also published their first essays about the legal status of Antarctica, including desirable arrangements for the future (Durdenevskij 1950; Kostritsyn 1951; Molodtsov 1954). By omitting any reference to the status of the Soviet Union either as party to the International Whaling Convention (which *had* been mentioned in the diplomatic Note of June 1950) or as a permanent member of the UN Security Council, they chose to base Soviet rights to participate in an Antarctic regime almost entirely on the Bellingshausen expedition. Kostritsyn, however, also deployed the 'common heritage of mankind' argument (CHM), 16 years before 1967, when one historian of the CHM wrongly supposes that such legal discussion first got under way (Taylor 2011).

None of the jurisprudents offered original historical insights or information. Their most interesting feature was that all three addressed themselves to the region, *Antarktika*, rather than the continent, *Antarktida*. Few, if any, Russians had set foot on the continent since Aleksandr Stepanovich Kuchin (1888–1913?) became the first to do so while serving as an oceanographer with Amundsen's South Pole Expedition (Simakova 2015).

Bibliographers and jurisprudents were doubtless seeking to make original contributions in their respective fields, but since such experts wrote no history apart from that of their own subjects no more such works will be considered.

*   *   *

In November 1971, as the golden age of Soviet Bellingshausen studies was drawing to a close, another theoretical analysis of geographical discovery was sent to press. It looked back, amongst other things, to Magidovich's treatment of the subject in 1949, when the revised assessment of the Bellingshausen expedition was being launched. The historian of exploration Naum Grigor'evich Fradkin (1904–1987) accepted an ascending hierarchy of chance encounter, tracking down a poorly reported landfall or other phenomenon, and full-blown discovery based on systematic exploration (Fradkin 1972: 38–48). As examples of the last two he cited Cook's verification of South Georgia and his discovery of Hawaii. He selected the Bellingshausen expedition as 'an example of scientific territorial discovery' [12] on the grounds of chronological priority in Antarctica, for which he cited but did not recapitulate the recent Soviet literature (the same: 47–48). He also asserted that James Clark Ross had acknowledged Bellingshausen and Lazarev as the first explorers to discover Antarctica, but he (Fradkin) failed to provide any evidence for that claim. All that Ross actually said was that Bellingshausen had discovered 'the southernmost known land' in 1821, an obvious reference to Peter I I. and the Alexander I Coast (J.C. Ross 1847 1: 188). That much has never been at issue, but it has no bearing on the dates of possible sightings of potentially 'continental' ice features a year earlier. It is inconceivable that Ross, who declared that the existence of Antarctica had not yet been established by any explorer including himself, should have saluted Bellingshausen as its discoverer. Nor did he in fact do so.

Fradkin made no attempt to explain how the sort of brief glimpse in poor visibility which he attributed to 28 January 1820, followed by the absence of any claim to have discovered anything during that passage through the ice fields, might qualify as a discovery, though he did make the valid point that initial discovery could not be expected to provide anything like complete knowledge of the territory in question. By focusing on the circumstances of discovery, rather than its essence, he also avoided tackling (despite quoting) Magidovich's awkward stipulation that the discoverer should in some sense arrive in or visit the place in question, as Cook had done on several occasions.

* * *

By coincidence, one of the last Soviet appearances of Bellingshausen's Remark (K. Markov and Others 1968) was approved by the censorship on 10 April 1968, the last day of the April plenum of the Central Committee of the Czechoslovak Communist Party, perhaps the high watermark of the Prague Spring before the Soviet invasion and the imposition of the Brezhnev Doctrine calling for the restoration of Communist orthodoxy and solidarity across the Warsaw Pact countries. The related domestic policies of the Brezhnev government may have been reflected in the falling away of research into the Bellingshausen expedition which was a feature of the later Soviet period (Chap. 10).[11] Before moving on to the final years of the Soviet Union, however, we must complete this survey of the golden age by examining the work of its two leading scholars.

Original Texts of Quotations

[1] ... Южный материк,

[2] ... необходимыѣ потребности...

[3] ...славному мореплавателю [нет главному]...

---

[11] The vigorous Soviet culture of the 1970s included sharp clashes over the social value of history (A. Yakovlev 1972). But, led by dissidents, conducted largely in *samizdat*, and focused on the recent past, it was unlikely to address the historiographic problems of a naval episode from the reign of Alexander I.

[4] *Признаковъ большой Южной земли, нигдѣ ни встречалъ, хотя большую часть плаванія имѣлъ за полярнымъ кругомъ и близь онаго...*

[5] ...сплошные льды и плохая видимость не позволили русским морякам увидеть материковые горы...

[6] ...основные звенья Антарктиды...

[7] ...открыли основные островные звенья Антарктиды...

[8] ...обледеневший низменный берег...

[9] Открытие Антарктиды русскими мореплавателями в 1819–1821 гг.

[10] ...во мгле и тумане...

[11] ... карта, составленная Брансфилдом, оказалась весьма неточной.

[12] В качестве примера научного территориального открытия...

REFERENCES[12]

Adamov, Arkadij. 1951. Bessmertnyj podvig russkikh moryakov [An Immortal Victory by Russian Seamen]. *Nauka i zhizn'* 1: 41–43.

Alejner, A.Z. 1958a. Osnovnye ètapy geograficheskogo issledovaniya Antarktiki [Basic Stages in Antarctic Exploration]. In *Antarktika: materialy po istorii issledovaniya i po fizicheskoj geografii*, ed. E.N. Pavlovskij and S.V. Kalesnik, 55–66. Moscow: Geografgiz.

———. 1958b. Geograficheskie naimenovaniya v Antarktike [Geographical Nomenclature in the Antarctic]. In *Antarktika: materialy po istorii issledovaniya i po fizicheskoj geografii*, ed. E.N. Pavlovskij and S.V. Kalesnik, 407–443. Moscow: Geografgiz.

Anon. 1821b. New Shetland. *Literary Gazette* 5: 691–692, 712–713, 746–747.

———. 1821c. New Shetland. *The Philosophical Magazine and Journal* 58 (280): 144–145.

———. 1946b. Edward Bransfield's Antarctic Voyage, 1819–1820, and the Discovery of the Antarctic Continent. *Polar Record* 4: 385–393.

Arsen'ev, V.A., and V.A. Zemskij. 1951. *V strane kitov i pingvinov [In the Land of Whales and Penguins]*. Moscow: Moscow Society of Naturalists.

———. 1954. *V strane kitov i pingvinov*. 2nd ed. Moscow: Moscow University Press.

---

[12] Key to archival references: SARN = State Archives of the Russian Navy, F = Fond, S = Series [*Opis'*], P = Piece [*Delo*], fo./fos = folio/s [*list/y*], v = verso [*oborotnoe*]. (The latter is necessary because verso pages were usually unnumbered.) Dates on manuscript documents are given in Universal Time (CE).

146   R. BULKELEY

Baird, G.H. 1821. Latitude and Longitude of Places in New South Britain. *The Edinburgh Philosophical Journal* 5 (9): 233.

Bellinsgauzen″, F.F. 1820a. Report to the Marquis de Traversay from Sydney, 20 April 1820. St Petersburg: SARN—F–166, S–1, P–660b, fos 239–245v.

———. 1820c. Report to the Marquis de Traversay from Sydney, 2 November 1820. St Petersburg: SARN—F–166, S–1, P–660b, fos 354–359.

Bellinsgauzen″, Kapitan″. 1821b. Vypiska iz″ doneseniya Kapitana 2 ranga Bellinsgauzena k″ Morskomu Ministru ot″ 8 Aprelya 1820 goda [O.S.] iz″ Porta Zhaksona [Extract from the Report of Junior Captain Bellingshausen from Port Jackson to the Minister of Marine, 8 April 1820 [O.S.]]. *Syn″ Otechestva*, April 23 1821 (O.S.), 69: 133–135.

———. 1821d. Karta Plavaniya Shlyupov″ Vostoka i Mirnago vokrug″ Yuzhnago polyusa v″ 1819, 1820 i 1821 godakh″ pod″ Nachal′stvom″ Kapitana Billensgauzena [Chart of the Voyage of Sloops *Vostok* and *Mirnyj* around the South Pole in 1819, 1820 and 1821 under the Command of Captain Bellingshausen]. St Petersburg: SARN—F–1331, S–4, P–536, fos 5–19.

Bellinsgauzen″, F.F. 1821e. Report to the Marquis de Traversay from Kronstadt, 5 August 1821. St Petersburg: SARN—F–203, S–1, P–826, fos 1–18v.

Bellinsgauzen″, Kap. 1823. Donesenie Kapitana 2 ranga Bellinsgauzena iz″ Porta Zhaksona, o svoem″ plavanii [The Report of Junior Captain Bellingshausen from Port Jackson, about His Voyage]. *Zapiski izdavaemyya Gosudarstvennym″ Admiraltejskim″ Departamentom″* 5: 201–219.

Bellinsgauzen″, Kapitan.″ 1831. *Dvukratnyya izyskaniya v″ yuzhnom″ ledovitom″ okeanye i plavanie vokrug″ svyeta v″ prodolzhenii 1819, 20 i 21 godov″ [Two Seasons of Exploration in the Southern Ice Ocean and a Voyage around the World, During the Years 1819, 1820 and 1821]*, ed. L.I. Golenishhev″-Kutuzov″, 2 vols. plus *Atlas*. St Petersburg: Glazunovs.

Belov, M.I. 1954. Vvedenie [Introduction]. In *Geograficheskij sbornik III*, 5–12. Moscow: Academy of Sciences.

———., ed. 1963b. *Pervaya russkaya antarkticheskaya èkspeditsiya 1819–1821 gg. i eyo otchyotnaya navigatsionnaya karta [The First Russian Antarctic Expedition 1819–1821 and its Official Navigational Chart]*. Leningrad: Morskoj Transport.

Bogoyavlenskij, G.P. 1955. *Russkie geografy i puteshestvenniki [Russian Geographers and Travellers]*. Moscow: State Library of the USSR.

———. 1963. *Fizicheskaya geografiya: bibliograficheskoe posobie poso uchitelej [Physical Geography: A Bibliographic Guide for Teachers]*. Moscow: Textbook Press.

Bolotnikov, N.Ya. 1953. Faddej Faddeevich Bellinsgauzen i Mikhail Petrovich Lazarev [Faddej Faddeevich Bellingshausen and Mikhail Petrovich Lazarev]. In *Russkie moreplavateli*, ed. V.S. Lupach, 183–209. Moscow: Military Press.

Bulkeley, Rip. 2008. Aspects of the Soviet IGY. *Russian Journal of Earth Sciences* 10: 1–17.

———. 2014a. *Bellingshausen and the Russian Antarctic Expedition, 1819–21.* Basingstoke: Palgrave.

Campbell, R.J. 2000. *The Discovery of the South Shetland Islands.* London: Hakluyt Society.

Cherevichnyj, I.I. 1963. *V nebe Antarktidy [In the Fog of Antarctica].* Moscow: Morskoj Transport.

Daly, J.C.K. 1991. *Russian Seapower and 'the Eastern Question' 1827–41.* Basingstoke: Macmillan.

Durdenevskij, V.N. 1950. Problema pravovogo rezhima pripolyarnykh oblastej (Antarktika i Arktika) [The Problem of the Legal Regime in Polar Regions (the Antarctic and Arctic)]. *Vestnik Moskovskogo Universiteta* 7: 111–114.

E.A. 1949. O prioritete russkikh v otkrytii Antarktiki [On Russian Priority in Discovering Antarctica]. *Morskoj flot* 4: 46–47.

Fradkin, N.G. 1972. *Geograficheskie otkrytiya i nauchnoe poznanie Zemli [Geographical Discoveries and Scientific Knowledge of the Earth].* Moscow: Mysl'.

Gould, R.T. 1941. The Charting of the South Shetlands, 1819–28. *The Mariner's Mirror* 27: 206–242.

Grigor'ev, S.G. 1937. *Vokrug yuzhnogo polyusa.* 3rd ed. Moscow: Textbook Press.

Guretskij, V.O. 1954. Russkie geograficheskie nazvaniya v Antarktike [Russian Geographical Names in the Antarctic]. *Izvestiya vsesoyuznogo geograficheskogo obshhestva* 86 (5): 457–465.

Hattersley-Smith, G. 1991. *The History of Place-Names in the British Antarctic Territory.* Cambridge: British Antarctic Survey.

Isakov, I.S., ed. 1950. *Morskoj atlas [Maritime Atlas].* Vol. 1. Naval General Staff: Moscow.

Kaminer, L.V., and Others. 1963. *Istoriya estestvoznaniya 1951–1956: literatura, opublikovannaya v SSSR [Bibliography of the History of Science in the USSR 1951–1956].* Moscow: Academy of Sciences.

Khvat, L. 1956. *Zagadochnyj materik [Mysterious Continent].* Moscow: Geografgiz.

Kostritsyn, B.V. 1951. K voprosu o rezhime Antarktiki [On the Question of an Antarctic Regime]. *Sovetskoe gosudarstvo i pravo* 3: 38–43.

Kotukhov, M.P. 1951. *Velikij podvig: otkrytie Antarktidy [A Great Victory: The Discovery of Antarctica].* Moscow: Ministry of the Navy—not seen.

———. 1955. *Velikij podvig: otkrytie Antarktidy.* 2nd ed. Moscow: State Geographical Press.

Kublitskij, Georgij. 1957. *Po materikam i okeanam [Across Continents and Oceans].* Moscow: Children's Press, Ministry of Education.

Lazarev", M.P. 1918. Letter 1 in: Pis'ma Mikhaila Petrovicha Lazareva k" Aleksyeyu Antonovichu Shestakovu v" g. Krasnyj Smolenskoj gubernii [Letters from

Mikhail Petrovich Lazarev to Aleksyej Antonovich Shestakov at Krasnyj in the Smolensk Gubernorate]. *Morskoj sbornik"* 403 (1): 51–66.

Lebedev, V.L. 1957. *Antarktika [The Antarctic].* Moscow: Geografgiz.

Magidovich, I.P. 1953. Entries for Bellingshausen and Lazarev in: *Russkie more-plavateli,* ed. V.S. Lupach, 473–578. Moscow: Military Press.

Markov, K.K., and Others. 1968. *Geografiya Antarktidy [Geography of Antarctica].* Moscow: Mysl'.

Markov, Sergej Nikolaevich. 1952. Otkryvatel' Antarktidy [Discoverer of Antarctica]. *Vokrug sveta* 1: 65.

McCannon, John. 1998. *Red Arctic.* Oxford: Oxford University Press.

Miers, J. 1820. Account of the Discovery of New South Shetland. *Edinburgh Philosophical Journal* 6: 367–380.

Molodtsov, S.V. 1954. *Sovremennoe mezhdunarodno-pravovoe polozhenie Antarktiki [The Situation of the Antarctic in Modern International Law].* Moscow: Gosyurizdat.

Moroz, V. 2001. *Antarktida: istoriya otkrytiya [Antarctica: The History of Discovery].* Moscow: Belyj gorod.

Nudel'man, A.V. 1959. *Sovetskie èkspeditsii v Antarktiku 1955–1959 gg. [Soviet expeditions to the Antarctic 1955–1959].* Moscow: Academy of Sciences.

———. 1960. *Sovetskie èkspeditsii v Antarktiku 1958–1960 gg.* Moscow: Academy of Sciences.

———. 1962. *Sovetskie èkspeditsii v Antarktiku 1959–1961 gg.* Moscow: Academy of Sciences.

———. 1965. *Sovetskie èkspeditsii v Antarktiku 1961–1963 gg.* Moscow: Nauka.

Osipov, K. 1950. *Kak russkie lyudi otkryli Antarktidu [How Russian People Discovered Antarctica].* Moscow: Geografgiz.

Ovlashhenko, Aleksandr. 2016. *Antarkticheskij renessans: provedenie pervykh kompleksnykh antarkticheskikh èkspeditsij i problema otkrytiya Antarktidy [Antarctic Renaissance: The Arrival of the first Combined Antarctic Expeditions and the Problem of the Discovery of Antarctica].* Saarbrücken: Palmarium.

Powell, G. 1822. *Chart of South Shetland Including Coronation Island, &c. From the Exploration of the Sloop Dove in the Years 1821 and 1822 by George Powell Commander of the Same.* London: Laurie. https://collections.rmg.co.uk/collections/objects/540915.html.

Purdy, John. 1822. *Memoir, descriptive and explanatory, to Accompany the New Chart of the Ethiopic or South Atlantic Ocean.* London: Laurie.

Ross, Captain Sir James Clark. 1847. *A Voyage of Discovery and Research in the Southern and Antarctic Regions During the Years 1839–43.* 2 vols. London: John Murray.

Ryabchikov, E. 1955. K zemle tajn! [To the Land of Secrets!]. *Ogonyok* 49: 18–19.

Samarov, A.A., ed. 1952. *M.P. Lazarev—dokumenty [M.P. Lazarev—Documents].* Vol. 1. Moscow: Ministry of the Navy.

Shherbakov, D.I. 1955. Zagadki Antarktidy [Mysteries of Antarctica]. *Ogonyok* 49: 17–18.

———. 1956. Nauchnye rezul'taty antarkticheskikh èkspeditsij [Scientific Results from Antarctic Expeditions]. In *Zagadochnyj materik*, ed. L. Khvat, 259–285. Moscow: Geografgiz.

Shvede, E.E. 1952. *Otkrytie Antarktidy russkimi moryakami [The Discovery of Antarctica by Russian Seamen]*. Moscow: Znanie.

———. 1960. Otkrytie Antarktidy russkimi moreplavatelyami v 1819–1821 gg. [The Discovery of Antarctica by Russian Navigators in 1819–1821]. In *Dvukratnye izyskaniya v yuzhnom ledovitom okeane i plavanie vokrug sveta v prodolzhenie 1819, 20 i 21 gg*, ed. E.E. Shvede, 9–52. Moscow: Geografgiz.

Simakova, Lyudmila. 2015. *Aleksandr Kuchin: Russkij u Amundsena [Alexander Kuchin: A Russian with Amundsen]*. Moscow: Paulsen.

Simonov, I.M. 1951. Shlyupy «Vostok» i «Mirnyj» ili plavanie rossiyan v Yuzhnom Ledovitom okeane i okolo sveta [Sloops *Vostok* and *Mirnyj*, or the Voyage by Russians in the Southern Ice Ocean and around the World]. In *Russkie otkrytiya v Antarktike*, ed. V.N. Sementovskij, 51–175. Moscow: Geografgiz.

———. 1955. Avtobiografiya I.M. Simonova (1848 g.) [Autobiography of I.M. Simonov (1848)]. *Istoriko-astronomicheskie issledovaniya* 1: 268–277.

Sokolov, A.V. 1950a. Obzor deyatel'nosti otdeleniya istorii geograficheskikh znanij i istoricheskoj geografii za period s oktyabrya 1947 g. po dekabr' 1949 g. [An Overview of the Work of the Historical Section from October 1947 to December 1949]. *Voprosy geografii* 17: 241–254.

———. 1950b. Obzor deyatel'nosti otdeleniya istorii geograficheskikh znanij i istoricheskoj geografii s dekabrya 1949 g. po aprel' 1950 g. [An Overview of the Work of the Historical Section from December 1949 to April 1950]. *Voprosy geografii* 20: 333–337.

———. 1953. Obzor deyatel'nosti otdeleniya istorii geograficheskikh znanij i istoricheskoj geografii s maya 1950 g. po yanvar' 1953 g. [An Overview of the Work of the Historical Section from May 1950 to January 1953]. *Voprosy geografii* 31: 274–285.

———. 1960. Obzor deyatel'nosti otdeleniya istorii geograficheskikh znanij i istoricheskoj geografii s fevralya 1953 g. po maj 1958 g. [An Overview of the Work of the Historical Section from February 1953 to May 1958]. *Voprosy geografii* 50: 238–252.

———. 1969. V moskovskom filiale geograficheskogo obshhestva SSSR [At the Moscow Branch of the Soviet Geographical Society]. *Voprosy istorii estestvoznaniya i tekhniki* 26 (1): 87–89.

———. 1970. Obzor deyatel'nosti otdeleniya istorii geograficheskikh znanij i istoricheskoj geografii s oktyabrya 1958 po dekabr' 1963 g. [An Overview of the Work of the Historical Section from October 1958 to December 1963]. *Voprosy geografii* 83: 173–189.

Sokolov, A.V., and E.G. Kushnarev. 1951. *Tri krugosvetnykh plavaniya M.P. Lazareva [The Three Circumnavigations by M.P. Lazarev]*. Moscow: Geografgiz.

Starosel'skaya-Nikitina, O.A., and Others. 1949. *Istoriya estestvoznaniya 1917–1947: literatura, opublikovannaya v SSSR [Bibliography of the History of Science in the USSR 1917–1947]*. Moscow: Academy of Sciences.

———. 1955. *Istoriya estestvoznaniya 1948–1950: literatura, opublikovannaya v SSSR [Bibliography of the History of Science in the USSR 1948–1950]*. Moscow: Academy of Sciences.

Taylor, Prue. 2011. Common Heritage of Mankind Principle. In *The Berkshire Encyclopedia of Sustainability. Vol. 3. The Law and Politics of Sustainability*, ed. Klaus Bosselmann, Daniel Fogel, and J.B. Ruhl, 64–69. Great Barrington, MA: Berkshire.

Tryoshnikov, A.F. 1963. *Istoriya otkrytiya i issledovaniya Antarktidy [A History of Discovery and Research in Antarctica]*. Moscow: State Geographical Press.

Uzin, S.V. 1950. *Zagadochnye zemli [Mysterious Lands]*. Moscow: Geografgiz.

Vadetskij, B. 1957. *Obretenie schast'ya [The Discovery of Fortune]*. 2nd ed. Moscow: Kirov Press.

Yablokov, A.V., and V.A. Zemsky, eds. 2000. *Soviet Whaling Data (1949–1979)*. Moscow: Centre for Russian Environmental Policy.

Yakovlev, A. 1972. Protiv antiistorizma [Against Antihistoricism]. *Literaturnaya gazeta*, November 15: n.p. http://left.ru/2005/15/yakovlev132.phtml.

Zubov, N.N. 1953. Russkie moryaki—issledovateli okeanov i morej [Russian Sailors—Explorers of the Seas and Oceans]. In *Russkie moreplavateli*, ed. V.S. Lupach, iii–xxxvii. Moscow: Military Press.

CHAPTER 9

Digging In, 2: Heavyweights (1960–1971)

The two leading experts on the Bellingshausen expedition in the 1960s were Vladimir Lebedev and Mikhail Belov. Lebedev's historical publications were confined to the Khrushhev Thaw. Belov, whose output on the expedition surpassed that of all other Russian commentators in both quality and quantity, produced some of his best work during the Khrushhev period, but continued to publish important contributions under Brezhnev. Not only did he have much more to say than Lebedev; he was also overtly more anti-Western and committed to the Soviet ideology, so probably had few difficulties with Brezhnev's efforts to re-establish it.[1]

Vladimir Lebedev

Vladimir L'vovich Lebedev was a physical geographer and oceanologist with no training or background in history. His work on the expedition focused mainly on the priority question, and most of it appeared in a new journal, *Antarctic: reports of the commission*. Although its members were never listed, the commission was an interdepartmental body first convened by the Academy of Sciences in June 1958 to support the Soviet Antarctic

---

[1] The original texts of translated quotations, indicated by numbers in square brackets, are provided at the end of each chapter. For the system of romanization used, the rendering of dates, and a few other matters, please consult the Apparatus at the end of the book.

© The Author(s), under exclusive license to Springer Nature   151
Switzerland AG 2021
R. Bulkeley, *The Historiography of the First Russian Antarctic Expedition, 1819–21*,
https://doi.org/10.1007/978-3-030-59546-3_9

Expedition (SAE) and the renamed *AANII* beyond the IGY, due to close at the end of the year. Similar arrangements already existed in other countries, reflecting the wide range of state interests in the Antarctic IGY. However the articles in *Antarctic* were not in fact reports from the interdepartmental commission, but academic contributions from natural scientists in various disciplines, as well as, for a few years, some on Antarctic history. From 1960, if not sooner, Lebedev was the commission's academic secretary, which gave him unfettered access to the pages of *Antarctic*.

The identity of the journal's first editor, the aerologist Viktor Antonovich Bugaev (1908–1974), reflected its official character. After leading the aerology section for the 3rd SAE (1957–1958) Bugaev became director of the Central Forecasting Institute in 1959, and when the Institute merged with the directorate of the Meteorological Service in 1963 he became director of the new entity, the Meteorological Centre or *Gidromettsentr*. Those institutions were all based in Moscow, where one of his main responsibilities was to report directly to the Council of Ministers. Outside the nuclear missile and space programmes few Soviet scientists can have been more closely connected with state power than Bugaev when the first, '1960' issue of *Antarctic* appeared in 1961.[2]

The most significant historical articles in *Antarctic* were by V. Lebedev and appeared between 1961 and 1964. From that point onwards the past was usually confined to palaeoclimatological studies, retrospective surveys of the work of the SAE, and one or two pieces on the origins of Antarctic place names. The last articles to deal with earlier exploration were a tribute to Amundsen on the centenary of his birth, and an account of the celebrations of the 150th anniversary of Bellingshausen's supposed discovery (Avsyuk and Kartashov 1971). Bugaev's successor, appointed in 1975, was a rather more civilian Leningrader, the glaciologist Grigorij Aleksandrovich Avsyuk (1906–1988), but his arrival had no effect on the journal's 'no history' policy, which seems to have been established by the early 1970s.[3]

[2] The journal was always published in the year after the cover date.

[3] The author has seen no issues of *Antarctic* after 1992, if such there be. The *AANII* has a long-standing interest in archaeological work, especially in the Arctic and usually in collaboration with other institutions. Historical research based on textual records is another matter. After the early 1970s no such contributions in *Antarctic* turned up in internet searches or in modern Russian polar literature, and in 2011 the author was informally advised that history was not part of the remit of the *AANII*.

On 15 June 1960 Lebedev launched a historical campaign for the express purpose of establishing once and for all, both at home and abroad, that Russians had been the first to sight the Antarctic continent (Anon. 1960).[4] His first short discussion of priority appeared in an English-language magazine (V. Lebedev 1960), and his second article in *Antarctic* was a bilingual exchange with a New Zealand journalist (V. Lebedev 1962). In both pieces he showed short facsimile extracts from the MSS of Bellingshausen's April 1820 report and Lazarev's letter to Shestakov for the first time in print.[5] He emphasized that they had not yet been studied by Western historians, and insisted that their authors were referring to a continent in the modern sense of the word. He also became the first modern commentator to publish and discuss Bellingshausen's Remark about finding no Antarctic land in 1820, which had last appeared in 1823 (V. Lebedev 1962: 156, 1963: 177; K. Markov and Others 1968: 13–21).

Lebedev began his first and most comprehensive *Antarctic* article by comparing the Antarctic voyages of Cook and Bellingshausen. He studied Cook's narrative in English, complaining that Russian translations had been inadequate, and explained that while Cook believed he had established the non-existence of any additional continent in the southern *temperate* zone, the English explorer remained convinced that a *polar* mainland existed, albeit beyond the reach of mariners in vulnerable wooden ships. Lebedev rebuked previous Soviet commentators for failing to understand the distinction. He also faulted a Soviet historical atlas for ahistorically showing Antarctica in every map of the region after Cook (V. Lebedev 1961: 10–11, 13). Not being a historian himself, however, Lebedev was unaware that the argument from heavy glaciation to the existence of land had been current for more than 150 years before it was taken up by Lomonosov (1763: 256), and mistaking a translation of a translation for an original work, he erroneously praised Andrew Kippis's (1725–1795) restatement of Cook's conclusions as coming from a Russian biographer (the same: 11, 7; see Kippis 1788: 62). Nor did he realize how much Russian authors in the early twentieth century, such as Shokal'skij, had taken from British and German predecessors.

---

[4] Conceivably he took this public stand shortly after being appointed to his post on the Antarctic Commission, but that could not be verified.

[5] Both documents were displayed in the Soviet Industrial Exhibitions at London and Paris in July and August 1961—whether in whole or in part is not known.

Lebedev's most valuable contribution was his thesis that Bellingshausen and Lazarev intended phrases like 'main of ice' and 'ice main' to refer to a geographical feature comprising at least a very high proportion of ice; in short, they meant what they said. Unfortunately his exegesis of this aspect of Bellingshausen came at the end of his main article about the expedition, after lengthy disagreements with foreign authors, some of which, such as his criticism of key passages in Debenham's translation, were well taken (V. Lebedev 1961: 22–23). Although he forced the language of his sources too far towards his modern understanding of Antarctica, Lebedev's interpretation was a significant achievement; the more so because he was unaware that people had been discussing the possibility of a vast marine Antarctic ice cap for 40 years before the expedition.

Throughout his historical work Lebedev claimed that the primary sense of 'main' (*materik*) in texts from the expedition was always 'continent'. As we saw in Chap. 3, however, the history of the word and the usage of the expeditioners were more complicated than that. Although we know that Buffon's theory was present on *Vostok* in the person of Simonov, because of the latter's remark about domes of ice formed like crusts on the sea around both Poles [1], Bellingshausen's usage is harder to pin down. Certainly his explanation of 'main ice' as 'ris[ing] ever higher, like mountain slopes, the closer they are to the South Pole' [2] shows that by 1824 at the latest he held something very like Buffon's conjecture about a massive Antarctic ice cap with no necessary connection to land. He may well have become acquainted with the idea during the voyage, but his punctilious character suggests that with 'main of ice' in his April 1820 report he was describing only what he had actually seen, namely a lofty main body of ice, with no visible termination, behind the seven large ice islands observed on 17 February 1820.

Without citing relevant sources such as contemporary dictionaries, Lebedev tried to make sense of the language used by Bellingshausen and Lazarev from the handful of examples in their texts.[6] He assumed, understandably, that they always meant the same thing by 'main ice', and that thing was some sort of continent. But his interpretation was ambivalent.

---

[6] Lebedev also consulted a note about Russian words for 'continent'. However its author had said nothing about their historical development or the exchange of ideas between Russia and other countries, and merely allowed that 'ice main' (*ledyanoj materik*) would be a reasonable modern term for the southern ice cap, if it became known that its lower surface never rose above sea level (Rodomanov 1959: 159).

On the one hand he shrewdly posited that, for Bellingshausen, a continent might be 'made up of ice' [3] (V. Lebedev 1961: 22). On the other hand he claimed that the Russian expedition had discovered and to some extent *understood* a continent similar to the one we know today, comprising rock largely covered by a massive ice shield. He insisted that Lazarev's '*main ice of extraordinary height*' had been what is now considered the 'underlying [or basic] shore' [4] of Antarctica (the same: 20). (The adjective has been used to distinguish between the rock relief and the glacial relief of Antarctica [Berlyant and Others 1981].) But crucially, he thought, Bellingshausen himself had had some inkling of the same thing. The ice coasts of Antarctica and the 'continent itself' [5] were not shown on the map of the voyage in the *Atlas* volume, according to Lebedev, only because other people did not understand the discovery and Bellingshausen took no part in preparing the book for press; in other words, they could have been shown even though they were not (the same: 22). That was a patently speculative contention, putting thoughts and knowledge into the minds of the expeditioners despite the lack of evidence for them.[7]

As explained in Chap. 7, Lazarev chose an unusual adjective, *l'dinnyj* ('pertaining to ice floes'), to describe the main which repeatedly barred their passage south. Most Soviet commentators assumed without comment that it was a slip of the pen for *l'dyanoj* ('made of ice'). Lebedev, however, interpreted the phenomenon observed by Lazarev as 'the underlying shore of the continent' purely on the basis of its size [6], which, besides stretching out of sight, must have been higher than the largest icebergs (V. Lebedev 1961: 20–21). Since the ice coast of that part of Antarctica could be as much as 900 m high, that must have been what Lazarev saw, he thought (the same: 22). But although size was the main feature of what Lazarev described, Lebedev's argument was a poor one, because Lazarev explained a few lines further on that he regarded 143 m as an exceptional height for an iceberg (Lazarev″ 1821: fo. 2v). So whatever he meant by 'an ice floe main', it would have been 'of exceptional height' if it were, say, over 140 m, or perhaps lower than that if he meant 'exceptional for such a massive, coast-like feature'. Nothing like 900 m need have been implied.

---

[7] Lebedev may well have written this article before seeing Bellingshausen's navigational chart, which, like other sources, did not show a continent, and had not been edited by anyone (Bellinsgauzen″ 1821d).

In general, Lazarev had mentioned nothing but ice south of the Circle apart from Peter I I. and the Alexander I Coast, which he called an island in the letter. If he saw the main as an agglomeration of ice floes, which by definition float, then his views were close to Buffon and Simonov, and not as described by Lebedev. Bellingshausen took much the same line, including his description of 17 February 1820. In his reflective, theoretical discussion, after explaining from his own observation that ice could be formed at sea, and then suggesting that it could be 'compressed or collected' [7] into very large bodies, one of which he had coasted for 300 nautical miles (555 km) (Bellinsgauzen" 1831, 2: 247), Bellingshausen went on to suppose that such ice extended beyond the Pole. He added his opinion that there were further islands and shoals 'touched' [8] by the ice cap near positions reached by the expedition (the same: 250). But he never said they might lie *under* it, and, since his reason for believing that further islands existed was that sea swallows (probably the Antarctic tern, *Sterna vittata*) observed in high latitudes did not feed at sea,[8] he cannot have been talking about land permanently concealed under ice. In short the conjectured touching was lateral, not vertical. Lastly, Russian has other adjectives, such as *ledenistyj*, *ledovityj*, and *obledennevshij*, for something covered in ice, but despite marvelling at the heavy glaciation of South Georgia, the South Sandwich Is., and their own discoveries, Lazarev and Bellingshausen never applied those words to land.

Lebedev also claimed that Bellingshausen had 'constantly' [9] used the word 'main' (*materik*) in his reports, plural (V. Lebedev 1961: 23). In fact, however, he used it in only one paragraph of one report, about the events of 17 February 1820, although his phrasing meant that the word occurred four times in that short passage (Bulkeley 2014a: 81).[9]

Despite his relatively open-minded approach, Lebedev provided little in the way of new arguments for priority. Formed in the authoritarian culture of the Soviet Union, he could no more conceive that Lazarev's

---

[8] Besides scavenging in the intertidal zone Antarctic terns also fish at sea but close inshore. In that sense Bellingshausen was right about them not being pelagic. Matters are complicated by the presence of the more pelagic Arctic tern (*Sterna paradisaea*) during the Antarctic summer. But Bellingshausen's thoughts about further land adjacent to the ice cap are revealed by his argument, not its biological accuracy.

[9] Technically, he also used it again, because all but the first paragraph of the April 1820 report was inserted at the beginning of his final report. But that was not so much another report as the second edition of an earlier one. Nor does it justify Lebedev's exaggeration, because he was unaware of its existence.

letter might contain factual errors than any of his colleagues. That made it his strongest card, and he quoted it repeatedly. Another argument on which he relied heavily was the claim that Bellingshausen's chief editor, Golenishhev-Kutuzov, had suppressed the explorer's references to a main.[10] That interference, he wrote, could be demonstrated by comparing the account of 17 February 1820 in the April report from Sydney with *Two Seasons*. In the latter, the earlier characterization of the coast as 'main' had allegedly been removed (V. Lebedev 1961: 15). A comparison of the texts, however, does not support Lebedev's reading (Table 9.1).

Bellingshausen gave a fuller account of the February 1820 probe in his book than in the original report, partly because he added some theoretical points. But both that change, and the replacement of 'main of ice' with the less technical 'main coast' in a work that would be available to the limited and poorly educated general public of those days, are the sort of decisions he could have taken himself. By surmising that the 'huge ice masses' had broken away from the main coast, Bellingshausen made it clear that the latter was also composed of ice. Lebedev's contention, that the word 'main' was expunged from this passage in the book, was therefore wrong. Part of his problem, evidently, was linguistic. Having pressed the word *materik* into meaning always and only 'continent', he could not understand that in the right context Bellingshausen might use *bereg* (coast, shore, (dry) land) as equivalent to *materik*, a parity which dictionaries confirm (Dal' 1863: 65).[11]

A year later Lebedev told his New Zealand correspondent that despite Bellingshausen's

> explanation of why he calls the ice continental («*matyoroy*»), [there were] no parts [in the book] where [he] called it so. Only the words «*matyoroy*» (continental) coast, mentioned once in the text of February 17(5), 1820, survived. (V. Lebedev 1962: 165)

---

[10] The view that Golenishhev-Kutuzov was biased in favour of Cook and therefore against Bellingshausen was expressed by several Soviet authors. Apart from anachronistically projecting their own approach to the subject onto early nineteenth-century Russia, it relied on ignoring the fact that G-K was lavish in his praise for the Russian expedition and led the struggle, at senior levels of the Imperial Navy, to get *Two Seasons* published (Bellinsgauzen" 1831, 1: Note from the Science Committee, n.p.).

[11] When Russian lookouts sighted land, they did not shout *Zemlya!*, *Susha!*, or *Materik!* They shouted *Bereg! Bereg!*, as Bellingshausen himself recorded.

Table 9.1   Bellingshausen's accounts of 17 February 1820

Report from Sydney, April 1820	Two Seasons, 1831
...between the 5th and 6th [O.S.], I reached latitude S 69°7′30″, longitude E 16°15′. There, beyond ice fields comprising small ice and [ice] islands, a main of ice was sighted, the edges of which had broken away perpendicularly, and which stretched as far as we could see, rising to the south like land. The flat ice islands that are located close to this main are evidently nothing but detached fragments of this main, since they have edges and upper surfaces which resemble the main.[10] (Bellinsgauzen″ 1820a: fo. 242v)	Eventually, at a quarter past three after noon, we saw a quantity of large, flat, high ice islands, beset with light floes some of which overlay one another in places. The ice formations to SSW join together into hilly, solid standing ice; its edges were perpendicular, forming coves, and its surface rose away to the South, to a distance whose limits we could not make out from the [main] cross-trees. ... Seeing that the ice islands had similar surfaces and edges to the aforementioned large ice formation, which lay before us, we concluded that those huge ice masses and all similar formations get separated from *the main coast* by reason of their own weight or from other physical causes and, carried by the winds, drift out into the expanse of the Southern Ice Ocean. ...[11] (Bellinsgauzen″ 1831, 1: 188–189— emphasis added)

That was disingenuous. Bellingshausen used the expression 'main coast' nine times in *Two Seasons*, usually in the sense of 'main land' (for which he preferred this expression, using 'main land' only once). On the occasion mentioned by Lebedev he used it for an ice coast, and when summarizing his ideas he thought that unusual usage warranted an explanation. That suggests that he saw the phenomenon only once. Lebedev, however, rejected Bellingshausen's testimony because he had assumed in advance that main ice was sighted more often.

The suppressed text argument was too vague to be much help with priority, because Bellingshausen had not used 'main' or anything like it when describing 28 January 1820 either in his original report from Sydney or in *Two Seasons*. Lebedev therefore used a different argument to bring Bellingshausen's account of that day more into line with the one in Lazarev's letter. It all came down to icebergs. According to Lebedev, 'Bellingshausen never used the word «icebergs» [*ajsbergi*] and always wrote «ice islands» instead' (V. Lebedev 1962: 166). As we saw in Chap. 3, the second point was mistaken. But it allowed him to treat Bellingshausen's sighting of icebergs on 28 January as a 'successful observation' [12], apparently because he also supposed that some sort of real hills were meant, rather than 'ice hills', as icebergs were starting to be called at the time (V. Lebedev 1961: 14).[12] This surprising reading of Lebedev is the only way to make sense of his view that Bellingshausen's description of icebergs surrounded by pack ice was a sighting of the continent, of equal weight with Lazarev's '*main ice of extraordinary height*' on that date. He repeated it a year later, together with the claim that Bellingshausen never called icebergs anything but 'ice islands' (V. Lebedev 1962: 155, 161).

It is all very well, in fiction, for a young radio operator to misunderstand a reference to icebergs (*ledyanye gory*) in the last message received from the famous Soviet pilot Sigizmund Aleksandrovich Levanevskij (1902–1937) (Muldashev 2013: 97). But for a leading Soviet specialist in Antarctic marine ice to have failed to understand the same phrase in the 1960s is scarcely believable. His contemporaries, after all, had no such problem (Krajner 1962: 203–204; Belov 1963a: 43; Ostrovskij 1966: 58, 61, 62, 72), and the phrase even appeared as a translation of 'iceberg' in an English-Russian oceanographic dictionary from the period (Gorskij and Gorskaya 1957: 92). Less believable still, Lebedev would have to have been unacquainted with Lomonosov's 'Thoughts on the origin of icebergs' (*ledyanye gory*), which was much discussed in the 1950s after a Russian translation (of the Swedish original) appeared in his Collected Works (Lomonosov 1763, 1953). But Lebedev could not have interpreted Bellingshausen's sighting of 'ice hills' on 28 January 1820 as he did, if he took Bellingshausen to be referring merely to icebergs.

The facts are, however, first, that Bellingshausen could not use the word *ajsbergi* because it was not a Russian word until about a century after

---

[12] Lebedev was not the first to put forward this interpretation of the 'ice hills' (*ledyanye gory*) in Bellingshausen's 1820 report. The most influential precedent was probably a note by Samarov (Samarov 1952: 420–421).

his voyage;[13] second, that he used the phrase *ledyanye gory* not only in his April report about 28 January but also in *Two Seasons* about the probe on 17–18 February 1820, where it appears twice; third, that he replaced 'hills' with 'islands', the more familiar Russian term for icebergs, when rewriting the 28 January passage for his final report (Bellinsgauzen″ 1821e: fo. 4—not available to Lebedev); and fourth, that in *Two Seasons* he said the 'ice hills' he tried to reach on 18 February resembled, and were probably a continuation of, the 'aforementioned' [13] large flat ice islands seen the day before (Bellinsgauzen″ 1831, 1: 190). All this makes it clear that for Bellingshausen ice hills were icebergs, not coastal hills comprising some proportion of rock as well as ice.[14]

There are only three ways to establish Bellingshausen's priority over Bransfield. The first is to rely solely on Lazarev's letter; but that is unconvincing, both because Bellingshausen was primarily responsible for reporting the results of the voyage and left more, and more substantial, accounts of it than anyone else, and because, as no Soviet commentator ever imagined, there were mistakes in the letter. The second is to bring Bellingshausen's version of 28 January 1820 closer to Lazarev's, either via the suppressed text argument, which is too vague for the purpose, or by forcing the sense of Bellingshausen's words, which is what Lebedev did by turning icebergs in an ice field into the coastal hills of an ice-covered continent. The third and potentially strongest option is to undermine Bransfield's claim to a sighting on 30 January, in which case Bellingshausen's undoubted sighting and reporting of a 'main coast' of ice on 17 February 1820 would be enough for priority, and 28 January could be set aside as the unlikely and unprovable possibility that it is.

Not surprisingly Lebedev advanced a version of the third argument as well as the second. He relied mainly on Gould's first article about Bransfield's voyage (1925), rather than the second (1941), which may not have been available in the Soviet Union. That was unfortunate, because Gould's second article cited more sources and gave a clearer explanation of those used in compiling the chart of Bransfield's estimated route, which

---

[13] The captioning of some untitled sketches by Mikhajlov with this word is an unwarranted anachronism (Petrova and Others 2012: 25—Russian edition).

[14] Although he described steep valleys choked with ice on South Georgia and some of the South Sandwich Is., Bellingshausen never referred explicitly to glaciers, the other possible meaning of 'ice hills'.

was also clearer.[15] Lacking the full picture, Lebedev highlighted Gould's initial reservation, that the name 'Trinity Land' might have been meant for the north coast of Trinity I. (as one of three possibilities), and disregarded Gould's statements on either side of that reservation, that 'There can be no doubt that he saw the northern extremity of the mainland', and that the account in the *Literary Gazette* 'appears to support the last theory', that the name was given to the mainland (R. Gould 1925: 224; V. Lebedev 1961: 18). Had Lebedev been able to consult Gould's later, more confident account of Bransfield's voyage (R. Gould 1941: 213–224), which contained such statements as: 'Bransfield's chart shows clearly that he sighted and coasted the northern shore of Trinity Peninsula' (the same: 223), and which published the chart in question separately, he might have hesitated before introducing the red herring of Trinity I., which was also deployed by Belov (below).

In his third *Antarctic* article Lebedev represented some Western authors, none of them historians, as endorsing Russian priority (V. Lebedev 1963). However one had simply praised Bellingshausen for reducing the area in which a polar continent might exist and for increasing its probability by observing '[hills of ice near] the Greenwich meridian' without a date and 'the mountains of Alexander Land' (Robin 1962—garbled English text restored from the Spanish edition); another put no date on the Russian probe and showed it exactly on the Greenwich meridian (Butler about 1958), whereas Bellingshausen's longitude for the approach on 28 January 1820 was 02°30′W, about 98 km from the meridian;[16] and the third, despite translating *materik* as 'mainland', rendered Bellingshausen's sighting that day as 'a mound-covered icefield stretching to the south' (Croome 1960). In the same article Lebedev wrongly supposed that Bellingshausen had speculated about possible islands or shoals *under* the ice cap rather than adjoining it, which is what he actually wrote (Bulkeley

---

[15] A complete journal of the voyage, containing numerous positions including those from which the Antarctic Peninsula was observed, has recently been discovered and published (Campbell 2000). That source, to which Soviet commentators had of course no access, shows that Gould's estimate of Bransfield's route was a fair approximation of the facts.

[16] Bellingshausen gave four different longitudes for this event, in his two reports, his book, and the track chart. The latter, which is the furthest west, has been followed here because Belov found that its measurements were usually the closest to modern values. Great circle distance also depends on latitude. Since that is hard to read off the *Karta*, the distance was calculated for 69°23′15″S, halfway between the values in the reports and the book. The longitude in *Two Seasons* puts the distance about 10km shorter.

2019). From that mistaken premise, Lebedev assumed without evidence that for Bellingshausen the Antarctic ice cap was made up of comparable amounts of sea and land (V. Lebedev 1963: 177). Two pages later he criticized the editors of the *UNESCO Courier* for saying that Bellingshausen had 'sighted mountains' (Anon. 1962: 17), whereas he only saw patches of bare rock on Alexander I Land after he had already (according to Lebedev) reported the existence of an ice continent. What they should have said, Lebedev suggested, was that he saw 'ice hills or ice slopes' [14] (V. Lebedev 1963: 179). Once again Lebedev was refusing to accept that 'ice hills' could be mere icebergs, in order to upgrade them into topographic components of the supposedly discovered continent. But in any case the rebuke to the *Courier* was pointless because it put no date on the sighting in question, so that it could just as well have been referring to the 1821 discoveries. To conclude this section, in his fourth and last historical article Lebedev presented a cogent case against Belov's dating theory (below) (V. Lebedev 1964).

Through his *Antarctic* articles, Lebedev became the first commentator to seek to understand the language used by the expeditioners, though only to a limited extent.[17] His difficulties followed a common pattern, which combined the overinterpretation of some words used in primary sources with inattention to others, such as Lazarev's 'ice floe main' or Bellingshausen's conjecture that further islands might be discovered near the polar ice mass, rather than under it. That was perhaps the inevitable result of trying to argue the case for priority more thoroughly than previous commentators, while at the same time being obliged by official requirements and his own declared intentions to extract the only outcome that he was prepared to accept from insufficient evidence. It was a pattern that would be repeated in the more comprehensive and impressive work of Lebedev's contemporary, Mikhail Belov.

---

[17] In recent years this essential analysis has been advanced by Tammiksaar and the present author (Bulkeley 2014a, 2019; Tammiksaar 2016; and this book), with Tammiksaar the first to have consulted a nineteenth-century Russian dictionary (Dal' 1881), of which several are now available online.

MIKHAIL BELOV

Historiographic Role

Before proceeding it will be useful to situate Belov's work on the expedition in a wider context. Sponsored by his harassed teacher A.I. Andreev, he made his debut in 1948 with a study of the seventeenth-century explorer Semyon Dezhnev (Belov 1948). It was part of a drive by Soviet scholars, in line with the cultural self-sufficiency message of late Stalinism but often with strong support from new evidence, to show that Russians had preceded foreigners in Arctic waters (Horensma 1991: 84–88, 90–96). There were close parallels between that historical campaign and the one around the Bellingshausen expedition that was announced in 1949, but had clearly been under way in the work of Andreev and Shvede for some years before that. (The Central Committee decision and consequent Geographical Society meeting [Chap. 6] did not initiate their work but rather gave or tried to give it a radical reset.) It was natural, therefore, for scholars like Andreev and Belov to move between the two campaigns. The main difference between north and south was that the big historical push on the Arctic came after the cult of heroic Soviet deeds in the region had passed its peak (McCannon 1998), whereas the timing was the other way around for the Antarctic.

By the 1960s, then, Belov was well versed in the ways of patriotic polar history, and clearly hoped to repeat for the Antarctic the upgrading of historical knowledge about the Artic in which he was still playing a leading role. He was the finest Russian historian to study the Bellingshausen expedition, and much of his work is a pleasure to read when compared with that of his colleagues. Furthermore his rejection in 1951 of the received tale of Stepan Andreev and in 1954 of the new, absolute priority version of the Bellingshausen expedition showed an independent turn of mind (Sokolov 1953: 279; Belov 1954). But perhaps because of overwork he was regrettably prone to error, such as misquotation and mis-citation, with those errors tending to yield shortcuts to desired conclusions. It is perhaps significant that the bibliography in his *magnum opus* on the Northern Sea Route contains no references to archival manuscripts. Although he was scrupulously accurate when describing his own discovery, the track chart, he could be distinctly cavalier with other documents, as we shall see. He was also more overtly political than Lebedev, occasionally letting fall comments along the lines of the debased residual Marxism of the day.

Publishing the Track Chart

Belov became interested in the expedition after discovering Bellingshausen's 15-sheet track chart (Belov 1963b—hereafter referred to as the *Karta*). That achievement was later celebrated in a very unusual literary work, part war story, part historiographic essay, and part political diatribe—Evgenij Petrovich Fyodorovskij's (1933–) *Fresh ocean breeze* (Fyodorovskij 1976). Using Belov's own account of the discovery and the record of his work in the 1950s to moderate Fyodorovskij's highly coloured version, the sequence of events appears to have been roughly as follows. While serving in the defence of Leningrad in the grim winter of 1943–1944 Lieutenant Belov learned that the Naval Archives had been stored in the basement of the Catherine Palace at Tsarskoe Selo, demolished by German forces before they retreated. Soon after that a friend found one of Bellingshausen's maps, not part of the track chart, and sent it to him. That in turn helped Belov to recognize the Bellingshausen track chart (Bellinsgauzen″ 1821d; see Fig. 3.1)[18] when in about 1950 he came upon it by chance, misfiled with Arctic maps in the Archives of the Central Cartographic Service, while working on his great four-volume history of the Northern Sea Route (NSR), a project sponsored by the Shipping Ministry through the Arctic Research Institute (*ANII*) of the Northern Sea Route Administration (NSRA) (Belov 1956–1969). Despite the fuss being made about the expedition at the time, and perhaps made cautious by the rough reception given to his Arctic maps discovery at the MHS, Belov kept the existence of the track chart to himself for some years and, as we have seen, did not accept the revision of Bellingshausen's achievement, from 1821 to 1820, until after 1954. (He may perhaps have mentioned the track chart at the LHS, especially after succeeding Andreev in the chair in 1959, but that is speculation.)

By the late 1950s Belov had completed the second of his NSR volumes (actually volume 3). After sending that to the publisher in January 1959 he simultaneously wrote a third, 760-page volume (1962) and prepared a facsimile of Bellingshausen's track chart (Bellinsgauzen″ 1821d),

---

[18] Belov listed several references to the track chart in early, now extremely rare catalogues. But despite providing archival references for other charts, he gave no location for the track chart. Nor did he or anyone else do so elsewhere before the recent 'virtual exhibition' on the Naval Archives website (Kondakova 2019). In 1963, that was probably because the archives of the Cartographic Service had only recently been transferred to the Naval Archives, in 1961.

supported by a collection of scholarly commentaries, for the same publisher, the Shipping Ministry's *Morskoj transport* press. By then, however, the renamed Arctic and Antarctic Research Institute (*AANII*) had been transferred from the NSRA to the Meteorological Service (*Gidromet*), and Belov had been appointed as the only historian in the editorial group at the *AANII* which was compiling the *Antarctic Atlas* on the basis of Soviet surveys and data exchanged between the IGY committees of several countries. Given that the Soviet public had from time to time been misinformed that Bellingshausen's 'expedition proceeded in immediate proximity to the Antarctic continent for hundreds of miles and described it with complete accuracy' [15] (Kublitskij 1949b: 144), the editors of the *Atlas* would have been anxious to support such assertions with maps recording Bellingshausen's major southings.

According to Belov details from two sheets of the track chart were published in 1954 and 1958, the latter an archival finding aid. From 1954, therefore, others besides Belov were aware of its existence (Kolgushkin and Maksimov 1958). However the discovery was not publicized until three years later, with an article in the national press by a naval oceanographer (Osokin 1961). Belov then made a similar announcement himself in a magazine of the Shipping Ministry (Belov 1961a), and gave a more detailed account, with illustrations, both in the *Bulletin of the Soviet Antarctic Expedition* and in the *Proceedings of the Geographical Society* (Belov 1961b, 1962). He and Osokin also made a joint presentation at the MHS (above). There may have been some rivalry between Belov and the senior naval cartographer Ivan Petrovich Kucherov (1912–1993), who published a description of the track chart in *Antarctic* at the time the *Karta* appeared (Kucherov 1963; see also Kucherov and Bogdanov 1962).[19]

Belov's preliminary articles need not be discussed in detail, because they were thoroughly recapitulated in his more extensive treatments. It should be mentioned, however, that the second one was bluntly antagonistic towards Western historians, with the single exception of Debenham, whom Belov spectacularly mis-cited by transferring a positive evaluation of

---

[19] The journal had double the print run of the book and the two men barely mentioned each other's work. As director of the Soviet Navy's Central Cartographic Department, Kucherov had led the cartographic team on the first SAE in 1955–1956 and then oversaw production of the Navy's first complete *Antarctic Pilot* in 1961. As a contender for ownership of Bellingshausen's track chart he outclassed Belov by a country mile.

17 February to 28 January 1820 (Belov 1962: 105; Debenham 1959: 45). That was, unfortunately, a taste of things to come.

The track of the expedition (effectively that of *Vostok*), from their approach to South Georgia on 26 December 1819 to crossing the latitude 53°15′S northwards on 14 February 1821, bound for Rio de Janeiro, was shown as a single line across the 15 sheets of the track chart, with small dated circles indicating observed or estimated noon positions. There was a substantial geographical and chronological interruption where the approach to Australia, two visits to Sydney, and the tropical phase between them, were omitted. The passage from Sydney to Macquarie I. and time spent lying off and on at Macquarie I. were also omitted when the chart moved on, on Sheet 9, to cover the second season. The track was copiously annotated with sea, wind, and other weather conditions, with observations of marine and avian species, and occasionally with other events. A key to the symbols and abbreviations was provided on Sheet 14 (Bellinsgauzen″ 1821d; Belov 1963b).

As Belov rightly declared, Bellingshausen's navigational chart of his two polar seasons provides a much needed day-by-day record to set beside that in the perhaps over-edited *Two Seasons*.[20] Besides reproducing the 15 sheets of the chart, Belov included four other maps from the expedition, a historical essay by himself, three analyses by other experts of Simonov's observational methods, Bellingshausen's survey methods, and the structures of HIMS *Vostok* and *Mirnyj*, graphological analyses, and a transcription by V.V. Kuznetsova of the noon-by-noon verbal and symbolic annotations on the chart, tabulated alongside information from *Two Seasons* and other sources. Although he explained that a complete analysis of the surviving records of the voyage was beyond the powers of any small group of researchers (Belov 1963a: 8),[21] Belov clearly intended his team to provide solid foundations for such work in the future.

As the collective product of nine scholars and a uniquely immediate, detailed, and authoritative source the *Karta* stands head and shoulders above the rest of the Bellingshausen corpus apart from Shvede's two

[20] Parts of *Vostok*'s track, in the vicinity of islands, had been shown in Bellingshausen's *Atlas* volume and reproduced by Novosil'skij and Debenham. However the *Atlas* coverage was limited in extent and showed none of the annotations for wind and sea conditions and other comments on the track chart.

[21] The epithets 'official' (*otchyotnaya*) and 'navigational' (*navigatsionnaya*) have been applied to the chart since it was published in 1963. Although descriptively accurate, they have no basis in the document itself, having been applied by Belov rather than Bellingshausen.

editions of *Two Seasons*. Its most important historiographic aspect was that it had a print run of only 500 copies, few of which survive intact. That was why, although it was regularly mentioned, few historians actually used it. Just eight years after it appeared, for example, two leading scholars mentioned it without citations but looked elsewhere when citing Belov's ideas (D. Lebedev and Esakov 1971: 385–386; Belov 1962).

Belov made his own significant historiographic comments. After deploring the fact that much of the data gathered by the expedition had never been published and had since been lost, he went on to express his regret that, despite having been published in the nineteenth century, Bellingshausen's April 1820 letter to de Traversay had not yet been republished, and his report of the same date had only been republished with cuts. This meant that Bellingshausen's own 'evaluation of the passage of the sloops through Antarctic waters' [16] had been withheld—a fairly transparent reference to Bellingshausen's superficially problematic Remark about having seen no signs of land in the far south in 1820 (Belov 1963a: 8, 10).

Belov opened his long essay in the *Karta* with an analysis of the maps themselves, explaining how they work, their history, and how exceptional they were by the standards of Russian naval cartography of the day. He provided a detailed comparison of the positional coordinates for the expedition given in the track chart, the *Atlas*, and the text of *Two Seasons*, pointing out that in the case of their Antarctic and sub-Antarctic discoveries, the positions in the book usually fell between those in the navigational chart and those in other maps, and that the positions in the chart were often closer than the others to modern measurements (taken from Soviet charts) (Belov 1963a: 13–19).

The Nautical Calendar

Next, Belov explained the noon-to-noon maritime calendar which was used at sea by the Imperial Navy at the time of the expedition, arguing that dates in *Two Seasons* and Bellingshausen's reports, as well as those in the track chart, should be read according to that calendar, in which events between noon and midnight were dated as the following civil day. His argument was unsound, for several reasons. First, there are very few externally dated p.m. events, such as eclipses, against which to check the expedition's written and published, as opposed to watchkeeping, usage, but for example Lazarev reported that he arrived off Port Jackson Heads on the night of 6(18) April 1820 (Bellinsgauzen" 1831, 1: 277) and entered Sydney Harbour the next morning on 7(19) April (Lazarev" 1821: fo. 2v), as confirmed by Governor Macquarie. Those dates should have been the same if his report, inserted in *Two Seasons*, was using the nautical calendar. Next, putting aside the marginal dates in *Two Seasons* as being of uncertain authorship, Bellingshausen often describes a morning event next to a new date within the text, but seldom an evening one. Furthermore Belov failed to explain passages in which the date appears to change overnight, or remains unchanged from morning into afternoon, in the manner of the civil calendar. A striking example from the final report (not seen by Belov) reads: '*from the 28th to the 29th at 4 in the morning in latitude 64°40′ longitude 126°40′* [17] (Bellinsgauzen" 1821e: fo. 12v).[22,23] Such wording was normal for the civil but impossible for the nautical calendar, in which an afternoon, evening, night, and following morning make up a single calendar day. After consulting the authority cited by Belov (Gamalyeya 1806–1808), the present author has concluded that Bellingshausen generally used the civil calendar in his reports and his book, if only for the convenience of civil servants and the general public,[24] but occasionally failed to transpose a nautical into a civil date, lapses which were not corrected in *Two Seasons* because he did not oversee its publication.

Applying his nautical calendar theory, Belov concluded that the expedition had discovered Antarctica on 27 January rather than 28 January 1820. The theory does not affect the priority question; indeed Belov, apparently unaware that the Royal Navy had switched to the civil calendar in 1805, backdated Bransfield's sighting as well as that attributed to Bellingshausen (below). But the 'Belov date' is a useful tracer for his influence on contemporaries and on later historians. The historian of geography Vasilij Alekseevich Esakov (1924–2016), for example, was an early and consistent convert. Thus he followed Belov in predating the discovery of Peter I I. by two days, one for the nautical calendar and another for the circumnavigation, and did so again by repeating the process for the

---

[22] Bellingshausen also showed the event graphically on the track chart, but without comment (Belov 1963b: Sheet 12, top right).

[23] The reference to an encounter with pack ice at a particular position, reached at a particular time on a particular night, rules out the possibility that this was a process or situation which began early on civil (and nautical) date 28 December (O.S.) and persisted for the rest of that civil day and on into the a.m. half of civil and nautical 29 December.

[24] Just as Cook, whom Bellingshausen much admired, had done in the narrative of his second voyage.

Alexander I Coast. That was unfortunate, because Belov had overlooked the fact that the latter was an a.m. event for which nautical and civil dates would have been the same, so that he only needed to predate by one day, for the circumnavigation (Esakov 1964: 64).

Belov's dating theory was not generally accepted; indeed, it attracted rather more public rejection than was usual for such work at that point in Soviet history. That led to an awkward moment in January 1970, when the Academy of Sciences celebrated the 150th anniversary of the supposed discovery of Antarctica in Leningrad on 28 January, and issued a medal to the same effect, whereas Belov dated the event to 27 January 1820 in his keynote address (Belov 1970a: 203). That year also saw the publication of another critique of Belov's dating theory, by one of his collaborators in the *Karta*, which claimed that the date of noon was the same in all three calendars—civil, nautical, and astronomical, whereas Belov held that nautical noon was dated a day later than the corresponding civil noon (Koblents 1970). Koblents concluded that there was no need to change the noon dates on the track chart. But the disagreement concerned only the reading of the chart. The narrative itself was not affected, because both men agreed that nautical p.m. events were dated one day later than the civil date, which was Belov's main point. The substantive issue, discussed above, is the extent to which surviving primary sources actually used the nautical calendar. Unlike the interpretation of Bellingshausen's ice vocabulary, however, the dating issue has few implications for the history of the expedition.

Analysing the Track Chart

Unlike Lebedev, Belov tended to assume that *materik* meant 'continent' in the modern sense without discussion.[25] Like Lebedev, he was unaware that well before Bellingshausen's expedition people were suggesting that a massive polar ice cap could be formed over the sea. Those two factors probably inclined him to misread the texts.

Belov's analytical flaws were especially evident when he tackled the thorny issues of discovery and priority. He began by arguing that what he took to be Lomonosov's view, that major icebergs could only be formed over land, reached Bellingshausen via Cook, rather than directly. He accepted, at first, that Lomonosov's book on metallurgy, in one of whose

---

[25] But see Belov (1971).

appendices he had argued from the occurrence of icebergs in relatively low southern latitudes to the existence of land further south (1763: 256), would not have been known to Bellingshausen. Mistaking Cook's naturalist Johann Reinhold Forster (1729–1798) for Cook himself,[26] Belov focused instead on Lomonosov's article about the origin of icebergs, which had been published in Swedish and was cited by Forster (Lomonosov 1763). That showed, he thought, that a version of Lomonosov's ideas had been taken on the Bellingshausen expedition, via the first Russian translation of Cook's voyage. On the subject of the generation of icebergs, however, Forster simply bracketed Lomonosov with Buffon and Crantz as sharing the view that 'the ice found in the ocean, is formed *near* lands, only from the fresh water and ice carried down into the sea by the many rivers in Sibiria, Hudson's Bay, etc.' (Forster 1778: 76—emphasis added).[27] That was by no means unfair to Lomonosov, who had emphasized the role of rivers, rather than glaciers, in the generation of icebergs in, and their transport from, fjords and estuaries, although icefall from glaciers into rivers was also mentioned. Belov took Lomonosov's view that land was involved in the formation of icebergs, so that the latter were a sign of the former, and overinterpreted it as saying that the ice which became icebergs had first to be accumulated '*on* "*main* land" ' [18] (Belov 1963a: 36—emphasis added). Significantly, he cited another Soviet author, rather than anything by Lomonosov, to support that reading. His assertion, that Cook (that is Forster) had been 'delighted' [19] with Lomonosov's ideas in particular (the same), was also baseless. Belov added that the resemblance of ideas between Cook/Forster and Lomonosov was further evidence of a connection, presumably because he was unaware that others had concluded there was a connection between land and icebergs long before either of them.

The important consideration here is not what Lomonosov thought, but Belov's hasty assumption, from merely seeing the name mentioned in

[26] Golenishhev-Kutuzov's translation of Cook's second voyage, consulted by Belov but not seen by this author, was from the French translation, probably in its larger six-volume format. Forster's scientific memoir, published separately in Britain, occupied the fifth and most of the sixth volumes of that edition, and his reference to Lomonosov appeared early in the fifth volume (Cook et Autres 1778).

[27] Forster may have seen Buffon's discussion of Antarctic icebergs in Book 1 of the *Histoire Naturelle*, which had suggested there might be a large south polar continent with rivers carrying ice to the sea. He was apparently unfamiliar with Buffon's alternative theory about a marine polar ice cap, published nearly 30 years later at the other end of the work.

a book from the Cook expedition, that Bellingshausen was thereby endowed with what Belov supposed to have been Lomonosov's ideas, whereas he should have looked more closely at the very little Forster actually said about them.[28] Two pages later, despite having just accepted that Bellingshausen had not known Lomonosov's book on metallurgy, Belov described the Russian navigator as convinced about the existence of an Antarctic 'continent of ice'[29] *by phrases in that book* together with the phrase 'continent, covered in ice' [20], allegedly from Cook (the same: 38).[30] But although Cook had concluded that any lands lying further to the south than he could reach must be 'a country doomed by nature ... to lie buried in everlasting snow and ice', neither he nor Forster ever called such land a continent outright, an idea which Cook treated as mere hypothesis on the same page (Cook 1777, 2: 231). Not for the last time, Belov was improving the text in front of him to suit his argument.

The fact that, alone of all Soviet commentators, Belov and Kuznetsova took the trouble to study Simonov's shipboard diary (Simonov″ 1822a), says much for their quality as historians. Unfortunately, however, Belov was even more creative with Simonov than he had been with Forster. Despite quoting Simonov's description of the ice caps as 'impenetrable domes [of ice] which cover the surface of the sea like crusts around both Poles' [21], Belov interpreted him as associating such an 'ice bastion' [22] with land merely because *other*, unnamed geographers had postulated land near the *North* Pole. In fact, Simonov mentioned land in the passage in question as just one possible origin for icebergs, with an ice bastion as an alternative, for which he posited no such connection (Simonov″ 1822a: 178–180; Belov 1963a: 38–39). As for Lazarev, instead of following the accurate transcription in his source Belov first rendered his 'ice floe main' as 'ice main' (the same: 38), then changed the noun as well from *materik*

---

[28] Nothing by Lomonosov was included in lists of books to be taken on the two parallel expeditions (Bellingshausen and Vasil′ev), though doubtless, if his study of the Northern Sea Route had been discovered and published 30 years earlier, it would have been.

[29] The phrase was 'main of ice' (*materik l′da*), but has been translated here to reflect what Belov understood by it.

[30] Unlike many of his contemporaries, or Belov for that matter, Forster was scrupulous about citing his sources. He specified that he had consulted Lomonosov's Swedish article, but he seems to have been unaware of the latter's Russian book on metallurgy. The citation was included in the French translation of Forster, so presumably also the Russian, which the author has not seen.

to *kontinent*, not once, but twice (the same: 39, 40).[31] (The reader might refer to the section about 'Ice Talk' in Chap. 3.)

Turning to Novosil'skij, Belov rightly drew attention to his concept of an 'ice wall' [23], though he missed the passage which most clearly stated the connection Novosil'skij made between such a phenomenon and land (Novosil'skij 1853c: 72). Despite having noted that Novosil'skij was writing some 30 years after the expedition, however, Belov failed to mention that in the interim J.C. Ross had discovered the 'Great Ice Barrier' which now bears his name in the vicinity of mountains and volcanoes, as was reported in the Russian press at the time (Anon. 1842: 30). He also wildly exaggerated the importance of Novosil'skij's four-part classification of ice (Novosil'skij 1853c: 86–87), claiming that it was the first ever made on the basis of observation, but had been ignored by foreign scientists (Belov 1963a: 39).[32] If Belov had taken his own sound advice about respecting the achievements of foreigners, however, he might have noticed that Cook had already listed and distinguished three of Novosil'skij's four types of sea ice (Cook 1777, 1: 42). That was followed by Scoresby's much fuller list of definitions (1818: 263–267), which was refined by himself (1820) and widely circulated before being assimilated into hydrographic literature as a list of 19 expressions (Evans 1824: 97–98). Even in Russia, Lomonosov's three-part classification, written in the 1760s and based on impeccably national observations, was published before Novosil'skij's (Lomonosov" 1847—not seen, but see Lomonosov" 1854: 77). A foreign expert would have to have been very lucky to come across Novosil'skij's pamphlet, given that no Russians paid any attention to it between 1870 and 1940, but even if they had, they were unlikely to have been impressed by its use of only four categories.

In the same parochial vein, Belov declared that Bellingshausen's observations of ice-blink were a great achievement, partly because, he claimed, it was a sure sign of the proximity of land (1963a: 40–41; Bellinsgauzen" 1831: 1, 188). That was little more than bluster. After more than a century of fairly intensive exploration of the Russian Arctic by Russians and others, and with ice-blink so well known across Europe (Crantz 1767: 32; Phipps 1774: 71) that it became a stock item for compendiums of general

---

[31] This, from the man who had recently rebuked Vladimir Lebedev for misplacing an accurate quotation from Bellingshausen by a single day (Belov 1962: 113).

[32] Although his rebuke to foreigners implied the existence of Russian scientists who *had* recognized Novosil'skij's classification, Belov did not identify any such people.

knowledge in Vienna (Schwab 1804: n.p.; Anon. 1804b: 286–287), it seems unlikely that Russians needed to be introduced to the phenomenon by Bellingshausen. It is true, however, that Fyodor Lütke used 'bright whitening' (*yarkaya nabel'*) for ice-blink in the 1820s (Litke 1828, 2: 121), whereas Bellingshausen used 'bright gleam' (*yarkij blesk"*) at about the same time,[33] which suggests that the concept had not yet settled down in Russian.[34] However both explorers regarded ice-blink as a predictor of 'continuous ice', rather than land, (below).

Continuous Ice

Belov's most significant discussion of ice was an argument to show that Bellingshausen's 'continuous ice' was equivalent to his 'hilly, solid stand-ing ice' for 17 February, the aim being to assimilate what was seen on 28 January 1820 to the scene observed three weeks later. Bellingshausen reported sighting continuous ice twice in his report on the first Antarctic season, several times in his report on the second (part of the final report), and about 40 times in *Two Seasons*. Perusal of those examples shows, first that widely different types of ice could be 'continuous', including small ice, pieces or hillocks (perhaps something like bergy bits), and icebergs, and second that the word most commonly associated with continuous ice was 'field', so much so that a field could be continuous, or become con-tinuous by closing up. At one point Bellingshausen explained 'continuous ice' as ice without even the smallest gap, in other words compact ice (1831, 2: 210). The first time he used the expression, continuous ice was the same thing as an ice field mentioned a few lines earlier (Bellinsgauzen" 1820a: fo. 241v), and he identified them again in the passage about 28 January 1820 in his book (Bellinsgauzen" 1831, 1: 171–172). In short

---

[33] As native German speakers, both men were probably translating the current expression in that language, *Eisblick*.

[34] In Britain, Scoresby's lucid discussion of the phenomenon, which distinguished between the ice-blink from field ice, pack ice, bay ice, and land, was in print before the expedition left Kronstadt (Scoresby 1818: 321–322). The earliest notice of Scoresby in Russia seems to have been a translation of a German account of his work (A.N. and M.G. 1823), but neither text could be consulted for this book. A translation of his book about the Greenland whale fishery appeared in 1825. Its dissemination of early glaciological ideas must have been limited, because although his terminology was translated, not always well, in the text, the appendix in which he had defined 22 varieties or aspects of ice, including ice-blink, was omitted (Skorezbi 1825).

the nearest thing to a synonym for his 'continuous ice' would have been 'continuous field', not 'hilly, solid standing ice', an expression which Bellingshausen used only once, and by which he seems to have meant something unique in his experience.

Apart from 'ice', Bellingshausen's commonest word for sea ice, not surprisingly, was 'floe' (*l'dina*). But floes could be of any size, from small to 'enormous' (*ogromnyj*), as high or even higher than 90 m above sea level; were sometimes mistaken for land; and were sometimes referred to on a par with ice islands. It was often possible to sail between or past them (Bellinsgauzen" 1831, 1: 152, 155). For that reason the small ones were extremely dangerous, but in the right conditions it was possible to heave to and send in boats to collect ice for drinking water, or penguins and other specimens. The word 'floe' occurs about 120 times in *Two Seasons*, but never in association with 'field' or 'continuous', except once as a synonym for a previously mentioned ice island (1: 155–156), and another time when Bellingshausen carefully described some continuous ice as made up of small horizontal floes (1: 193). In at least two places the presence of floes near the ships was contrasted with an ice field further off (1: 177; 2: 179). Overall, and despite the word being used very widely, 'floe' signalled ice conditions that were the opposite of continuous ice or an ice field, because the former could be navigated but the latter could not. That means that he used both 'floe' and 'ice field' differently from modern usage, in which floes are the main components of ice fields, but consistently with his tendency to equate ice fields with continuous ice.

Sea ice was shown on the track chart in varying shades of blue, with dark blue reserved for large ice islands and extensive fields of continuous ice, for which dark blue was mottled across a background of light blue (Fig. 3.1). But Bellingshausen did not show on the chart all the ice he mentioned in *Two Seasons*, for example, on the afternoon of 29 January 1820. One consideration may have been the complexity of the track and the number of annotations sometimes required, as, for example, between 27 January and 5 February 1820. As to whether ice was ever shown beyond the limits of possible observation, the author has compared the depiction of ice fields barring further passage south with sections showing the expedition proceeding through relatively open water with icebergs on either beam, but only out to a certain distance. The conclusion is that, if that distance was exceeded when depicting southerly and impassable ice fields, it was not by very much.

Belov drew attention to a small number of annotations along the track that were made with something like a fine indelible pencil, instead of the ink used elsewhere. The graphologist S.N. Balk concluded that they were added by Bellingshausen himself (Belov 1963b: 62). All but one of them read 'Sighted land' [24]. The exception, entered for 28 January 1820, read 'Sighted continuous ice' [25]. Belov interpreted that as evidence that Bellingshausen saw the encounter with an impassable ice field, a short distance SW from 69°21'28"S 2°14'50"W, as a significant approach to the ice main, or continent as Belov liked to call it, on a par with that on 17 February 1820 (Belov 1963a: 40). To make that reading work, however, he chose only some references to continuous ice for the wished-for status of continental discovery.[35] He achieved that selection by associating the references to continuous ice for 28 January, 2 and 17 February 1820, and 20 January 1821, with ice-blink, even though, as he pointed out himself, ice-blink was only recorded on the third and fourth of those days. Nothing daunted, he declared that ice-blink was the key to understanding the weather conditions on all four days. On those occasions continuous ice had been encountered when the ships themselves were under low cloud in murky weather, 'but at the same time it was clear over the continent' [26] (the same).

There were several things wrong with that argument. First and most obvious, Belov had not found a feature common to all four dates; instead, he had imposed his understanding of ice-blink onto two days where it was not recorded, in particular onto 28 January 1820, for which he was seeking to interpret Bellingshausen's 'continuous ice' annotation. Putting words into Lazarev's mind, he also interpreted the latter's recollection, that the evening of that day was briefly fine before the weather turned bad again, as a reference to fine weather 'over the continent' [27] at a time which Bellingshausen recorded as murky weather between noon and midnight. All that Lazarev actually said was that it was a very fine evening [28], by implication both for the ships and for the scene before them, thus nothing like a distant ice-blink observed in bad weather (Lazarev" 1821: fo. 2). Second, if continuous ice with real or alleged ice-blink had merited this special annotation from Bellingshausen for 28 January 1820, why did he not repeat it for 17 February 1820 and 20 January 1821, when that combination was actually recorded, as he did with sightings of land? Third,

---

[35] There would otherwise have been too many to be plausible, including some references to small ice, and in one case to a body of continuous ice itself in motion.

not only were there numerous other references to continuous ice, but ice-blink was observed on two of them, at about 66°S on 26 February 1820 and about 65°S on 9 January 1821, which were not on Belov's list.[36] Fourth, Bellingshausen stated more than once that, like Cook, he regarded ice-blink as a reliable sign of continuous ice (1831, 2: 212–213). That he did not associate it with land is clear from his remark, for 26 February 1820, that the ice field indicated by ice-blink to the south was at a position through which Cook had sailed (1831, 1: 200). Last, he once observed ice-blink over some 40 large flat ice islands, spread across the sea to the *north* of the ships (2: 212). Overall, then, Belov seems to have manufactured an occurrence of ice-blink, wrongly associated with land, for 28 January 1820 purely in order to assimilate continuous ice with land and thus secure the wished-for Russian priority. Whether he was simply unfamiliar with the variety of situations in which ice-blink occurs, or deliberately pretended that it occurs only over land, is impossible to tell.

There is, furthermore, another possible but less dramatic explanation for Bellingshausen's single 'continuous ice' annotation on the chart at 28 January 1820. Ice conditions were not usually annotated along the track, presumably because they were illustrated instead, but they were not explained in the symbols key either. Bellingshausen may have added his 'continuous ice' annotation, therefore, to explain what was meant by the larger, less intuitively obvious, areas mottled with dark blue ink.[37] A reasonable objection is that, in that case, he should have done so the first time such a patch was shown, on Sheet 1, rather than on Sheet 2, the third time.[38] However Sheet 1 had to show a southerly and northerly track close together and both passing through loose ice and icebergs on the west side of the first large field of continuous ice (about 170 km long), and Bellingshausen's assistants had already (on Belov's estimate of the timing) made an ice-related annotation at that point, about collecting ice (for water) with the ships' boats. There were also the usual annotations about

---

[36] Bellingshausen was not far from land at the first position, though too far to see it, which reflects well on Cook if he was further south at that meridian. The second position mentioned here was far from land.

[37] As mentioned above, continuous ice was shown as an agglomeration of ice islands with lighter patches between them, rather like frogspawn (Figs. 3.1 and 3.2). So it was not that hard to read, with additional separate ice islands shown nearby. But Bellingshausen was a careful man.

[38] The second patch had to be broken across Sheets 1 and 2, so that Bellingshausen may have seen that too as an inconvenient place to explain the convention he was using.

sea and weather conditions etc., some of which had to be entered on the east side of that ice field, away from the two tracks, because the latter were so crowded. Bellingshausen may have preferred, therefore, to insert a one-off explanatory annotation beside the third continuous field, where it so happened there was vacant space on the chart for him to do so. This reading of his 'continuous ice' annotation also explains why he did not repeat it wherever such ice was shown on the chart, including the few places where it was observed with ice-blink, such as 17 February 1820 and 20 January 1821, whereas he annotated 'Sighted land' wherever appropriate.

Sheet by Sheet

The rest of Belov's long essay in the *Karta* was a sheet-by-sheet exposition of the track chart. The usual falsehood, about Cook declaring unreservedly that the South Sandwich Is. were a mainland, or continent in Belov's usage, was produced for Sheet 1. Sheet 2, which covered 28 January 1820, was discussed for five pages, about a third of this part of the essay. This time subtleties were dispensed with. The misquotation of Lazarev's 'ice floe main' as 'ice continent' was repeated (Belov 1963a: 45). The weather was overcast with snow, but at a distance recently calculated as between 42 and 38 miles (between 78 and 70 km)[39] the expedition had been 'in immediate sight of the Antarctic coast' [29] (the same: 44), despite which a few pages earlier something like ice-blink had been needed, at least by Belov, to establish the proximity of land. As for dark blue areas on the chart, they represented 'shelf ice and "main ice of extraordinary height"' [30] (the same: 45), this time with no mention of ice-blink as an additional requirement. That implied that both the long ice field to the east and south-east of the South Sandwich Is., and the ice field seen on 20 January 1821 and judged to be at a position through which Cook had sailed, had also been intended by Bellingshausen to represent main ice, a highly improbable corollary. Next, by treating the third dark blue ice field, on Sheet 2, as a sighting of shelf ice at the very least, Belov located the 1820 ice front further north than modern measurements to bring it within 1.5 miles (2.4 km) of the expedition, who must therefore have seen it (the same: 45–46). As was noted in Chap. 8, that form of argument is known as the fallacy of arguing in a circle.

---

[39] The conversion for nautical miles.

Still on Sheet 2, Belov coolly backdated Bellingshausen's 'solid standing ice' from 17 February to 28 January 1820, before prolonging it to 1 February, shown on the chart as a second encounter with the same ice field (the same: 47). Turning to the actual probe of 17 February on Sheet 3, Belov upgraded that too by citing Bellingshausen as his source for 'an approach to within 1.5 miles [2.4 km] of a barrier ice shelf' [31] (the same: 48). No such distance or language, however, appears in any primary source. For Sheet 4, Belov quoted a passage in which Bellingshausen summarized Cook accurately as referring to 'impassable ice' [32]. After first quoting this, Belov then paraphrased it as 'firm ice fields' [33], shown inaccurately inside quotation marks as if also from Cook (the same: 49). He then reverted to his blanket equation of 'continuous' with 'solid standing ice' (the same). However, perhaps because that was inconveniently sweeping, he replaced 'continuous' with a different adjective, 'compact' (*plotnyj*), used nowhere by Bellingshausen, when discussing Sheet 10 (the same: 51). The dating error with the discovery of 'Alexander I Land', mentioned above, was repeated for Sheet 14 (the same: 53). Instead of consulting the findings of modern Antarctic explorers, Belov turned back to Krusenstern, writing more than a century earlier, as his authority for stating that Alexander Land was part of the mainland (the same: 54).

Belov used Bellingshausen's names for the South Shetlands when describing the brief Russian survey, but without commenting on whether they should be acknowledged outside Russia. He remarked that the chart did not support an American claim that Nathaniel Palmer had provided Bellingshausen with useful information, including a map (the same: 54–55; Bulkeley 2015a). But no record of their conversation would have been entered on the chart; a similar event, the visit of sealers at South Georgia, was not even noted.

Belov closed his account of the track chart by highlighting its 'politico-scientific significance' [34], giving as his reason its superior quality by comparison with Russian and, he thought, foreign charts and atlases of the day (the same: 55–56). The one does not, however, follow from the other. He listed ten approaches by the expedition to the ice or rock coasts of Antarctica, with proximities of between 1.5 and 52 miles (between 2.4 and 96 km). The fallacious reasoning which gave him the first of those was discussed earlier. A minor slip concerned the distance from which the Alexander I Coast had been observed. After first citing Bellingshausen's estimate of 40 (nautical) miles (74 km) (Bellinsgauzen" 1831, 2: 254;

Belov, as before: 54), Belov accidentally replaced it with 14 miles (26 km)—the proximity achieved at Peter I I. (the same: 56).

The supplements by Belov's colleagues will not be analysed here, but A.L. Larionov's account of the dimensions and structures of the two ships, using archival documents wherever possible, was an important contribution (Larionov 1963). His conversion of miles to kilometres shows that he, at least, was using the nautical mile, as did Bellingshausen. It is hard to be certain, but Belov seems never to have explained that. Kuznetsova's collation of the annotations on the navigational chart with other accounts of the voyage was another impressive achievement (Belov 1963b: 65–120). Bellingshausen's annotations of sightings were of course included. Another aspect of the tabulation was less straightforward. For 28 January, 1 February, and 17 February 1820 the question-begging phrase 'ice coast' [35], which does not occur on the chart, was added in square brackets. Apart from that the other approaches proposed by Belov were not singled out. Kuznetsova also showed her independence by not listing any of Bellingshausen's (or Belov's) references to ice-blink in the column for events, and by not adjusting the dates in her table to the nautical calendar.

The Antarctic Atlas

Throughout the 1960s Belov was simultaneously completing his monumental four-volume history of the NSR and publishing on early Antarctic history. As the sole historian in the team producing the Soviet *Antarctic Atlas*, he was responsible for 16 maps showing the history of cartography, 32 maps on eight sheets showing the history of exploration, and a further sheet with two details from the navigational chart alongside facsimile passages from Lazarev's letter and Bellingshausen's April 1820 report (Bakaev and Others 1966).[40] Bellingshausen's Antarctic circumnavigation was shown alongside others in Map II of Sheet 7. The expedition's track in January 1820 (O.S.) was shown as passing down the west side of a feature later named the Jutulstraumen Ice Tongue, but designated the

---

[40] For the maps showing the history of exploration until 1963, Belov used a common template outlining the semi-permanent ice edge of the continent, including the Jutulstraumen Ice Tongue, though the line showing the estimated minimum summer distribution of seasonal 'ice', undefined, varied slightly between those maps. The method had been condemned as ahistorical by V. Lebedev a few years earlier.

Bellingshausen Ice Tongue in the Soviet Union (Kruchinin 1965: Fig. 7),[41] then back out again and in to the tip of the same ice promontory. That was quite unlike Sheet 2 of Bellingshausen's navigational chart, where the two probes of 28 January and 1 February reach similar latitudes.

Bransfield's voyage was also shown, together with those of Palmer and Powell, 'in the region of the Antarctic Peninsula' [36] in an inset map at 1:20,000,000 on Sheet 7. Apart from the omission of Deception I., the rendering of Bransfield's anticlockwise course around the Bransfield Strait (so designated on Sheet 24) was close to that estimated by R. Gould, the probable source. However the only date given for Bransfield was a mistakenly nautical '30 January 1820' near Joinville and d'Urville Islands. It implied that Bransfield had been close to Tower I. and the Peninsula the previous day, 30 January CE, but without positively saying so.[42]

Belov's reasonably accurate treatment of Bransfield in the maps volume of the *Atlas* contrasts sharply with the version of that historic voyage which he offered in a history of Antarctic exploration for the companion, text volume three years later (Belov 1969). To appreciate the problem we need to look first at the account published in the *Literary Gazette* in November 1821. It is generally accepted that the *Gazette* was quoting and paraphrasing the journal of one of Bransfield's three midshipmen, Thomas Main Bone, since lost:

> On the 29th, a glimpse was caught of a very high mountain due north; and on the 30th, a small group of islands, extending S.E. to E. by S. ... the horizon very hazy, which opened and shut occasionally, offering to our view an unknown coast ... 'At noon, our latitude ... was 63°16′, and longitude ... 60°28′W.' They now ... steered southward, and seemed to be running from land; but at three o'clock in the afternoon, ..., the haze clearing, they very unexpectedly saw land to the S.W.; and at four o'clock were encompassed by islands spreading from N.E. to E. ... land was undoubtedly seen in latitude 64°, and trending to the eastward. ... A round island was called Tower

---

[41] Today, SCAR's Composite Gazetteer accepts only the Norwegian name, Trolltunga, for this ice tongue, without national alternatives. The modern Russian *Atlas*, however, retains the Soviet name (Kuroedov and Others 2005: 48). Ironically, a major calving event in 1963 had probably reduced Trolltunga to a stump before the Soviet name was bestowed. As a result, Soviet and other scientists had very little time to study this feature as it had been shortly before that date. What was actually there in 1820 is of course impossible to tell.

[42] The name 'Trinity Land' was not used in the *Atlas*, but the tip of the Peninsula was designated 'Trinity Peninsula' on several of its maps, most clearly in Map VII on Sheet 24.

Island, latitude 63°29′, longitude 60°34′, and the land Trinity Land, in complement [thus] to the Trinity Board. (in Campbell 2000: 88)

Belov seems to have consulted both (Gould 1925), which mentioned other sources, and this account from the *Gazette*, republished in 1946, before writing as follows:

On 29 January 1820 the English saw land outlined on the horizon—at first in the shape of a high mountain and then as a mountainous ridge. They called the largest island they could get close to 'Trinity Land'. (The modern name is Trinity Island.) According to Bransfield's observation, it lay at 63°10′S 60°28′W, which corresponds to the location of the offshore Trinity Island, separated from the Antarctic Peninsula by a wide gulf. Taking a north-western course along the strait, E. Bransfield saw land on the horizon, but was unable to approach it on account of the ice. ...

... Researchers have been obliged to settle for a summary of the ship's documents and an 1822 map of 'New South Shetland' created by the cartographer K. [presumably Robert] Laurie.[152] A legend on that map shows a coastline south of 'Trinity Land', but it is quite impossible to confirm that it is Antarctica [37]. (Belov 1969: 42)

By overlooking Bone's 'due north', which shows that at that point Bransfield was inside the Strait looking at a mountain on the South Shetlands (the 'group of islands'), Belov conflated the first part of the excerpt with the rest. He ignored 'steered southward' and the four hours (at least) that elapsed between the noon position of the *Williams*, south of Deception I., and the discovery of Tower I. and Trinity Land. Instead, with one digit mistranscribed, he transferred the ship's noon position onto what was discovered,[43] ignoring the position for Tower I. given in both his sources. In fact he ignored Tower I. altogether. He needed to do so, because it is close to the Peninsula and its latitude, which gives that proximity, was measured quite well by Bransfield and his officers. Instead of following the *Gazette*, Belov picked up on Gould's remark in 1925 (not repeated in 1941) that Trinity I. was one possible candidate for Bransfield's Trinity Land, a speculation which Gould immediately discounted as inconsistent with 'land ... undoubtedly seen in latitude 64°, and trending to the eastward', as reported in the *Gazette*. There is no land at Bransfield's noon

---

[43] The Russian text could be read as 'he lay' rather than 'it lay', but the rest of the sentence shows that Belov was referring to the island, not Bransfield.

position, nor did Bone say there was, though the land seen to SW before they sighted Tower I. may perhaps have been Trinity I., which unknown to Gould, Lebedev, or Belov was given other names as well, and was not known to be separate from the Peninsula until the 1870s (Hattersley-Smith 1991, 2: 573).

The Trinity I. suggestion would not have mattered if Belov had followed the map which he published in 1966, showing Bransfield's track NE along the Peninsula and then passing Joinville and d'Urville islands before turning away north to discover the Elephant and Clarence group. Instead, having left the *Williams* somewhere in the middle of the strait, Belov proceeded to turn her NW, which would have taken Bransfield back towards the south-western end of the South Shetland Is. The context suggests, however, that that was probably another careless mistake, writing 'north-western course' instead of 'north-eastern'. If that is right, Belov was prepared to acknowledge that Bransfield saw land while coasting the tip of the Peninsula, but resisted the obvious implication, that he saw and named part of the Peninsula as Trinity Land, by leaving Bransfield further north and moving the land away to 'the horizon'—once again without support from any sources. Bransfield's chart, available to Belov in (R. Gould 1925), certainly showed part of the Peninsula as 'supposed' due to fog rather than ice, but other annotations, that Trinity Land was 'partly covered with snow' and that further along the coast they saw 'High mountains covered with snow', demonstrated that from time to time the Peninsula was in plain enough view, as Novosil'skij had accepted more than a century earlier.

Ignoring inconvenient testimony from on board the *Williams*, Belov preferred to claim that an 1822 chart of 'New South Shetland' did not support a sighting of Antarctica (Belov 1969: 42). It is hard to follow his reasoning, first because he *discussed* an 1822 chart of the South Shetlands, presumably (Powell 1822), but *cited* an 1821 'Chart of the World' also published by R. Laurie[44]; second because Gould described an 1831 revised version of (Powell 1822), on which, unlike the original, part of Bransfield's track had been added (not seen); and third because (Powell 1822) shows the legend 'Palmer Land' south of a coastline, not a coastline south of the

---

[44] Confusion is worse confounded, if a later version is any guide, because Purdy's 1821 *Chart of the World*, updated to 1824 and published by R. Laurie, also shows 'Palmer's Land' as a coast on the south side of the Bransfield Strait (Purdy 1824).

legend 'Trinity Land',[45] which was how Belov described what (one hopes) he had seen. Both (Purdy 1824) and (Powell 1822) gave clear representations of land on the south side of the Bransfield Strait for any but the most obdurate sceptic, as too did (Weddell 1825), which Belov consulted for the next section of his narrative, and from which he may, with his usual licence, have transferred the legend 'Trinity Land' to whichever chart he meant, given that neither of the two possible charts actually used that name.

The section on the Bellingshausen expedition in Belov's historical survey preceded that on Bransfield and Palmer. It added little new to his essay in the *Karta* six years earlier. The usual mistakes and fudges were still in evidence. Belov replaced his previous sound opinion, that Bellingshausen's Remark about finding no signs of land in the first season had been an overall 'evaluation of the passage of the sloops through Antarctic waters' [38], with the prevarication that it had been limited to the search for land directly south of the South Sandwich Is. Since the Remark came at the end of the April report and explicitly referred to the whole season of exploration which had just been described, there were no grounds whatsoever for that interpretation. On the same page Belov repeated his habitual misquotation of the word *kontinent* from both Bellingshausen and Lazarev (Belov 1969: 40). Despite the fact that just three years earlier maps in the *Atlas* had shown Alexander Land as an island (Sheets 16–17, 19–20), the nineteenth-century opinion of Krusenstern, as a 'Russian navigator' [39], was once again dusted off to support its mainland status (the same: 41). At the South Shetlands, Belov claimed Rozhnov/Gibbs I. and others (unnamed) for Bellingshausen in 1821 (the same; see also Bellinsgauzen" 1831 *Atlas*: Sheet 62). As was said in Chap. 8, Rozhnov/Gibbs was a legitimate attribution because Bransfield, working in fog in 1820, had only recorded 'O'Brien's Is.' (Eadie, O'Brien and Aspland) a short distance to the west of Rozhnov/Gibbs, whereas the *Atlas* volume of *Two Seasons* showed the latter as well.

In summing up, Belov repeated his usual 'ice continent' misquotation and elevated sightings of the aurora australis to the status of 'data' [40] (the same: 41), but rightly praised the expedition for the quantity of

---

[45] Of the large number of maps of the Bransfield Strait, from between 1820 and 1850, that were collected by W.H. Hobbs, all showed the legend below the coastline rather than above. Most showed 'Palmer's Land' (the case being made by Hobbs), but those by Weddell, or by French or German cartographers, used the legend 'Trinity Land' (Hobbs 1939a). Only Ross put the legend physically above the coastline, but that was because his chart was drawn with south at the top (J.C. Ross 1847, 2: opp. p. 328).

geographical and oceanographic information that had been obtained. He drew attention to such of the expedition's scientific observations as were published, but without a citation his improbable claim that Bellingshausen's magnetic data (G.[auss] 1840) influenced the work of the Ross expedition, which sailed before they were published, was empty talk. The still more extraordinary claim that Ross had republished them in the Leipzig general knowledge weekly *Pfennig-Magazin* was a complete fabrication (Belov 1969: 41, 116), because first, the articles in question were (partly) about Ross, not by Ross, and second, neither Bellingshausen nor his magnetic measurements were even mentioned. Nor was Bellingshausen's estimated position for the South Magnetic Pole (SMP) published in Germany then, as Belov also alleged.

Misquoting to the end, Belov explained the failure of their contemporaries to understand Bellingshausen and Lazarev (the latter unpublished!), when they referred to an 'ice continent' or 'continent of ice', as due to the backwardness of nineteenth-century glaciology (the same: 42)—not that he seems to have studied the latter at all closely. From the many foreign tributes to the expedition he selected: (a) Gravelius' translation of *Two Seasons*, which he wrongly supposed to have been complete;[46] (b) the 1822 British chart discussed above, no candidates for which showed any Russian information, such as Peter I I. and the Alexander I Coast, since at that point, given that Simonov's university lecture was not published (in Russian) until late 1822, they would have needed to extract it from a Russian periodical (Anon. 1821a); and (c) an 1867 'maritime atlas' by Ravenstein. The necessary cross-reference to Map И on Sheet 5 of the 1960 maps volume was carelessly omitted.[47]

The Anniversary Lecture

Belov's 1970 anniversary lecture was his most political treatment of the subject, as well as the one in which he stretched his modernizing reinterpretation of sources to its limits. The discussion of Cook was unusual, to say the least. During the Khrushhev Thaw, just like V. Lebedev, Belov

---

[46] It was complete in manuscript, but the Dresden Society for Earth Sciences could not afford to publish the whole thing (Mill 1903: 151).

[47] Besides the discoveries of January 1821 that map showed 'Eis [ice] 1820' for Bellingshausen between 30° and 35°E longitude, through which the expedition passed at about 65°S and without meeting continuous ice. It appears to come from the third edition of Meyer's general atlas (Ravenstein 1867—not seen).

accepted that Cook had concluded there probably was an ice-girt continent near the South Pole [41] (with the 'probably' inserted by Belov)
(Belov 1963a: 6). That went unchanged in 1970, but Belov still managed
to engineer a confrontation by claiming that Bellingshausen had coun-
tered Cook's 'groundless' [42] view with his own 'new geographical theory' [43] about an ice continent (Belov 1970a: 203). Nothing much
wrong with that, apart from the combative language, the absence of
Buffon, and Cook's having argued that 'The excessive cold, the many
islands and vast floats of ice, all tend to prove that there must be land to
the South …' (Cook 1777, 2: 239), which was anything but 'groundless'
reasoning. But as we shall see Belov was also keen to claim that
Bellingshausen and Lazarev had envisaged a continent every bit as terrestrial as the one Cook believed in (the same: 205), which meant that the
opposition between them earlier in the lecture had been entirely contrived.
Another novelty served up by Belov was the claim that Cook thought
South Georgia was a 'Land' rather than an island, such that the Russian
expedition needed to set the record straight (the same: 203). The facts
that Cook named it 'the Isle of Georgia', called it that six times in his narrative, included a chart showing that it was an island, and both before and
after surveying its NW coast took a good look along the SE one,[48] were
brazenly ignored (Cook 1777, 2: 212, 217, 218).[49] Small wonder that the
stale canard, about Cook having no idea that Sandwich Land was a group
of islands, was added to the mix.

On the positive side of Belov's presentation, Lomonosov was hailed as
the first person ever to place the existence of a southern continent on a
scientific basis because, in his book on metallurgy, he had cited the occurrence of large icebergs, at about 53°S between Tierra del Fuego and the
meridian of the Cape of Good Hope, as evidence that:

> there are remote islands and main land covered with much unremitting
> snow, and a great[er] proportion of the earth's surface is burdened with
> them around the South Pole, than in the North [44]. (Lomonosov"
> 1763: 256)

---

[48] Relying on Cook as he usually did, Bellingshausen surveyed only the SW coast without
looking at the NE one.

[49] Cook did mention that they had taken the island 'for part of a great continent' at first,
but went on to explain that he had demonstrated that it was an island after all (the same: 217).

Lomonosov was almost certainly referring to recently published maps (Buache 1740, 1763),[50] but unfortunately for Belov's thesis first Behrens (1738: 50) and then Buffon had published this idea before Lomonosov. As mentioned above, in the 'Theory of the Earth' with which he opened the *Histoire Naturelle*, Buffon referred to 'enormous icebergs which detach themselves from polar continents' [45] (Buffon and Daubenton 1749: 72). There might be a continent in the Antarctic as large as Europe, Asia and Africa combined, he suggested, and went on, after referring to Buache's 1740 map, to explain that:

> These icebergs must come from lands within & near the southern Pole, & one can surmise that they follow the course of several large rivers which water those unknown lands,... [46] (the same: 215–216)

In short, 'The more ice, the more land' [47], as de Brosses summarized it, also before Lomonosov (1756: 60). Lomonosov read Buffon in 1762 (Berkh" 1828: 290), and published his article on icebergs a year later. So much for his supposed priority in the matter.

As for the Antarctic zone being colder than the Arctic, that too was suggested well before Lomonosov (thus enabling Buffon to disagree with it in 1749 before changing his mind 29 years later):

> Because of ye much greater cold and ye Seas being more frozen toward ye South than North Pole, discoveries have not been made so far to ye Southward as to ye Northwd. but Open Seas are never known to be frozen, only ye borders near ye Land, through ye great quantity of fresh Water brought from ye Land: Whence it may resonably be concluded that there lies much more Land, tho' less discovered about ye South Pole than the North Pole. (Senex 1725)

When the brief passage in Lomonosov's study of metallurgy is considered alongside his article on icebergs, discussed above, it becomes clear that on the question of the existence of land near the South Pole he was synthesizing the ideas of his day, rather than coming up with something

---

[50] The second item here was Buache's English version of a map which he presented to the Académie Royale in 1757. The original was published in colour in the *Histoire de l'Académie Royale* (1757 pl. XI opp. *Mém.* 202—not seen but readily available to Lomonosov), but the small greyscale image available on the internet is difficult to read because the map overlays the Arctic on the Antarctic.

new.[51] Belov had failed to appreciate that Lomonosov was very well informed by and connected with his peers across Europe.

Despite having shown the voyages of Bransfield and Palmer on a map in the *Atlas* in 1966, as he approached the end of his 1970 address Belov went even further in his misrepresentation of their achievements than he had done in the 1969 *Atlas* text volume. The sealers were presented as the agents of British and American aspirations for control of the Antarctic. According to their 'tales', they had seen some sort of body of land to the south of their camps [48]. Reluctant to concede priority to their Russian rival, they gradually and quietly announced that they had discovered the southern continent (Belov 1970a: 206). But even if the sealers, while combing the area for new sealeries, had improbably gained access to information which had only been published in Russian at that point, they would have found in it repeated statements that no land was sighted in 1820 after the South Sandwich Is. (Bellinsgauzen" 1821b, c; Simonov" 1822a, b). No such rivalry for priority, therefore, could possibly have motivated even the most polyglot and cloak-and-dagger sealer.

Avoiding any allusion, however slight, to Bransfield's survey with a team of naval officers, Belov focused on Palmer and John Davis (1784–18??) (1970a: 206–207). Palmer was said to have sighted the Low Islands (thus) and the South Orkneys, rather than the Peninsula. The second attribution is generally accepted, but Belov had no grounds for diverting Palmer to Low I., given that it lies WbSW from Deception I., whereas his course on 17 November 1820 was 'S by E½' according to his log (Hinks 1940: 422), the relevant parts of which had been published 30 years earlier. (The Low I. red herring was also pointless, because Belov needed only the date to refute claims that Palmer saw the mainland first.)[52] Next, Belov translated a substantial extract from Davis's log, but dismissed the possibility that he was the first person to land on the mainland (on 7 February 1821), partly because he had given no longitude, but also because this had been erroneously claimed as the discovery of Antarctica *by someone else* (Stackpole

---

[51] Lomonosov was himself a Pomor (Arctic Russian), and his real contributions to early glaciology were his detailed observation of varieties and behaviours of ice, supported with local words like *padun* (ice fall) and *toros* (ice foot, but applied to ice fringing large bergs as well as shorelines), and the inclusion of inland ice-fall in the generation of icebergs, albeit without renouncing the rivers theory.

[52] There are ominous resemblances between the western detour which Belov imposed on Palmer and the 'north-western' detour which he previously imposed on Bransfield (Belov 1969: 42), but which has been interpreted here as a slip of the pen (above).

1955). All in all, he concluded, even if the sealers had discovered 'this or that … microscopic geographical object' [49] a year or two before Bellingshausen, their achievements would have borne no comparison with the scope and thoroughness of the Russian expedition (the same: 207).

Belov closed his anniversary address with a well-taken tribute to Bellingshausen's magnetic observations which he also published separately (Belov 1966b, 1970b).[53] He repeated the claim that subsequent geomagneticians and explorers had been indebted to Bellingshausen in the matter of the SMP. That was certainly true for Duperrey, who used Bellingshausen's measurements to calculate different coordinates for the Pole from those which the latter sent Gauss in a letter, but which remained unpublished in the nineteenth century (Duperrey 1841).[54] But Duperrey had last taken his own measurements in the 1820s and far from the Antarctic. By contrast J.C. Ross and his geomagnetician Edward Sabine were using thousands of new measurements of the three main elements in the earth's magnetic field (intensity, declination, and inclination), taken close to the SMP and collated with simultaneous readings from across the world, so probably felt little need to consult 202 measurements of a single element taken by Bellingshausen.[55]

*   *   *

Unlike Vladimir Lebedev, Belov made very substantial contributions to knowledge of the expedition by publishing the track chart and describing the magnetic observations, and would do so again in 1974 (below). His

[53] For want of the relevant Soviet journal, the author was obliged to rely on the English translation of the second item, published in Israel. In that version, Belov mentioned that he had not been able to find a single reference to Bellingshausen's list of declinations in works on terrestrial magnetism. Two sentences later he declared that 'Bellingshausen's magnetic work in the Antarctic was highly valued by his contemporaries' (1970b: 246).

[54] Belov seems to have missed Duperrey's acknowledgement to Bellingshausen, since he mentioned another estimate for the SMP from Duperrey but not this one, made in 1837 from declinations he can only have taken from *Two Seasons*, rather than from those in (G. [auss] 1840), which were four times more numerous. Duperrey also gave different coordinates for J.C. Ross's determination of the SMP from those given by Belov.

[55] One has to wonder whether Belov was able to consult the dense volumes of magnetic and other data from mid-nineteenth-century voyages, such as Sabine (1850–1853); Wüllerstorf-Urbair and Others (1861–1875). Even Bellingshausen's (privately funded and better-equipped) contemporary in the Russian Navy Kotzebue had published more data more promptly (Kotzebue 1821: 221–228).

arguments for priority, however, resembled Lebedev's work by overinterpreting or neglecting the texts, which are all we have to go on. His more varied and sometimes more subtle reasoning tended to draw him into more serious misconceptions than those of Lebedev, such as when considering the possible role of Lomonosov. On the basis of the analysis presented here, the judgement of one modern commentator, that Belov's work overall was 'scrupulous ... detailed and cogent' [50], can no more be accepted than the same author's unsupported guess that Belov and his contemporaries worked with logbooks and journals from the voyage [51], none of which in fact survive (Tsuker 2015: 45).

To the best of this author's knowledge, no Soviet or Russian historian ever offered an explanation for the scarcity of substantial research into the Bellingshausen expedition during the last 25 years of the Soviet Union, to which we now turn.

Original Texts of Quotations

[1] ... техъ не проницаемыхъ сводовъ, кои на подобіе коры покрываютъ морскую поверхность у обоихъ полюсовъ, ...

[2] Огромные льды, которые по мѣрѣ близости къ Южному полюсу подымаются въ отлогія горы, называю я матерыми...

[3] ... сложенного льдом ...

[4] ... «матерой лед чрезвычайной высоты» был коренным берегом Антарктиды и ничем иным.

[5] Ледяных берегов Антарктиды и самого материка нет также на карте, ...

[6] ...увиденный Лазаревым «матерой лед чрезвычайной высоты» был коренным берегом Антарктиды ...

[7] ...таковые спертіеся или сплоченные льды...

[8] ...касаясь...

[9] ...постоянно...

[10] ...съ 5 на 6 число [ст.ст.] дошелъ до широты S. 69°, 7′, 30″, долготы О. 16°, 15′. Здесь за льдеными [так] полями мелкаго льда, и Островами виденъ Материкъ льда. коего края отъ ломаны перпендикулярно, и Которой [так] продолжался по Мѣрѣ на-шего зрѣнія, возвышаясь къ Югу подобно берегу. Плоскіе льдяные острова, близь сего материка находящіися [так], ясно

показывають, что онѣ суть отломки сего материка; ибо имѣють края и верьхную поверьхность подобную материку.

[11] …наконецъ въ четверть четвертаго часа по полудни, увдѣли множество большихъ, плоскихъ, высокихъ льдяныхъ острововъ, затертыхъ плавающими мелкими льдами, и мѣстами одинъ на другомъ лежащими. Льды къ SSW примыкаются къ льду гористому, твердо стоящему; закраины онаго были перпендикулярны и образовали заливы, а поверхность возвышалась отлого къ Югу, на растоянiе, предѣловъ котораго мы не могли видѣть съ салинга. …

Видя льдяные острова, поверхностью и краями сходные съ поверхностью и краями большаго вышеупомянутаго льда, предъ нами находящагося, мы заключили, что сiи льдяныя громады и всѣ подобные льды, отъ собственной своей тяжести, или другихъ физическихъ причинъ, отдѣлились отъ матераго берега, вѣтрами отнесенные, плавають по пространству Ледовишаго Южнаго Океана; …

[12] …успешные наблюдения…

[13] … сколько можно подойти къ дальнимъ льдянимъ горамъ, …
Далѣе представлялись льдяныя горы, подобныя вышеупомянутымъ, и вѣроятно, составляють продолженiе оныхъ.

[14] … а ледяные горы или ледяные склоны …

[15] Экспедиция проходила в непосредственной близости от Антарктического материка на протяжении сотен миль и совершенно точно описала его…

[16] …оценк[а] плавания шлюпов в антарктических водах…

[17] …*съ 28 на 29е утромъ въ 4е часа въ широте 64°40′ долготе 126°40′*…

[18] … на «матёрой земле» …

[19] … понравилась Куку.

[20] … континент, покрытый льдами …

[21] … льдяному оплоту.

[22] … техъ не проницаемыхъ сводовъ, кои на подобiе коры покрывають морскую поверхность у обоихъ полюсовъ, …

[23] … ледяная стѣна …

[24] *Увидѣли берегъ.*

[25] *Увидѣли сплошной ледъ.*

[26] … а над континентом в это время было ясно.

[27] … над континентом …

[28] … *въ прекрасный тогда вечеръ* …

[29] … в пепосредственной видимости антарктического берега.

[30] … условно обозначавшей шельфовый и «матёрой лед чрезвычайно высоты», …

[31] … подошли … к барьеру шельфового ледника на расстояние 1,5 мили.

[32] … непроходимые льды …

[33] … «твердые льды»,

[34] Политико-научное значение…

[35] [Ледяной берег].

[36] …в районе Антарктического Полуострова.

[37] 29 января 1820 г. англичане увидели на горизонте силуэт земли—сначала в виде высокой горы, а затем горного хребта. Самый большой остров, к которому им удалось подойти на близкое растояние, они назвали «Землей Тринити». (1. Современное название о. Тринити.) По вычислению Брансфилда, он находился на 63°10′ ю. ш. и 60°28′ з. д., что соответствует местоположению прибрежного острова Тринитн, отделенного от Антарктического полуострова большим проливом. Следуя по проливу северо-западным курсом, Э. Брансфилд на горизонте увидел землю, но из-за льда пройти к ней не мог.…

…Исследователям приходится довольствоваться пересказом судовых документов и картой «Новой Южнной Шетландии» 1822 г., автором которой был картограф К. Лори [152]. На этой карте к югу от «Земли Тринити» условно показано побережье, но утверждать, что это Антарктида, никак нельзя.

[38] …оценк[а] плавания шлюпов в антарктических водах…

[39] …высказивания русских мореплавателей…

[40] …получены данные о…полярном сиянии…

[41] Главный вывод Кука заключался в том, что у Южного полюса, возможно, лежит ледяной континент…

[42] …несостоятельное…

[43] …новая географическая теория…

[44] …въ большемъ отдаленіи островы и матерая земля многими и несходящими снѣгами покрыты, и что большая обширность земной поверхности около южнаго полюса, занята оными, нежели въ сѣверѣ.

[45] …ces glaces énormes qui se détachent des continens des poles,…

[46] Ces glaces doivent venir des terres intérieures & voisines du pole austral, & on peut conjecturer qu'elles suivent le cours des plusieurs grands fleuves dont ces terres inconnues sont arosées,...

[47] Plus il y a de terre plus il y a de glace:...

[48] ...из рассказов которых явствовало, что к югу от своих становищ они видели какую-то большую землю.

[49] ...какого-то...объекта микросопической величины...

[50] ...скрупулезном изучении...Наиболее подробное и убедительное доказательство...

[51] ...изучении...судовых журналов и штурманских расчетов...

References[56]

A.N., and M.G., trans. 1823. Izvestie ob″ otkrytiyakh″ kap. Villiyama Skorresbi na vostochnom″ beregu zapadnoj Grenlandii [A Report on Captain William Scoresby's Discoveries on the West Coast of Greenland]. *Severnyj Arkhiv″* 7 (13): 50–59, 7 (14): 189–200, 7 (15): 240–258—not seen.

Anon. 1804b. Fortsetzung des Aufsatzes: Das nordliche Amerika [North America—Continued]. *Skizzen in Kupfern geografisch-historisch-artistisch-ökonomischen Inhalts* 36, Vienna: 284–288.

———. 1821a. Dal'nyejshiya svyedeniya o plavanii otryada, sostoyashhago iz″ shlyupov″ Vostoka i Mirnago, pod″ komandoyu Kapitana Bellinsgauzena [Further Information about the Voyage of the Squadron Comprising Sloops Vostok and Mirnyj, Commanded by Captain Bellingshausen]. *Russkij Invalid* 196: 786–788.

———. 1842. Otkrytie novago materika v Yuzhnom″ polusharii [Discovery of a New Mainland in the Southern Hemisphere]. *Syn″ Otechestva* 1: 24–32.

———. 1960. Russia "Discovered Antarctica". *The Times*, June 16: 10.

———. 1962. Trail Blazers of Antarctica. *UNESCO Courier* 1: 17.

Avsyuk, G.A., and S.N. Kartashov. 1971. 150 let otkrytiya Antarktidy russkoj èkspeditsiej [150 Years Since the Discovery of Antarctica by a Russian Expedition]. *Antarktika—doklady kommisii* 1968: 5–12.

Bakaev, V.G., and Others. 1966. *Atlas Antarktiki [Atlas of the Antarctic]*. Vol. 1. Moscow–Leningrad: Geodesic and Cartographic Department.

[56] Key to archival references: SARN = State Archives of the Russian Navy, F = Fond, S = Series [*Opis'*], P = Piece [*Delo*], fo./fos = folio/s [*list/y*], v = verso [*oborotnoe*]. (The latter is necessary because verso pages were usually unnumbered.) Dates on manuscript documents are given in Universal Time (CE).

Behrens, C.F. 1738. *Der wohlversuchte Südländer: Reise um die Welt [The Much Sought After Southerner: A Circumnavigation].* Leipzig: Self.

Bellinsgauzen″, F.F. 1820a. Report to the Marquis de Traversay from Sydney, April 20, 1820. St Petersburg: SARN—F–166, S–1, P–660b, fos 239–245v.

Bellinsgauzen″, F.F. 1821e. Report to the Marquis de Traversay from Kronstadt, August 5, 1821. St Petersburg: SARN—F–203, S–1, P–826, fos 1–18v.

Bellinsgauzen″, Kapitan″. 1821b. Vypiska iz″ doneseniya Kapitana 2 ranga Bellinsgauzena k″ Morskomu Ministru ot″ 8 Aprelya 1820 goda [O.S.] iz″ Porta Zhaksona [Extract from the Report of Junior Captain Bellingshausen from Port Jackson to the Minister of Marine, April 8 1820 [O.S.]]. *Syn″ Otechestva*, April 23 1821 (O.S.), 69: 133–135.

———. 1821c. Vypiska iz″ pis″ma Kapitana Bellinsgauzena k″ Morskomu Ministru ot″ 8 Aprelya 1820 goda [O.S.] iz″ Porta Zhaksona [Extract from Captain Bellingshausen's letter to the Minister of Marine from Port Jackson, April 8 1820 [O.S.]]. *Syn″ Otechestva*, April 23 1821 (O.S.), 69: 135–137.

———. 1821d. Karta Plavaniya Shlyupov″ Vostoka i Mirnago vokrug″ Yuzhnago polyusa v″ 1819, 1820 i 1821 godakh″ pod″ Nachal′stvom″ Kapitana Billensgauzena [Chart of the Voyage of Sloops *Vostok* and *Mirnyj* around the South Pole in 1819, 1820 and 1821 under the Command of Captain Bellingshausen]. St Petersburg: SARN—F–1331, S–4, P–536, fos 5–19.

———. 1831. *Dvukratnyya izyskaniya v″ yuzhnom″ ledovitom″ okeanye i plavanie vokrug″ svyeta v″ prodolzhenii 1819, 20 i 21 godov″ [Two Seasons of Exploration in the Southern Ice Ocean and a Voyage around the World, During the Years 1819, 1820 and 1821],* ed. L.I. Golenishhev″-Kutuzov″, 2 vols. plus *Atlas.* St Petersburg: Glazunovs.

Belov, M.I. 1948. *Semyon Dezhnev 1648–1948.* Moscow: GlavSevMorPut.

———. 1954. *Vvedenie [Introduction].* In *Geograficheskij sbornik III,* 5–12. Moscow: Academy of Sciences.

———. 1956–1969. *Istoriya otkrytiya i osvoeniya severnogo morskogo puti [History of the Discovery and Conquest of the Northern Sea Route].* 4 vols. Moscow: Morskoj Transport.

———. 1961a. Zabytaya karta [A Forgotten Chart]. *Vodnyj Transport* 21 (138): 3.

———. 1961b. Otchyotnaya karta pervoj russkoj antarkticheskoj èkspeditsii [The Official Chart of the First Russian Antarctic Expedition]. *Byulleten′ Sovetskoj Antarkticheskoj Èkspeditsii* 31: 5–14.

———. 1962. Shestaya chast′ sveta otkryta russkimi moryakami (Novye materi-aly…) [The Sixth Continent Discovered by Russian Seamen—New Sources…]. *Izvestiya vsesoyuznogo geograficheskogo obshhestva* 94 (2): 105–114.

———. 1963a. O kartakh pervoj russkoj antarkticheskoj èkspeditsii 1819–1821 gg. [The Maps of the First Russian Antarctic Expedition 1819–1821]. In *Pervaya russkaya antarkticheskaya èkspeditsiya 1819–1821 gg. i eyo otchyotnaya navigatsionnaya karta,* ed. M.I. Belov, 5–56. Leningrad: Morskoj Transport.

————. 1963b. *Pervaya russkaya antarkticheskaya èkspeditsiya 1819–1821 gg. i eyo otchyotnaya navigatsionnaya karta [The First Russian Antarctic Expedition 1819–1821 and its Official Navigational Chart]*. Leningrad: Morskoj Transport.

————. 1966b. Slavu pervogo tochnogo vychisleniya mestopolozheniya yuzhnogo magnitnogo polyusa anglichanin Ross dolzhen razdelit's Bellinsgauzenom [The Englishman Ross should Share the Glory of the First Accurate Determination of the South Magnetic Pole with Bellingshausen]. *Nauka i Zhizn'* 8: 21–23.

————. 1969. Istoriya otkrytiya i issledovaniya Antarktiki [History of the Discovery and Exploration of Antarctica]. In *Atlas Antarktiki [Atlas of the Antarctic]*, ed. E.I. Tolstikov, vol. 2, 35–97. Moscow–Leningrad: Geodesic and Cartographic Department.

————. 1970a. Otkrytia ledyanogo kontinenta [The Discovery of the Icy Continent]. *Izvestiya vsesoyuznogo geograficheskogo obshhestva* 102 (3): 201–208.

————. 1970b. The First Maritime Magnetic Survey around Antarctica. In *Problems of Polar Geography*, 244–252. (Trans. of *Trudy Arkticheskogo i Antarkticheskogo nauchnogo issledovatel'skogo instituta*, 185). Jerusalem: Israel Program for Scientific Translations.

————. 1971. Comment by Professor M.I. Belov. *Polar Record* 15 (99): 890–891.

Berkh″, V.N. 1828. Dopolnenie k zhizneopisaniyu M.V. Lomonosova [Supplement to the Biography of M. V. Lomonosov]. *Moskovskij Telegraf* 21 (11): 289–314.

Berlyant, A.M., T.F. Vasil'ev, and I.A. Suetova. 1981. Prostranstvennye soötnosheniya lednikovogo i korennogo rel'efa Antarktidy [Spatial Relationships between the Glacial and Basic Relief of Antarctica]. *Antarktika—doklady kommisii 1980*, n.p.

de Brosses, C. 1756. *Histoire des Navigations aux Terres Australes [A History of Voyages to Southern Lands]*. Vol. 1. Paris: Durand.

Buache, Philippe. 1740. *Carte des Terres Australes [Map of Southern Lands]*. Paris: Académie Royale des Sciences. https://www.davidrumsey.com/luna/servlet/detail/RUMSEY~8~1~299835~90070885:Carte-des-Lieux-ou-les-Differentes-.

Buache, [P.]. 1763. Geographical and Physical Observations, Including a Theory of the Antarctic Regions, and the Frozen Sea which They are Supposed to Contain. Includes: Chart of the Antarctic Polar Circle, Opp. p. 32. *Gentleman's Magazine* 33: 32–36.

de Buffon, M. le Comte, and L.-J.-M. Daubenton. 1749. *Histoire naturelle générale et particulière*. Vol. 1. Paris: Imprimerie Royale.

Bulkeley, Rip. 2014a. *Bellingshausen and the Russian Antarctic Expedition, 1819–21*. Basingstoke: Palgrave.

————. 2015a. The Bellingshausen-Palmer Meeting. *Polar Record* 51 (2): 212–222.

———. 2019. Bellingshausen's 'Mountains': The 1820 Russian Sighting of Antarctica and Bellingshausen's Theory of the South Polar Ice Cap. *Polar Record* 55 (6): 392–401.

Butler, Raymond A. About 1958. *The Age of the Sealer. Map 2. Antarctica.* Washington, DC: US Antarctic Programs.

Campbell, R.J. 2000. *The Discovery of the South Shetland Islands.* London: Hakluyt Society.

Cook, James. 1777. *A Voyage Towards the South Pole, and Round the World; Performed in His Majesty's Ships the Resolution and* Adventure, *in the Years 1772, 3, 4, and 5.* 2 vols. London: Strahan and Cadell.

Cook, Jacques, et Autres. 1778. *Voyage au Pôle Austral et autour du monde [A Voyage Towards the South Pole, and Round the World].* Vol. 5. Paris: Merigot.

Crantz, David. 1767. *The History of Greenland.* Vol. 1. London: Society for the Furtherance of the Gospel.

Croome, Angela. 1960. Geophysics and Space Research. *Discovery* 21 (9): 401–402.

Dal', V.I. 1863. *Tolkovyj slovar' zhivago velikorusskago yazyka [Reference Dictionary of the Greater Russian Language].* Vol. 1. Moscow: Semyon.

Dal', Vladimir. 1881. *Tolkovyj slovar' zhivago velikorusskago yazyka [Reference dictionary of the Greater Russian language].* Vol. 2. St Petersburg–Moscow: Vol'f.

Debenham, F. Frank. 1959. *Antarctica.* London: Herbert Jenkins.

Duperrey, L.I. 1841. Notice sur la position des pôles magnétiques de la terre [The Position of the Earth's Magnetic Poles]. *Bulletin de la Société de Géographie* 16: 314–324.

Esakov, V.A. 1964. Otkrytie Antarktidy i issledovaniya èkspeditsii F.F. Bellinsgauzena i M.P. Lazareva [The Discovery and Exploration of Antarctica by the Bellingshausen–Lazarev Expedition]. In *Russkie okeanicheskie i morskie issledovaniya v XIX—nachale XX v*, 53–69, eds. V.A. Esakov, A.F. Plakhotnik, and A.I. Alekseev. Moscow: Nauka. https://flot.com/publications/books/shelf/explorations/7.htm.

Evans, John. 1824. *Revision and Explanation of the Geographical and Hydrographical Terms.* Bristol: Rose.

Forster, John R. 1778. *Observations Made During a Voyage Round the World.* London: Robinson.

Fyodorovskij, E. 1976. Svezhij veter okeana [Fresh Ocean Breeze]. *Iskatel'* 3: 1–74.

G.[auss]. 1840. Abweichungen der Magnetnadel, beobachtet vom Capitaine Bellingshausen in den Jahren 1819–1821 [Magnetic Declinations Observed by Captain Bellingshausen in 1819–1821]. *Beobachtungen des magnetischen Vereins im Jahre* 1839: 117–119.

Gamalyeya, P.Ya. 1806–1808. *Teoriya i praktika korablevozhdeniya [Theory and Practice of Navigation].* St Petersburg: Naval Press.

Gorskij, N.N., and V.I. Gorskaya. 1957. *English-Russian Oceanographic Dictionary.* Moscow: State Publishing House for Technical and Theoretical Literature.

Gould, R.T. 1925. The First Sighting of the Antarctic Continent. *Geographical Journal* 65: 220–225.

———. 1941. The Charting of the South Shetlands, 1819–28. *The Mariner's Mirror* 27: 206–242.

Hattersley-Smith, G. 1991. *The History of Place-Names in the British Antarctic Territory.* Cambridge: British Antarctic Survey.

Hinks, Arthur R. 1940. The Log of the *Hero. Geographical Journal* 96 (6): 419–430.

Hobbs, William Herbert. 1939a. The Discoveries of Antarctica within the American Sector, as Revealed by Maps and Documents. *Transactions of the American Philosophical Society* 31 (1): 1–71.

Horensma, Pier. 1991. *The Soviet Arctic.* Milton Park, Abingdon: Routledge.

Kippis, Andrew. 1788. *The Life of James Cook.* Vol. 2. Basle: Tourneisen.

Koblents, YaP. 1970. Schislenie vremeni na russkom flote i khronologiya sobitij pervoj russkoj antarkticheskoj èkspeditsii [The Reckoning of Time in the Russian Navy and the Chronology of Events in the First Russian Antarctic Expedition]. *Byulleten' Sovetskoj Antarkticheskoj Èkspeditsii* 80: 5–23.

Kolgushkin, V.V., and P.R. Maksimov. 1958. *Opisanie starinnykh atlasov, kart i planov XVI, XVII, XVIII vekov i pervoj poloviny XIX veka [A Description of Old Atlases, Maps and Plans from the 16th to the First Half of the 19th Century].* Leningrad: Hydrographic Department, Soviet Navy—not seen.

Kondakova, O.N. 2019. *K 200-letiyu otkrytiya Antarktidy Èkspeditsiej F.F. Bellinsgauzena i M.P. Lazareva na shlyupakh "Vostok" i "Mirnyj" [For the Bicentenary of the Discovery of Antarctica by F.F. Bellingshausen and M.P. Lazarev on Sloops Vostok and Mirnyj].* St Petersburg: Russian Naval Archives. https://rgavmf.ru/virtualnye-vystavki/k-200-letiu-otrritiya-antarktidi.

Kotzebue, O. 1821. *Entdeckungsreise in die Süd-See und nach der Berings-Strasse, zur Erfahrung einer nordöstichen Durchfahrt [Voyage of Discovery to the Pacific Ocean and through the Bering Strait, to find out a North-East Passage].* Vol. 1. Weimar: Hoffman.

Krajner, N.P. 1962. P.A. Kropotkin o proiskhozhdenii valunov [P.A. Kropotkin on the Origin of Boulders]. *Trudy Instituta Istorii Estestvoznaniya i tekhniki* 42 (3): 195–211.

Kruchinin, Yu.A. 1965. *Shel'fovye ledniki Zemli Korolevy Mod [Ice Tongues of Queen Maud Land].* Leningrad: Gidromet.

Kublitskij, G. 1949b. *Otkryvateli Antarktidy [The Discoverers of Antarctica].* Moscow–Leningrad: Children's Press, Ministry of Education.

Kucherov, I.P. 1963. Navigatsionnye karty Antarktiki, sostavlennye v èkspeditsii Bellinsgauzena–Lazareva v 1819–1821 gg. [Navigational Charts of the Antarctic, Created During the Bellinshausen–Lazarev Expedition 1819–1821]. *Antarktika—doklady kommisii 1962* (3): 153–165.

Kucherov, I.P., and K.A. Bogdanov. 1962. Svidetel' nauchnogo podviga i geroizma [Testimony to a Heroic Scientific Achievement]. *Priroda* 5: 89–91.

Kuroedov, V.I., and Others, ed. 2005. *Atlas Okeanov: Antarktika [Atlas of the Oceans: the Antarctic]*. St Petersburg: Ministry of Defence and AANII.

Larionov, A.L. 1963. Korabli Pervoj russkoj antarkticheskoj èkspeditsii—shlyupi «Vostok» i «Mirnyj» [The Ships of the First Russian Antarctic Expedition—Sloops *Vostok* and *Mirnyj*]. In *Pervaya russkaya antarkticheskaya èkspeditsiya 1819–1821 gg. i eyo otchyotnaya navigatsionnaya karta*, ed. M.I. Belov, 128–142. Leningrad: Morskoj Transport.

Lazarev″, M.P. 1821. Pis′mo Mikhaila Petrovicha Lazareva k″ Aleksyeyu Antonovichu Shestakovu, 24 sentyabrya 1821 [A letter from Mikhail Petrovich Lazarev to Aleksyej Antonovich Shestakov, September 24, 1821 [O.S]]. St Petersburg: SARN—F–315, S–1, P–775, fos 1–6v.

Lebedev, D.M., and V.A. Esakov. 1971. *Russkie geograficheskie otkrytiya i issledovaniya [Russian Geographical Discoveries and Exploration]*. Moscow: Mysl′.

Lebedev, V. 1960. Who Discovered the Antarctic. *Soviet Union* 127: 52.

Lebedev, V.L. 1961. Geograficheskie nablyudenya v Antarktike èkspeditsij Kuka 1772–1775 gg. i Bellinsgauzena–Lazareva 1819–1821 gg. [Geographical Observations in the Antarctic by the Cook, 1772–1775, and Bellingshausen–Lazarev, 1819–1821, Expeditions]. *Antarktika—doklady kommisii 1960* (1): 7–24.

———. 1962. Kto otkryl Antarktidu? (otvet za pis′mo g-na Ternera) [Who Discovered Antarctica? A reply to Mr Turner's Letter]. *Antarktika—doklady kommisii 1961* (2): 153–166.

———. 1963. Reshenie spornykh voprosov Antarkticheskoj istorii na novoj osnove [A Solution for Contentious Issues in Antarctic History on a New Basis]. *Antarktika—doklady kommisii 1962* (3): 176–179.

———. 1964. O raznom tolkovanii nekotorykh mest iz dokumentov èkspeditsii Bellinsgauzena i Lazareva [On Alternative Interpretations of Some Passages in Documents from the Bellingshausen–Lazarev Expedition]. *Antarktika—doklady kommisii 1963* (4): 170–174.

Litke, Fyodor. 1828. *Chetyrekratnoe puteshestvie v″ Syevernyj Ledovityj Okean″ [Four Seasons of Exploration in the Northern Ice Ocean]*. St Petersburg: Naval Press.

Lomonosov, M.V. 1953. Mysli o proiskhozhdenii ledyanykh gor v severnykh moryakh [Thoughts on the Origin of Icebergs in Northern Seas] in *Polnoe sobranie sochinenii* [Complete Collected Works], vol. 3: 447–459. Moscow and Leningrad: Soviet Academy of Sciences.

Lomonosov″, M.V. 1763. *Pervyya osnovaniya metallurgii, ili rudnykh″ dyel″ [Elements of Metallurgy, or Mining]*. St Petersburg: Imperial Academy of Sciences.

Markov, K.K., and Others. 1968. *Geografiya Antarktidy [Geography of Antarctica]*. Moscow: Mysl′.

McCannon, John. 1998. *Red Arctic*. Oxford: Oxford University Press.

Mill, Hugh Robert. 1903. Bellingshausen's Antarctic Voyage. *Geographical Journal* 31: 150–159.

Muldashev, Ernst R. 2013. *Propavshee zoloto Levanevskogo [The Lost Gold of Levanevskij]*. Moscow: Molma.

Novosil'skij, P.M. 1853c. *Yuzhnyj polyus": iz" zapisok" byvshago morskago ofitsera [The South Pole: From the Memoirs of a Former Naval Officer]*. St Petersburg: Vejmar".

Osokin, Cap S. 1961. Nauchnyj podvig russkikh moryakov [A Scientific Victory by Russian Seamen]. *Krasnaya zvezda* 227: 5.

Ostrovskij, B.G.B. 1966. *Lazarev*. Moscow: Young Guard.

Petrova, Evgenia, and Others, ed. 2012. *Pavel Mikhailov 1786–1840: Voyages to the South Pole*. St Petersburg: Palace.

Phipps, C.J. 1774. *A Voyage towards the North Pole*. London: Nourse.

Powell, G. 1822. *Chart of South Shetland Including Coronation Island, & c. From the Exploration of the Sloop Dove in the Years 1821 and 1822 by George Powell Commander of the Same*. London: Laurie. https://collections.rmg.co.uk/collections/objects/540915.html.

Purdy, John. 1824. *A Chart of the World on Mercator's Projection*. London: Laurie. https://texashistory.unt.edu/ark:/67531/metapth193446/.

Ravenstein, L. 1867. *Meyer's Hand-Atlas der neuesten Erdbeschreibung in 100 Karten [Meyer's Portable Atlas]*. Hildburghausen: Bibliographisches Institut—not seen.

Robin, G. de Q. 1962. Discovery, Exploration, Adventure & Courage. *UNESCO Courier* 1: 14–20.

Rodomanov, B.B. 1959. O ponyatiakh «materik», «kontinent» i «chast' sveta» [On the Meanings of *materik, kontinent* and *chast' sveta*]. *Izvestiya vsesoyuznogo geograficheskogo obshhestva* 91 (2): 159–160.

Ross, Captain Sir James Clark. 1847. *A Voyage of Discovery and Research in the Southern and Antarctic Regions During the Years 1839–43*. 2 vols. London: John Murray.

Sabine, Edward. 1850–1853. *Observations Made at the Magnetical and Meteorological Observatory at Hobarton, in Van Diemen island, and by the Antarctic Naval Expedition*. 3 vols. London: Hobart Observatory.

Samarov, A.A., ed. 1952. *M.P. Lazarev—dokumenty [M.P. Lazarev—Documents]*. Vol. 1. Moscow: Ministry of the Navy.

Schwab, Jakob Friedrich, ed. 1804. *III Naturlehre. Das Meer. [Natural History Part 3—the Oceans]*. *Oestereichischer Toleranz-Bote*. Vienna: Nehm.

Scoresby, William, Jr. 1818. On the Greenland or Polar Ice. *Memoirs of the Wernerian Society* 2: 261–388.

———. 1820. *An Account of the Arctic Regions*. Vol. 1. Edinburgh: Constable.

Senex, Iohn. 1725. *A Map of the World*. London: Self. https://nla.gov.au/nla.obj-230683748/.

Simonov", Prof i Kav. 1822a. Plavanie shlyupa *Vostoka* v" Yuzhnom" Ledovitom" Morye [The Voyage of the Sloop *Vostok* in the Southern Ice Ocean]. *Kazanskij vyestnik"* 4 (3): 156–165, 4 (4): 211–216, 5 (5): 38–42, 5 (7): 174–181, 6 (10): 107–116, 6 (12): 226–232.

———. 1822b. *Slovo o uspyekhakh" plavaniya shlyupov"* Vostoka *i* Mirnago *okolo svyeta i osobenno v" Yuzhnom" Ledovitom" morye, v" 1819, 1820 i 1821 godakh" [An Address about the Results from the Voyage of the Sloops Vostok and Mirnyj around the World and Especially in the Southern Ice Ocean, in 1819, 1820 and 1821].* Kazan: Kazan University Press.

Skorezbi, Ml Villiam". 1825. *Podennyya zapiski o plavanii na syevernyj kitovyj promysl" [Journal of a Voyage to the Northern Whale Fishery].* St Petersburg: Naval Press.

Sokolov, A. 1953. Obzor deyatel'nosti otdeleniya istorii geograficheskikh znanij i istoricheskoj geografii s maya 1950 g. po yanvar' 1953 g. [An Overview of the Work of the Historical Section from May 1950 to January 1953]. *Voprosy geografii* 31: 274–285.

Stackpole, Edouard A. 1955. *The Voyage of the Huron and the Huntress.* Mystic, CT: Marine Historical Association.

Tammiksaar, E. 2016. The Russian Antarctic Expedition under the Command of Fabian Gottlieb von Bellingshausen and its Reception in Russia and the World. *Polar Record* 52 (5): 578–600.

Tsuker, Viktor. 2015. Vydayushhijsya vklad moryakov rossii [A Notable Contribution from Russian Seamen]. *Vesti morskogo Peterburga* 1: 44–45.

Weddell, James. 1825. *A Voyage Towards the South Pole, Performed in the Years 1822–24.* London: Longman and Others.

Wüllerstorf-Urbair, Bernhard, and Others. 1861–1875. *Reise der österreichischen Fregatte Novara um die Erde in den Jahren 1857, 1858, 1859 [Circumnavigation by the Austrian Frigate Novara in 1857, 1858, 1859].* 21 vols. Vienna: K. Gerold's Sohn.

CHAPTER 10

Standing Pat (1966–1991)

Brezhnev

In February 1966 the magazine *Novyj Mir* published an article by the war veteran Emil′ Vladimirovich Kardin (1921–2008) in which he questioned some treasured tales from Russian history, including the generally accepted beliefs that the cruiser *Aurora* had fired the shot which launched the October Revolution in 1917, and that 28 *Panfilovtsy* had sacrificed their lives to destroy a German tank column in 1941 (Kardin 1966). On 10 November 1966 the Politburo of the Central Committee of the Soviet Communist Party held a meeting to discuss the state of Soviet ideology and decide on steps to remedy any falling away of public commitment to Communist aims and values. The Party's General Secretary, Leonid Il′ich Brezhnev (1907–1982), delivered a diagnosis of the problem in unequivocal terms:

> That which is held most sacred and best in the hearts of our people is being subjected to analysis in certain productions, in our journals and other publications. Thus some of our writers venture to suggest (and they get published) that supposedly there never was a shot from the *Aurora*, that it was only a blank and so on, that there never were 28 Panfilovtsy, there were

R. Bulkeley, *The Historiography of the First Russian Antarctic Expedition, 1819–21*,
https://doi.org/10.1007/978-3-030-59546-3_10

fewer of them; they almost go so far as to claim that Klochkov[1] never existed and never voiced his slogan that 'Moscow is behind us and there is nowhere for us to retreat'. They even dare to insinuate libels against the October Revolution and other historic steps in the heroic history of our Party and of our Soviet people [1]. (Petrov and Èdel'man 1997: 149)

The events to be described in this chapter transpired in the cultural context created both by Brezhnev's ideological project and by its failure, and indeed by the slow motion failure of the Soviet project as a whole, despite the relative economic prosperity of the 1960s and 1970s. Thus on the one hand Bellingshausen's Remark, that litmus test for some degree of honesty and openness in accounts of the expedition, disappeared completely until the twenty-first century; on the other hand occasional expressions of dissent from the official version, that the expedition had discovered Antarctica on 28 January 1820, continued to be heard. It is worth bearing in mind that the revised assessment of the expedition was still only 15 years old when Brezhnev became First Secretary of the Party, as the office then was, in 1964.[2]

KUZNETSOVA

After 1965 the stream of Soviet research into the Antarctic achievements of the expedition began to dry up, apart from additional contributions from Belov and his *Karta* colleague V.V. Kuznetsova. The decline came on quite sharply in the journal *Antarctic* but less so elsewhere. Thus the *Bulletin of the Soviet Antarctic Expedition*, a Leningrad-based periodical, continued to accept original treatments of the expedition for a few years. In 1967 the *Bulletin* published an article by Kuznetsova in which she reported that she had discovered Bellingshausen's final report, a document which Belov had considered lost and Lebedev non-existent.[3] Summarizing its contents, she pointed out that Bellingshausen had revised most of his longitude calculations during the homeward voyage, but she

---

[1] According to the received version of the story, Company Commissar Vasilij Klochkov had led the *Panfilovtsy* in their heroic combat.

[2] The original texts of translated quotations, indicated by numbers in square brackets, are provided at the end of each chapter. For the system of romanization used, the rendering of dates, and a few other matters, please consult the Apparatus at the end of the book.

[3] Due to regrettable ignorance at the time, the author failed to credit Kuznetsova with the discovery when he published the final report for the first time (Bulkeley 2014a: 102–123).

laid perhaps more stress on the final report's account of the first season than it deserves, since apart from longitudes it was almost a literal transcription of the original report from Sydney in April 1820. Her most interesting contention was that the Marquis de Traversay had suppressed Bellingshausen's discovery of main ice (*materoj lyod*) in his summary of the reports (filed alongside the final report) because he considered it of no use to Russia (Kuznetsova 1967). But given that no one else, before Debenham, paid any attention to Bellingshausen's description of the 'main coast' (*materoj bereg*) of ice that was sighted on 17 February 1820, Kuznetsova was hardly justified in pinning so much blame on the ill and overworked Minister (Bulkeley 2014a: 99, 242).

For the supposed 150th anniversary in 1970 the *Bulletin* published a short survey both of the expedition and of modern Soviet Antarctic stations which did not include any new findings or insights in respect of the former (Aver'yanov and Koblents 1970). A few months later they published Koblents's critique of Belov's dating theory, mentioned above. From that point the Antarctic fog bank which had rolled over Bellingshausen in *Antarctic* moved on to engulf the *Bulletin*. Nor did any other periodical come forward to replace them.

The treatment of Kuznetsova's 1967 article perhaps reflects the advent of the Brezhnev administration. Ten years earlier, or even five, such an archival discovery would have been headline news and her description of a substantial new primary source would have appeared in one of the journals of the Soviet Academy of Sciences, as did her other work. But in 1967 it appeared only in the house magazine of the SAE and was not taken up by anyone else. The hypothesis that something had begun to deter research into the Bellingshausen expedition is made more plausible by other indicators. A year later Kuznetsova published an article in the *Proceedings of the Geographical Society* in which she reported her discovery of an officer's journal from the parallel northern expedition, commanded by Vasil'ev. In it she mentioned that she was:

> Busy collecting material in the Fonds of the Soviet Naval Archives and the manuscripts division of the State Public Library, in preparation for the publication of a collection focused on the expedition to *both* poles … [2] (Kuznetsova 1968: 237—emphasis added)[4]

---

[4] In referring to the 'double polar venture' as a single expedition, Kuznetsova was following the usage of naval officials at the time.

However the Bellingshausen/Vasil'ev collection announced by Kuznetsova never saw the light of day, despite the fact that a similar collection of archival documents from other expeditions was published a few years later (Shur 1971).

Kuznetsova may have envisaged, as documents suitable for a new collection of primary sources from the 'double polar venture', first her own recent discoveries; next, Simonov's still unreprinted shipboard diary, which she had used while working on the *Karta*; then perhaps the full text of Bellingshausen's report from Sydney in April 1820, which had only been reprinted with cuts in 1952; and possibly also the covering letter to de Traversay which accompanied it, both of which Belov had singled out as in need of republication. Like the last two items, the newly discovered final report included Bellingshausen's Remark about finding no signs of a polar landmass in 1820, a paragraph whose unveiling might have been perceived as undermining a historic step in the heroic history of Russia, to borrow Brezhnev's language. But given that Lebedev had published the Remark several times in recent years, Kuznetsova may have supposed that it was now established as historical fact. Whether for its political unacceptability, which can only be an unsupported conjecture, or for some other reason, the projected sourcebook never appeared.

In 1974 Kuznetsova and Belov co-authored a sequel to Kuznetsova's previous articles about the double expedition, their final contribution. In it they discussed the political goals of the project as revealed in early planning, and expressed regret that the original proposal, drafted by the veteran Arctic explorer Captain Gavrijl Andreevich Sarychev (1763–1831) and still in the archives, had never been published (Belov and Kuznetsova 1974). On the face of it, Sarychev's memorandum would have been eminently suitable for inclusion in the sourcebook once projected by Kuznetsova. On the other hand the fact that, in their opinion, it showed the double expedition in the role of a vehicle for Russian imperialism in the Pacific may perhaps have weighed against publication from the point of view of the authorities.

General books and articles about Antarctica, including popular albums of travel and exploration which covered the Bellingshausen expedition, continued to appear in the 1970s and 1980s. Their treatments of the expedition were usually unoriginal. Meanwhile work from archival historians about other early Russian maritime exploration was also being published (Goncharova 1973; Tumarkin 1978; Sopotsko 1978), including slightly later an article discussing further documents from Vasil'ev's northern squadron (Tumarkin 1983). A collection of archival documents about the history of Russian America was published in 1979 (Khlebnikov 1979). Evidently, whatever lay behind the neglect of the Bellingshausen expedition during this period, there was no lack of appetite for maritime history as such on the part of the scholars themselves.

Markov

Just one other piece of original research into the expedition appeared between 1966 and 1989. As the only text in that period to address the priority question by re-evaluating the evidence, it was an exceptionally bold example of dissent from the official tale of Russian priority in 1820. Its author was the poet and historian of exploration Sergej Nikolaevich Markov (1906–1979), who once again avoided any mention of the approved 'discovery day' (28 January 1820), as he had done 21 years earlier (S. Markov 1952):

> On 15 January 1820 [O.S.] the sloops *Vostok* and *Mirnyj* crossed the Antarctic Circle for the first time. In the following days both ships were near the actual continent of Antarctica in the neighbourhood of today's Princess Martha Land. The murky weather prevented them from seeing the unknown coast. ... Moving on a few days we note that on 5 February 1820 [O.S.] the ships of Bellingshausen and Lazarev came very close to the South Polar continent for a third time. But Bellingshausen was still not sure whether he had seen the real coast of the continent or endless ice hillocks [3]. (S. Markov 1982 [1973]: 245–246)[5]

Markov described the expedition's voyage in the South Pacific, visits to Australia, and the joyful discovery of Peter I I. during their second Antarctic season. Then:

> At last, on the morning of 17 January [1821, O.S. and s.d.], in sunny, cloudless weather, with sea swallows circling overhead—foretelling the proximity of land, the mariners caught sight of a snowbound coast with steep mountains and cliffs.
> This was Antarctica! It stretched far to the south, and there seemed to be no end or limit to its white coast.

[5] The text in the posthumous second edition was identical with that in the first, of which the author has seen only an unpaginated, electronic version.

Thus occurred the discovery of the Alexander I Coast. Russians had become the first people to reach the edge of south polar land [4]. (the same: 247–248)

There was nothing covert or gnomic, of the sort sometimes found in dissenting Soviet texts, about this rejection of the 1949 revision of the Bellingshausen expedition. Although he was largely self-taught, having left school at the age of 14 to care for younger siblings, Markov surely knew what he was doing. As a young man he travelled widely in Siberia and the Soviet Far East before being imprisoned for counter-revolutionary activities in 1932—the 'Siberian brigade' affair. After Maxim Gorky intervened to secure his release he worked as a journalist in Archangel'sk until the outbreak of war in 1941. Meanwhile his poems began appearing in various magazines and newspapers. His first books were travelogues, followed by a tribute to the explorers Przheval'skij and Miklukho-Maklaj in 1944, after which his output was divided almost equally between poetry and the history of exploration. By 1942, at the age of 36, his library of exploration and general Pacific history contained more than 3000 items. Although he joined the Geographical Society he seems to have had little to do with the political or intellectual establishment.

Markov's *Eternal Footsteps*, which included this account of the Bellingshausen expedition, was a comprehensive survey of exploration and geographical knowledge, and was evidently intended to be his last word on the subject. It was written very much in the early Soviet mode, in which authors were expected to study their topic rigorously but then summarize their findings in clear prose, accessible to a mass readership. For 113 brief chapters Markov did just that, ending with a note about his own extensive research in Russian archives. For the Bellingshausen chapter he was careful both to mention that he was using Old Style dates and to specify his sources, two practices which his more academic counterparts might have done well to imitate. The sources are revealing: the first editions of Bellingshausen's book and Novosil'skij's *South Pole* pamphlet (1853c), and the manuscript of Simonov's last, unfinished account of the voyage (1951), which he had studied in the archives of the University of Kazan and discussed with its senior librarian, M.K. Andreev. As secondary sources he listed only minor articles by Belov (1962) and Shvede (1962) and two general anthologies of exploration, thus none of the leading proponents of Russian priority from 1949 onwards, such as D. Lebedev and

A. Grigor'ev. Shvede's piece, in particular, had been one of that author's most diffident treatments of the priority claim:

> The Russian expedition first came close to the continent of Antarctica on 16 January 1820, during their first 'thrust' aimed at penetrating southwards, and *we count that day* as the date on which it was discovered. Visibility conditions, however, were not good enough … [5] (Shvede 1962: 426— emphasis added)

Elsewhere Markov mentioned the second edition of Berg's history of geography (1949b), another distinctly lukewarm treatment of the priority question. All this showed, and was probably meant to show, that he had gone back to some of the primary accounts of the voyage before reaching his own conclusion, which was that the expedition discovered Antarctica, in the shape of the Alexander I Coast (typically reverting to Bellingshausen's original name), not in 1820 but in 1821.

Markov's work had faults. He located the Blue Mountains in Tasmania and Vostok Island (now in Kiribat) south of Fiji, trotted out the old falsehoods about Cook's disbelief in Antarctica and Bellingshausen's priority in naming the South Shetlands, and left Bransfield right out of the picture. Nor did he explain why he disregarded Bellingshausen's reports and Lazarev's letter, published archival documents which suggested that something significant had been achieved in 1820. Nevertheless his single-handed attempt to turn the clock back to the pre-1949 assessment of the expedition was remarkable. For all that, it had no discernible effect. If there is any truth in the hypothesis that from the late 1960s academic historians began to avoid the subject of Bellingshausen's Antarctic achievements, for whatever reason, then Markov's heretical interpretation, buried inside an encyclopedic popular survey of exploration from Snorri Sturluson to Przheval'skij, may not even have been noticed. That would explain how it came to be republished twice in the Soviet period, in 1982 and 1990.[6]

---

[6]For an appreciation of Markov's work, with no reference to his apostasy over Bellingshausen, see Markova (1985). For a tribute to his thorough research methods, see Mogil'nitskij (2006).

A New Atmosphere

As time passed, treatments of the expedition with any vestige of originality became increasingly rare. Two illustrated booklets showing plans of *Vostok* and *Mirnyj* for model-makers in DOSAAF, a nationwide paramilitary association for friends of the armed forces, should be mentioned (Luchininov 1973a, b). They were probably based on (Larionov 1963), but provided more, and clearer visual information than Larionov, so that they remain useful today.[7]

Broadly speaking, late Soviet Antarctic literature fell into two categories. The first were scientific reports published in the *Papers* of the Soviet Antarctic Expeditions, in the SAE's *Bulletin*, or elsewhere, which did not concern themselves with the history of exploration. And second came more popular treatments such as the memoirs of expeditioners and journalists or articles in general knowledge magazines (Gusev 1972; Dubrovin 1976; Rybakov 1976; Somov 1978; Fedoseev 1979; Pasetskij 1981; Zotikov 1984a; Myasnikov 1986; Strugatskij 1986; Dubrovin and Preobrazhenskaya 1987; Grushinskij and Dralkin 1988; Tsigel'nitskij 1988; Bardin 1989). Most, but by no means all, of the second group mentioned the Bellingshausen expedition,[8] but without adding anything to what had been published in the 1950s and 1960s and without following Markov down the path of dissent. Not surprisingly memoirs by expeditioners often reflected the internationalist spirit of Antarctic science (Zotikov 1984b).

With the passage of time, mildly nonconformist treatments began to appear. Despite its 'discovery' title, for example, (D. Shherbakov 1976) did not record such an achievement, let alone its date, in its brief text.[9] Popov confined his remarks to the history of the South Shetland Is., where he had worked himself, much of which he got wrong. He noted Bellingshausen's visit to the group in 1821 but said nothing about events elsewhere (L. Popov 1979: 3–4). Kozlovskij accepted that the probes of 28 January and 17 February 1820 had both encountered 'a barrier of shelf

---

[7] To take one very small detail, by comparing Luchininov's version of *Vostok*'s original plans with a sketch by Mikhajlov (Petrova and Others 2012: 6), one can see that the standard continuous horizontal stowage of hammocks had to be modified into a succession of slanting rows, the obvious reason being that polar voyages required thicker bedding rolls.

[8] Zotikov's book was a typical, and well-written, 'non-mentioner'.

[9] This D. Shherbakov was probably a different author from the D.I. Shherbakov mentioned in Chap. 8, who died in 1966.

ice', [6] but refrained from calling either of them the discovery of Antarctica (Kozlovskij 1988: 6). The contrast between some of these treatments and those which had preceded them was especially striking in the case of V. Lebedev, who no longer mentioned Bellingshausen (V. Lebedev 1975). It was also typical of the last decade of the Soviet Union that a year after the death of Brezhnev one of the country's most popular authors and a serving naval officer, Viktor Viktorovich Konetskij (1929–2002), felt able to publish a mocking anecdote about having been force-fed with the cult of Bellingshausen as a naval cadet in the last years of Stalin:

> I should mention that in those days every examination paper in any subject—be it astronomy or ship theory—was obliged to include the question: 'Who discovered Antarctica?'. And even lubbers in the preparatory classes knew the answer. But the ringleader of the 'no-groaners' didn't even know that much. During an electrical engineering test, along with serious questions about motors and generators he was presented with the sacramental 'Who discovered Antarctica?'. The ringleader dried up. Despite their sworn enmity towards the no-groaners, even lads from the rival sect of the 'bonquistas' were whispering 'Bellingshausen!'. But the fellow misheard them, and answered 'Munchausen!' [7]. (Konetskij 1983: 17)

The book has been republished twice this century.

Although it qualifies neither as a memoir nor as a contribution to general knowledge, Aleksandr Kondratov's *Atlantises of Five Oceans* is of some interest because, 14 years after Markov, Kondratov followed the latter's example by abandoning the 1949 revision in favour of the earlier view that the expedition discovered Antarctica in January 1821 (Kondratov 1988 [1987]: 87–88).[10] Both Kondratov's and Kozlovskij's books appeared during the era of *glasnost'* in the closing years of the Soviet Union, although the effects of the policy are perhaps more visible elsewhere in the Antarctic literature (Kalenikin 1989) than in references to the Bellingshausen expedition.

---

[10] The citation is from the second edition, but unpaginated online copies of the 1987 edition have the same text.

MITIN AND ARISTOV

Two authors who deserve notice for their relatively full recapitulations of previously published accounts of the expedition were Georgij Pavlovich Lemeshhuk and Rear Admiral Lev Ivanovich Mitin (1925–1998), and it is perhaps significant that apart from encyclopedia entries they are almost the only commentators from this barren period listed in the bibliography on the *Maritime Encyclopedia* website, oddly enough alongside S. Markov (Anon. n.d.-a). A digital version of Lemeshhuk's book is widely available and his very readable account has probably influenced large numbers of Russians alive today (Lemeshhuk 1984: 36–64). Mitin's articles were less informative, because they were (usually) quite short and aimed primarily at publicizing the 1982–1983 Antarctic expedition by two research vessels of the Black Sea Fleet, rather than describing the distant past (Mitin and Dorogokupets 1983, 1984). Two fictional treatments by Ivan Ivanovich Firsov (1926–) were also widely read (Firsov 1983, 1988).

The only item from the *glasnost'* period to exhibit historical scholarship on a par with Andreev, Shvede, Belov, or Kuznetsova was an edition of Simonov's 1822 *Address* together with his unfinished but substantial account of the voyage, prepared by the renowned archivist of the University of Kazan, Vyacheslav Vasil'evich Aristov (1937–1992) (Simonov 1990a, b). Although both works were already known, Aristov made a valuable contribution by providing definitive texts and the only regret is that he kept his annotations so brief. (He also included the draft of a previously unknown letter which Simonov sent from Sydney in 1820.) But for his early death Aristov might have gone on to shed light on one of the most puzzling aspects of the expedition, Simonov's failure to analyse and publish most of his data. However he did report that there are no further autograph materials by Simonov at Kazan or elsewhere for researchers to study. Aristov avoided the awkward question of Bellingshausen's priority by leaving it to the veteran polar explorer and author Tryoshnikov to say the usual things in a foreword. In his own introduction Aristov echoed the evasions of Berg and Shvede: 'as a result [of the expedition] Antarctica was discovered'; 'an expedition which was deemed to have discovered ... Antarctica' [8] (Aristov 1990: 8, 10). Not surprisingly, he neither drew attention to the absence of anything about seeing an ice cliff on 28 January 1820 in the accounts by Simonov that he was presenting (Simonov 1990a:

23, b: 134),[11] nor placed those passages alongside the corresponding entry in Simonov's diary for that day (Simonov″ 1822a, pt 4: 181).

Aristov's editions of Simonov were awkwardly packaged with Admiral Mitin's fullest account of his 1982–1983 expedition. Mitin's other contribution to the book, an assessment of Simonov's work on the Bellingshausen voyage, is more pertinent to this investigation (Mitin 1990). It was a serious but lamentably deficient attempt, which failed even to mention, let alone analyse, Simonov's main report on his measurements of latitude and longitude (Simonov″ 1828). Mitin also disregarded Simonov's summary of the work undertaken, an obvious place to start (Simonov″ 1822c), and several other publications. Perhaps by way of distracting attention from his neglect of Simonov's published accounts of the voyage, Mitin referred to magnetic observations that were taken by others (warrant officers, according to Shvede), and later measurements taken by Simonov at Kazan (Mitin 1990: 293). The trick probably succeeded with many readers. Eventually, outdoing even the usual cosmetic treatment of the expedition's scientific work, Mitin declared that Simonov had surpassed Darwin, at least in the range of his investigations (the same: 308). Such adulation depended on avoiding the issue which Belov had raised as long ago as 1963 in the *Karta* (a work which Mitin did consult), namely Simonov's failure to publish some of his geographical, most of his meteorological, and all of his stellar measurements, despite explicit instructions to observe and record the southern stars which he acknowledged in 1822. Rather than tackle such a thorny question, Mitin padded out his assessment of Simonov still further by revisiting the priority issue, on which Simonov's work has little to offer, and by inserting several more pages about his own expedition, as if a 74-page chapter in the same book were not enough.

The problem with evaluating Mitin's overblown representation of Simonov, however, is that apart from the obvious omissions it is impossible to tell whether it was entirely his own doing, or whether he was constrained by some sort of official or unofficial policy which frowned on the raising of awkward questions about the expedition, a policy which may perhaps have operated since the late 1960s. With no firm evidence for such guidance from the state, beyond the decline of historical work on the expedition during the last 20 years of the Soviet Union, and in the context

---

[11] The highly sociable Simonov was more concerned to mention that *Mirnyj*'s officers came across to *Vostok* for dinner that evening.

of *glasnost'*, the gagging hypothesis is less plausible than the alternative, namely that Mitin gagged himself.

Lastly on this period, a few works were published about the Pacific phase of the expedition (for example Kabo and Bondareva 1974). As explained above they do not fall within the scope of this survey, because they were not part of the expedition's central, Antarctic historiography.

* * *

From 1949 until the end of the Soviet period dissent and indifference towards Bellingshausen's supposed discovery of Antarctica in January 1820 persisted to varying degrees. A possible explanation for this somewhat unexpected finding is that for more than two decades after 1949 the discovery claim was maintained primarily for purposes of foreign policy, which may have meant that unanimous conformity was not required from commentators, provided that a reputable cadre were prepared to support it and any dissent was not flagged up as outright rebuttal. Possible causes of the trend towards a more widespread conformity with the official view, which seems to have begun in the late Soviet period before reaching its peak in the present century, will be discussed in Chaps. 11 and 12.

ORIGINAL TEXTS OF QUOTATIONS

[1] Подвергается критике в некоторых произведениях, в журналах и других наших изданиях то, что в сердцах нашего народа является самым святым, самым дорогим. Ведь договариваются же некоторые наши писатели (а их публикуют) до того, что якобы не было залпа «Авроры», что это, мол, был холостой выстрел и т. д., что не было 28 панфиловцев, что их было меньше, чуть ли не выдуман этот факт, что не было Клочкова и не было его призыва, что «за нами Москва и отступать нам некуда». Договариваются прямо до клеветнических высказываний против Октябрьской революции и других исторических этапов в героической истории нашей партии и нашего советского народа.

[2] Занимаясь сбором материала для готовящегося к изданию сборника, позвященного эксдедции к двум полюсам, в фондах ... ЦГАВМФ и рукописном отделе Государственной публичной библиотеки ...

[3] 15 января 1820 года шлюпы «Восток» и «Мирный» впервые пересекли Южный полярный круг. На следующие сутки оба корабля находились близ самого материка Антарктиды, около теперешней Земли Принцессы Марты. Пасмурная погода не позволила рассмотреть неведомый берег. ... Забегая несколько вперед, скажем, что 5 февраля 1820 года корабли Беллинсгаузена и Лазарева в третий раз подошли вплотную к Южному полярному материку. Но Беллинсгаузен еще не был уверен в том, видит ли он действительный берег материка или нескончаемые ледяные бугры.

[4] Наконец утром 17 января при солнечной и безоблачной погоде, когда в небе кружились морские ласточки—предвестники близости суши, с кораблей увидели снежный берег с крутыми горами и скалами.
Это была Антарктида! Она простиралась далеко к югу, и белому берегу, казалось, не было ни конца ни края.
Так совершилось открытие Берега Александра I. Русские люди первыми достигли окраины южно-полярной суши.

[5] К материку Антарктиды русская экспедиция впервые близко подошла 16 января 1820 г., во время своего первого «покушения» проникнуть на юг, и этот день *мы считаем за дату его открытия*. Условия видимости, однако, были недостаточно хорошими, ... [разрядка наша]

[6] ...барьером шельфого ледника,...

[7] А надо заметить, что в те времена на любом экзамене по любому предмету—и в астрономию, и в теорию корабля—обязательно вставляли в билет вопрос: «Кто открыл Антарктиду?» И ответ знали уже салаги подготовительных курсов, но вождь «ни-бум-бумов» и этого не знал. На экзамене по электротехнике попадается ему среди разных сериесных двигателей и генераторов сакраментальное: «Кто открыл Антарктиду?» Вождь завял. Ребята из племени «долб-долбов» хотя и враждовали с «ни-бум-бумами», но подсказывают: «Беллинсгаузен!» Парень недослышал и отвечает: «Мюнхгаузен!»

[8] ...об экспедиции ... в результате которой была открыта Антарктида ... ... экспедици[я], которой было суждено открыть ... Антарктиду...

REFERENCES[12]

Anon, n.d.-a. *Bellinsgauzen Faddej Faddeevich [Bellingshausen, Faddej Faddeevich]*. Morskaya Èntsklopediya.

Aristov, V.V. 1990. Zhizn' i deyatelnost' I.M. Simonova [The Life and Work of I.M. Simonov]. In *Dva plavaniya vokrug Antarktidy*, ed. T.Ya. Sharipova, 8–14. Kazan: Kazan University Press.

Aver'yanov, V.G., and YaP. Koblents. 1970. 150-letie otkrytiya Antarktidy i sovetskie issledovaniya v Antarktike [150th Anniversary of the Discovery of Antarctica, and Soviet Antarctic Research]. *Byulleten' Sovetskoj Antarkticheskoj Èkspeditsii* 77: 5–18.

Bardin, Vladimir. 1989. *V gorakh i na lednikakh Antarktidy [On the Mountains and Glaciers of Antarctica]*. Moscow: Znanie.

Belov, M.I. 1962. Shestaya chast' sveta otkryta russkimi moryakami (Novye materialy…) [The Sixth Continent Discovered by Russian Seamen—New Sources…]. *Izvestiya vsesoyuznogo geograficheskogo obshhestva* 94 (2): 105–114.

Belov, M.I., and V.V. Kuznetsova. 1974. Pervonachal'nyj proekt russkoj èkspeditsii v Yuzhnyj i Severnyj ledovityj okeany [The Initial Proposal for a Russian Expedition to the Southern and Northern Ice Oceans]. *Izvestiya vsesoyuznogo geograficheskogo obshhestva* 106 (6): 491–497.

Berg, L.S. 1949b. *Ocherki po istorii russkikh geograficheskikh otkrytii [Towards a History of Russian Geographical Discoveries]*. 2nd ed. of (Berg 1929). Moscow: Academy of Sciences.

Bulkeley, Rip. 2014a. *Bellingshausen and the Russian Antarctic Expedition, 1819–21*. Basingstoke: Palgrave Macmillan.

Dubrovin, L.I. 1976. *Chelovek na ledyanom kontinente [Man in the Icy Continent]*. Leningrad: Gidromet.

Dubrovin, L.I., and M.A. Preobrazhenskaya. 1987. *O chem govorit karta Antarktiki? [What does the Map of the Antarctic Tell Us?]*. Leningrad: Gidromet.

Fedoseev, I.A. 1979. F.F. Bellinsgauzen. *Voprosy istorii estestvoznaniya i tekhniki* 67–68: 122–123.

Firsov, Ivan. 1983. *I Antarktida, i Navarin [Both Antarctica and Navarino]*. Yaroslavl': Verkhnevol'sk Press.

———. 1988. *Polveka pod parusami [Half a Century Under Sail]*. Moscow: Mysl'.

Goncharova, N.N. 1973. Khudozhnik krugosvetnoj èkspeditsii 1819–1822 godov E. Korneev [E. Korneev, Artist with the Circumnavigating Expedition of 1819–1822]. *Izvestiya vsesoyuznogo geograficheskogo obshhestva* 105 (1): 67–72.

---

[12] Key to archival references: SARN = State Archives of the Russian Navy, F = Fond, S = Series [*Opis'*], P = Piece [*Delo*], fo./fos = folio/s [*list/y*], v = verso [*oborotnoe*]. (The latter is necessary because verso pages were usually unnumbered.) Dates on manuscript documents are given in Universal Time (CE).

Grushinskij, N.P., and A.G. Dralkin. 1988. *Antarktida [Antarctica]*. Moscow: Nedra.

Gusev, A.M. 1972. *Ot Èl'brusa do Antarktidy [From Elbrus to Antarctica]*. Moscow: Soviet Russia.

Kabo, V.P., and N.M. Bondareva. 1974. Okeaniskaya kollektsiya I.M. Simonova [I.M. Simonov's Oceania Collection]. *Sbornik muzeya antropologii i Etnografii* 30: 101–111.

Kalenikin, Sergej. 1989. Pyl' v Antarktide [Dust in Antarctica]. *Smena* 17: 12–15.

Kardin, E.V. 1966. Legendy i Fakty [Legends and Facts]. *Novyj Mir* 2: 237–250.

Khlebnikov, K.T. 1979. *Russkaya Amerika v neopublikovannykh zapiskakh [Russian America in Unpublished Memoirs]*. Leningrad: Nauka.

Kondratov, A.M. 1988. *Atlantidy pyati okeanov [Atlantises of Five Oceans]*. Leningrad: Gidromet.

Konetskij, Viktor. 1983. *Tretij lishnij [Three's a Crowd]*. Leningrad: Soviet Author.

Kozlovskij, A.M. 1988. *Vokrug tol'ko lyod [Icebound]*. Leningrad: Gidromet.

Kuznetsova, V.V. 1967. Novye dokumenty pervoj russkoj antarkticheskoj èkspeditsii [New Documents from the First Russian Antarctic Expedition]. *Byulleten' Sovetskoj Antarkticheskoj Èkspeditsii* 66: 5–11.

———. 1968. Novye dokumenty o russkoj èkspeditsii k severnomu polyusu [New Documents on the Russian Expedition to the North Pole]. *Izvestiya vsesoyuznogo geograficheskogo obshhestva* 100 (3): 237–245.

Larionov, A.L. 1963. Korabli Pervoj russkoj antarkticheskoj èkspeditsii—shlyupi «Vostok» i «Mirnyj» [The Ships of the First Russian Antarctic Expedition—Sloops *Vostok* and *Mirnyj*]. In *Pervaya russkaya antarkticheskaya èkspeditsiya 1819–1821 gg. i eyo otchyotnaya navigatsionnaya karta*, ed. M.I. Belov, 128–142. Leningrad: Morskoj Transport.

Lebedev, V.L. 1975. Ot kuda v more led? [Where does Sea Ice Come From]. *Khimiya i zhizn'* 2: 35–38.

Lemeshhuk, G.P. 1984. *Iz goroda na Neve: moreplavateli i puteshestvenniki [From a City on the Neva: Navigators and Travellers]*. Leningrad: Lenizdat.

Luchininov, S.T. 1973a. *Shlyup Vostok [Sloop Vostok]*. Moscow: DOSAAF.

———. 1973b. *Shlyup Mirnyj [Sloop Mirnyj]*. Moscow: DOSAAF.

Markov, Sergej Nikolaevich. 1952. Otkryvatel' Antarktidy [Discoverer of Antarctica]. *Vokrug sveta* 1: 65.

———. 1982. *Vechnye sledy*. Moscow: Sovremennik.

Markova, G. 1985. Tikhookanskaya kartoteka Sergeya Markova [Sergej Markov's Pacific Card Index]. *Al'manach bibliofila* 19: 107–124.

Mitin, L.I. 1990. Sovremennaya otsenka nauchnoj deyatel'nosti I.M. Simonova v èkspeditsii 1819–1821 gg. [A Modern Evaluation of the Scientific Work of I.M. Simonov on the Expedition]. In *Dva plavaniya vokrug Antarktidy* T.Ya. Sharipova, 280–312. Kazan: Kazan University Press.

216   R. BULKELEY

Mitin, Rear Admiral L., and Lt. Cmdr S. Dorogokupets. 1983. Novaya vstrecha s «Terra australis» [A New Encounter with *Terra australis*]. *Vokrug sveta* 6: 28–32, 7: 39–41.

———. 1984. Marshrutom pervootkryvatelej [Along the Track of the First Discoverers]. *Nauka i zhizn'* 4: 144–151.

Mogil'nitskij, Valerij. 2006. Vechnye sledy Markova [Markov's Eternal Footsteps]. *Vesti Saryarki*, March 14. 9: 6. http://old-site.karlib.kz/semenov/vechnye_sledy_markova.pdf.

Myasnikov, V.F. 1986. *Puteshestvie v stranu belogo sfinksa [A Journey to the Land of the White Sphinx]*. Simferopol': Tavriya.

Novosil'skij, P.M. 1853c. *Yuzhnyj polyus″: iz″ zapisok″ byvshago morskago ofitsera [The South Pole: From the Memoirs of a Former Naval Officer]*. St Petersburg: Vejmar″.

Pasetskij, V.M. 1981. Russkie geograficheskie otkrytiya i issledovaniya pervoj poloviny XIX veka [Russian Geographical Discoveries and Exploration in the First Half of the 19th Century]. *Voprosy istorii* 12: 98–108.

Petrov, N., and O. Èdel'man. 1997. Novoe o sovetskikh geroyakh [A New Perspective on Soviet Heroes]. *Novyj Mir* 6: 140–151.

Petrova, Evgenia, and Others, ed. 2012. *Pavel Mikhailov 1786–1840: Voyages to the South Pole*. St Petersburg: Palace.

Popov, L.A. 1979. *God v Antarktike [A Year in the Antarctic]*. Moscow: Nauka.

Rybakov, S.N. 1976. *Zhivaya Antarktika [The Living Antarctic]*. Leningrad: Gidromet.

Shur, L.A. 1971. *K beregam Novogo Sveta [To the Shores of the New World]*. Moscow: Nauka.

Shvede, E.E. 1962. F.F. Bellinsgauzen. In *Lyudi russkoj nauki: geologiya i geografiya*, ed. I.V. Kuznetsov, 419–431. Moscow: Physics and Mathematics Press.

Simonov, I.M. 1951. Shlyupy «Vostok» i «Mirnyj» ili plavanie rossiyan v Yuzhnom Ledovitom okeane i okolo sveta [Sloops *Vostok* and *Mirnyj*, or the Voyage by Russians in the Southern Ice Ocean and around the World]. In *Russkie otkrytiya v Antarktike*, ed. V.N. Sementovskij, 51–175. Moscow: Geografgiz.

———. 1990a. Slovo ob uspekhakh plavaniya shlyupov «Vostok» i «Mirnyj» okolo sveta i osobenno v Yuzhnom Ledovitom more, v 1819, 1820 i 1821 godakh [An Address about the Results from the Voyage of the Sloops *Vostok* and *Mirnyj* around the World and Especially in the Southern Ice Ocean, in 1819, 1820 and 1821]. In *Dva plavaniya vokrug Antarktidy*, ed. T.Ya. Sharipova, 18–40. Kazan: Kazan University Press.

———. 1990b. «Vostok» i «Mirnyj» [*Vostok* and *Mirnyj*]. In *Dva plavaniya vokrug Antarktidy*, ed. T.Ya. Sharipova, 46–248. Kazan: Kazan University Press.

Simonov″, Prof. i Kav. 1822a. Plavanie shlyupa *Vostoka* v″ Yuzhnom″ Ledovitom″ Morye [The Voyage of the Sloop *Vostok* in the Southern Ice Ocean]. *Kazanskij vyestnik″* 4 (3): 156–165, 4 (4): 211–216, 5 (5): 38–42, 5 (7): 174–181, 6 (10): 107–116, 6 (12): 226–232.

Simonov", Prof. 1822c. Kratkij otchyot" [A Brief Report]. *Kazanskij vyestnik"* 3 (10): 98–107.

Simonov", I. 1828. *Opredyelenie geograficheskogo polozheniya myest" yakornago stoyaniya shlyupov" VOSTOKA i MIRNAGO u.m.ò. [A Determination of the Geographical Location of the Anchorages of Sloops Vostok and Mirnyj etc.].* St Petersburg: Department of Education.

Somov, Mikhail. 1978. *Na kupolakh zemli [On the Domes of the Earth].* Leningrad: Lenizdat.

Sopotsko, A.A. 1978. Vakhtennye zhurnaly korablej V.I. Beringa [V.I. Bering's Logbooks]. *Izvestiya vsesoyuznogo geograficheskogo obshhestva* 110 (2): 164–170.

Strugatskij, V. 1986. *Podvig na Polyuse kholoda [Victory at the Pole of Cold].* Leningrad: Lenizdat.

Tsigel'nitskij, I.I. 1988. *V morya studenye ukhodyat korabli [And Ships Move Out to Bitter Seas].* Leningrad: Gidromet.

Tumarkin, D.D. 1978. Materialy pervoj russkoj krugosvetnoj èkspeditsii kak istochnik po istorii i ètnografii Gavajskikh ostrovov [Materials from the First Russian Circumnavigating Expedition as a Source for the History and Ethnography of the Hawaiian Islands]. *Sovetskaya Ètnografiya* 5: 68–84.

———. 1983. Materialy èkspeditsii M.N. Vasil'eva—tsennyj istochnik po istorii i ètnografii Gavajskikh ostrovov [Materials from M.N. Vasil'ev's Expedition—A Valuable Source for the History and Ethnography of the Hawaiian Islands]. *Sovetskaya Ètnografiya* 6: 48–61.

Zotikov, Igor'. 1984a. *Za razgadkoj tajn Ledyanogo kontinenta [Unlocking the Secrets of the Icy Continent].* Moscow: Mysl'.

———. 1984b. *Ya iskal ne ptitsu kivi [I was not looking for Kiwis].* Leningrad: Gidromet.

CHAPTER 11

The Triumph of 'Fact' (1992–2020)

A Changing Context

Russian treatments of the Bellingshausen expedition in the modern, post-Soviet period are hard to survey for several reasons, which include the interruption to the dissemination of journals abroad, the shortage of bibliographic sources, the rise of the internet, and the advent of work published in the Russian diaspora. Diaspora publications are mentioned here if they were probably intended for at least some distribution in Russia. But the other, less tractable obstacles mean that this period will probably be described less thoroughly than earlier ones.[1]

During the remainder of the 1990s, or the presidency of Boris Yeltsin, Russian scientific work on the Antarctic continued to appear despite the many institutional and personal problems of the times. Much as before most scientists steered clear of historical commentary, even in semi-popular works (Kotlyakov 1994). There were signs of life from maritime or partly maritime historians (Dmitriev 1993–1994; Favorov 1994; N. Bolkhovitinov 1997), but apart from conventional encyclopedia entries nothing that dealt with the Bellingshausen expedition; in those days, Russian historians were doubtless more concerned to reassess the recent than the remote past

---

[1] The original texts of translated quotations, indicated by numbers in square brackets, are provided at the end of each chapter. For the system of romanization used, the rendering of dates, and a few other matters, please consult the Apparatus at the end of the book.

© The Author(s), under exclusive license to Springer Nature Switzerland AG 2021
R. Bulkeley, *The Historiography of the First Russian Antarctic Expedition, 1819–21*,
https://doi.org/10.1007/978-3-030-59546-3_11

of their country. One has to turn to the diaspora to find an example of competent, though not innovative, work (Govor 1995). Popular treatments of the subject were also rare (Shmatkov 1998), but in 1996 Rusakov's 1903 pamphlet on exploration, with its ill-informed glorification of Bellingshausen, was reissued in facsimile by a government publishing house, since when it has remained in print.

As Vladimir Putin succeeded Yeltsin with the turn of the century there were few signs of any immediate change. Most commentators either were not, or chose not to present themselves as, historians (Fyodorov 2000), and a new *Antarctic Atlas* retold the usual story, this time without misrepresenting Bransfield because he was not mentioned (Kuroedov and Others 2005: 5, 8). Popular articles also repeated the established version (Udintsev 2002; Volkov 2003), in one case (Volkov) reviving the groundless claim that Bellingshausen visited and named the South Shetlands before Smith and Bransfield. An illustrated textbook, approved by the Ministry of Education, repeated the Soviet tale of first discovery for the children of the new democracy. Unfortunately for them it is riddled with factual errors, such as placing Cook's death in New Zealand or calling William Smith an American (Moroz 2001: 3, 4). Children are still required to learn about Bellingshausen's priority today (Sergeeva 2003; Anon. n.d.-b). One such teaching aid informs students that Lazarev, rather than Bellingshausen, discovered Antarctica and the South Pole was first conquered in 1980 (Anon. n.d.-c). A collection of voyages repeated the usual story, complete with a bogus quotation from Bellingshausen's instructions, two absentee German naturalists instead of one, alleged discovery of the South Shetlands etc. (Lubchenkova 2001: 33–41). But a hint that things could be done better, whether or not they ever would be, was provided by Fyodorovskij's novel *Bellingshausen* (2001). Although Fyodorovskij stuck to the orthodox version, his careful and thorough presentation showed something hardly seen since Belov, a real interest in the voyage as a whole.

A more promising development this century has been the publication of several scholarly studies in maritime history, most of which were preceded by the usual articles in academic journals, and some of which address the history of science or exploration (Kozlov 2003—not seen; Unkovskij 2004; V. Smirnov 2005, 2006; Grebenshhikova 2010; Simakova 2015). A fundamental tool for maritime historians, the Imperial Russian Navy List from 1696 to 1917, is being republished by Atlant in St Petersburg. Relatively lightweight formats such as the anthology and the memoir have also received a makeover (Norchenko 2003; Utusikov 2012; Lajba 2015).

Some items in this literature discuss the early career of Bellingshausen or the later career of Lazarev (Kopelev 2010; Chernousov 2011).

'RUSSIAN ANTARCTICA'

Before describing the reprints of Bellingshausen's *Two Seasons* which have appeared in recent years, some of the cultural and political obstacles to renewed scholarly investigation of the expedition should be pointed out. The first is the widespread but as we have seen mistaken belief that the event has already been exhaustively studied by historians and nothing more needs to be said (Tsuker 2015). The phrase 'as is well known' (*kak izvestno*) scores over 21,000 internet hits when coupled with 'Bellingshausen' (see for example R. Potapov 2012: 89). That complacency is reflected in the tally of only two new primary sources relating to the expedition discovered by Russian scholars between 1967 (Kuznetsova) and 2018: Simonov's draft letter, published by Aristov as long ago as 1990, and a collection of documents on Bellingshausen's subsequent financial affairs at the Russian State Historical Archives, which were kindly pointed out to the present author by Dr Valentin Smirnov, since appointed director of the Naval Archives. The 'virtual exhibition' of archival documents that was posted on the Naval Archives website in 2019 (below) is a welcome relief after that long drought, but the first scholarly article to use material from it is not by a Russian (Bulkeley 2021).

Another factor has been the rise of the mighty but fickle internet itself. While a rich if sometimes frustrating or illusory resource, the internet is better suited to providing summaries of knowledge or belief than to the exegesis of complex and nuanced argument. A recent article about the expedition on the Naval Archives website, by the distinguished naval historian Captain Sergej Prokof'evich Siryj, was a good example of the scope and limits of such work (Siryj 2010).

The internet has enabled some related social developments. One is the seemingly insatiable exploitation of Antarctica as a theatre for fantasy, often projected onto the recent past (Grejg 2008; Telitsyn 2013). The repetitiveness of much of this material appears to be no bar to its saleability. The implication that Antarctica is something like Atlantis or, in Russian culture, America, a place that anyone is free to say anything they like about within the global entertainment industry, may not be widely shared, but it is real. In a recent example a news report about the discovery of a virus in the Greenland ice cap resurfaced on the *Pravda* website some four years later, by which time it had been transferred to Antarctica and linked on no grounds whatsoever to the enduring myth about a Nazi base allegedly established in Antarctica in the 1940s (M. Walker 1999; Novikova 2003).

Doubtless the rest of Europe and North America contribute as much to the fantasy life of Antarctica as does Russia. But another internet and media development in respect of the Bellingshausen expedition is peculiarly Russian. It amounts to a cult of manifest (and profitable) national destiny under the slogan 'Russian Antarctica'. In its mildest version, this social advocacy movement takes the form of harmless role-playing on the popular *VK* social network, where a child can appear in fancy dress as the 'Crown Princess of Marie Bird Land'. On the *Moj mir* network, a small 'Russian Antarctica' group was set up in about 2013 to exchange tourist photographs and general chat. In 2015 enthusiasts in Ekaterinburg set up the not-for-profit foundation 'Russian Antarctica 2020' to prepare for the bicentenary of the expedition (Chernykh 2015), but their activities have been largely confined to low-key reunions of Antarctic veterans. The Twitter hash-tag *Antarktidanasha* ('Antarctica is ours') received only one tweet in 2017 and again in 2018, in one case from someone who could not see the point of the phrase, and it was not popular on *VK* either. The account has since been deactivated. One proponent of the position, clearly unaware of (Campbell 2000), claims that since no logbooks have survived nothing can be known about Bransfield's voyage (Kudryashov 2018). That is a self-defeating argument, since it would also disqualify Bellingshausen, if ever Russian commentators could bring themselves to accept the fact that no logbooks or officers' journals have survived from his voyage, instead of alleging the opposite (Tsuker 2015: 45 [1]; Anon. 2019 [2]).

The promotional film *Russian Antarctica—21st Century* was released on *VK* and elsewhere in January 2015, and had considerable impact at the time. It received its national premiere on the *Russia 1* television channel a year later, on 28 January ('discovery day') 2016, where it was billed as not only describing the work of modern scientific expeditions but also as posing the question of how Russian sovereignty over Antarctica might best be secured (Anon. 2016).[2] One of the contributors, Valerij Vladimirovich

---

[2] A new three-part documentary, *Antarctica—a Trek to Three Poles*, was shot in Antarctica by director Valdis Pel'sh during the 2018–2019 season with support from the Russian Geographical Society, and released by Russia 1 on 14 January 2020 for the bicentenary of Bellingshausen's supposed discovery of Antarctica. It has a total length of almost three hours and is now on YouTube. The cinematography and presentation are impressive, but the historical passages continue to support the discovery claim in the usual terms.

Lukin (1946–), is an eminent polar oceanologist and expedition leader with some 40 years of experience on Arctic and Antarctic drift stations. He has been deputy director of the *AANII* since 2003, and has attended more than 50 international meetings within the ATS. His regular appearances in the Russian media have made him a virtual Mr Antarctica with the public, and his political views have been on record for some time. In 2001, in response to government discussions about securing national interests in Antarctica, Lukin remarked that it would be impossible to build an ideal society in Antarctica while Russia itself was going to the dogs (*naperekosyakh*) (Lukin 2001). The revealing aspect of that comment was that, apart from a single, rather bizarre dystopic novel (Batchelor 1983), no one outside Russia has ever supposed it might be possible to build such a society. Five years later, however, Lukin replied to questions about Russia's position in Antarctica with a balanced and informative survey of the ATS (Lukin 2006). Seven years later still, in an interview for Russia's Council on International Relations, Lukin emphasized Russia's interests in upholding the ATS and conducting space-related and other advanced research, while focusing on fisheries on the economic side (Lukin 2013; see also Lukin 2018). Between those two contributions, however, he reverted to chauvinistic mode for an interviewer from *Rossiya kul'tura*, citing the 1939 Soviet Note to Norway (Chap. 5) as proof that Russia holds reserved rights to the entire continent on the grounds of first discovery. The interviewer paraphrased him to the effect that Antarctica would be part of Russia today if only the Empire had established such institutions as a governor's residence and a post office, and concluded that 'Russian Antarctica lives on' [3] in the minds of such veterans (Anon. 2010). The 2015 film (above) took much the same position. Elsewhere, an interesting map, which juxtaposes the Bellingshausen expedition with modern Russian research stations, was recently posted on the 'Defending Russia' website to support the claim that Antarctica belongs to the Russian people.[3]

At its highest pitch, on 17 January 2018 a prolific but unidentified blogger on the Aftershock news website presented the usual priority claim as an unsupported headline and concluded: 'Let us remember that Crimea is ours, the Arctic is ours, and Antarctica will also be ours' [4].[4] However the movement has also been satirized in the Russian edition of *Forbes* magazine (S. Medvedev 2015).

---

[3] https://defendingrussia.ru/inf/rossija_antarktida-2126/#&gid=1&pid=1
[4] https://aftershock.news/?q=node/606964&full

OFFICIAL POLICY

Turning to Russia's official position, the *AANII*'s annual expedition almost failed in 1992 when the decision to maintain the Antarctic research programme was left dangerously late in the year (Decree 1992). From that low point, however, and having convinced the government of the geopolitical significance of Antarctica for Russian national interests (below), the renamed Russian Antarctic Expeditions (RAE) were able to maintain a basic five-station presence through the 1990s, albeit with fewer personnel than in Soviet times. In 1997 financial restraints were imposed but a set level of expenditure was guaranteed; whether that was delivered after the rouble devaluation of 1998 is unclear. In 2001 (shortly after Vladimir Putin became president for the first time) prospects for the RAE began to improve, according to Lukin. In 2005 it was decided to reopen three additional Soviet-era stations and to build a new Antarctic research vessel; the existing *Progress* station would be developed as 'the capital of Russian Antarctica' in Lukin's words (Timofeev 2006; Order 2005).[5] For the International Polar Year of 2007–2008 Russia put more effort into the Arctic than the Antarctic, in line with the aims of the programme. The headline achievement in the Antarctic was the installation of a money-saving *Vaisala* automatic weather station (AWS) near another mothballed station, *Molodyozhnaya* in East Antarctica. The Finnish manufacturer, which has undertaken a nationwide modernization programme for *Rosgidromet*, made the surprising claim that this was 'the first automatic weather station—ever—in Antarctica' (Vaisala 2010).[6] By about 2011, Lukin estimated, the RAE budget had returned to a level comparable with the SAE in 1985 (Lukin 2013). The new oceanographic vessel *Akademik Tryoshnikov* carried out her first Antarctic cruise in 2012–2013. The three old stations mentioned in 2006 were visited in 2007 and 2008 but found to be in very poor condition. By 2011 one of them, *Soyuz*, was described by Lukin as in use (or destined) for 'seasonal fieldwork' [5] (Lukin 2011), and by 2017 the other two, *Russkaya* and *Leningradskaya*, were accorded the same status (Danilov and Mart'yanov 2017). In 2014 plans to replace *Russkaya* with a completely new structure, including a *Vaisala* automatic anemometer and a ground station for Russia's GLONASS navigational satellite network, were reported to be at an advanced stage (Mart'yanov

---

[5] None of the details for which Timofeev cited the Order appear to be there.
[6] For the real history of Antarctic AWS since the 1980s see Lazzara and Others (2012).

2014). The successful installation of a new station will be of great importance for Russian prestige, because while other Antarctic players, including Belgium, Brazil, Britain, China, Germany, South Korea, and the United States, have been erecting new state-of-the-art research stations around the continent, the only new Russian building so far this century has been a small wooden church. According to a recent news report, *Russkaya* station was due to be reoccupied in February 2020 and then, after a comprehensive makeover, to commence operations in 2022.[7]

The strategic perspective which informs such determined struggles against adversity has been clearly stated by Lukin, by officials at the Ministry of Nature (*MinPrirody*), and elsewhere. It includes considerations of prestige, and of the unique value of Antarctica for scientific investigations of climate change and the development of space technology. But, in a clear echo of Berg's address to the Geographical Society in 1949, the dominant theme is economic. The Protocol on Environmental Protection (PEP) to the Antarctic Treaty is not expected to survive its 2048 review without radical changes which will affect the entire ATS. After that point an intense competition for the resources of the region, including water, oil, gas, metals and seafood, and made increasingly viable by climate change, is anticipated (Timofeev 2006; Kolesnikova 2007; Sergeev 2016; see also Fox 2014). It is worth recalling that for centuries the Russian economy has focused on resource extraction to a greater extent than most other technologically advanced countries. It seems at least plausible, therefore, that one basic goal of the RAE since the 1990s has been to establish and maintain the grounds for an eventual claim to a disproportionate share in the resources of the continent.[8]

The supposed fact that Russian explorers were the first to discover the continent, in January 1820, seems to be viewed as sufficient basis for a *prima facie* claim. While it is also true that official policy has been founded on a commitment to pursue Russian interests in the region in line with international law, the latter is believed to favour a Russian claim on the grounds of first discovery. Commitment is also given to the ATS, but it is subordinated to the wider long-term goal (Order 2010). That is not to say that Russia expects its claim to Antarctic resources to succeed outright

[7] https://iz.ru/935967/nataliia-mikhalchenko/na-sviazi-antarktida-zachem-rf-raskonserviruet-stantciiu-russkaia
[8] A recent study of the economic potential of the Antarctic refers to the general desirability of protecting the region, but without mentioning the PEP (Korzun 2009).

after 2048, but the available evidence suggests an intention to put it forward, at least as a bargaining position. The parallel presentation of bold talk about 'Russian Antarctica' alongside more measured official pronouncements, sometimes in the same news report (Usov 2010), is compatible with this interpretation. Lukin himself, active both as lobbyist and as senior government scientist, has embodied just such a, perhaps deliberate, ambiguity.

Commentators have remarked that modern Russian nationalism includes a powerful nostalgia for both Soviet and imperial episodes of the national past, and cares much for glory and little for truth, the more so since its conception of the future is poorly defined (Ostrovsky 2015: 233–240). Former President Medvedev's heavy-handed 'Commission to counter attempts to falsify history to the detriment of Russia's interests' is no longer with us, but its creation was part of a wider anti-intellectual movement, in which calls for a partial revival of the Stalin–Zhdanov campaign against 'internationalism' and 'kowtowing to the West' have been heard (Vdovin 2008). The recent *Panfilovtsy* incident, described in the Preface, confirms that such attitudes still prevail with senior members of the Russian government. Meanwhile in Antarctica itself the economic handicaps experienced by the RAE's scientific programme may have foregrounded the supposed priority of the Bellingshausen expedition as the USP for Russian policy and ambitions in the region.

In 2014 Lukin published a comprehensive account of Russian Antarctic policy in English, with the proviso that it was not an official document (Lukin 2014). While on the surface it appeared to support the ATS, on closer reading it exhibited several of the features noted above. They included: Russian priority; the right, supposedly shared with the United States, to claim the whole continent and adjacent waters; the financial hardships of the Russian Antarctic programme; a characterization of Soviet and Russian Antarctic expeditions as prospectors, largely concerned with the economic dividends of their work (the word 'resources' occurs 26 times); and a determination to maintain or enlarge the scale of Russian Antarctic fisheries—currently the only extractive industry in the region—against any obstruction from the PEP regime. Lukin explained that Russia prefers to confront 'the exacerbated contradictions in the ATS' (2014: 218), where it has an effective veto, rather than see the Treaty replaced with a multilateral, bizonal treaty for the Arctic and Antarctic under UN auspices, which would reduce Russian influence and might also restrict Russian activities in the north. He did not, however, refer to the review of the PEP in 2048 or other long-term aspects of the Antarctic Treaty. In his latest contribution, after reaching back to 1966 for a Soviet plan to ring Antarctica with research stations, Lukin concludes that in addition to their scientific work Russian stations have 'worked up a large quantity of legally normative documents which regulate our country's activities in the Antarctic' [6] (Lukin 2020: 65, 68).

Lastly under this heading, a report broadcast by the national television company *VGTRK* on 30 November 2019 featured interviews with Foreign Minister Sergej Lavrov and former Ambassador Oleg Khlestov in which both men emphasized the role of the Antarctic Treaty in barring the exercise of national territorial claims, although the commentary regretted its inapplicability to the continental shelf, seen as a potentially abundant repository of mineral resources.[9] According to Lavrov the Treaty was under no foreseeable threat, but once again the PEP was not mentioned.

RECENT HISTORICAL TREATMENTS

Lukin

The social and political background surveyed above may explain why recent historical scholarship has focused on the six decades of the SAE/RAE (Savatyugin and Preobrazhenskaya 1999–2009; Lukin and Others 2006), to the relative neglect of the Bellingshausen expedition, whose priority is seen as firmly established, but perhaps also as better not reopened for discussion. Apart from work which touches on other phases in the careers of Bellingshausen or Lazarev, or which concerns only the history of Russian America or other topics within Pacific Studies (above), post-Soviet domestic Russian publications on the expedition can almost be counted on the fingers of one hand. In 2005 Lukin wrote a historical article for the business and current affairs magazine *Vlast'*. Despite taking his lead from Lazarev's familiar statement about January 1820, Lukin focused to an unusual extent on the exploration of the South Shetlands by Bellingshausen and others, including the Russian encounter with Nathaniel

---

[9] https://www.youtube.com/watch?v=NZYhB4yn_tg. In view of the support for the Treaty expressed by the interviewees, it is unfortunate that the person posting this on YouTube saw fit to add the caption, in English: 'The White Continent Belongs to Russia!' The original 'hook' for this item was the new monument to be erected at Bellingshausen Station in the South Shetland Is., a project which had been promoted by *VGTRK* (below).

Palmer on 5 February 1821. He accepted that English names for the group had preceded Russian ones, and perhaps for that reason he did not claim that the Russian names were given on the spot (Lukin 2005: 77). However he displaced the Bellingshausen–Palmer meeting from its actual location near Deception I. to the vicinity of King George I., at the other, eastern end of the main chain. Given that the facts of the meeting have long been recognized in Russia (Berg 1951), then unless it was due to carelessness such a revision may have been intended to represent Bellingshausen as conducting his rapid survey unaided by a local informant.

Lukin's most political revision of the past came when he tried to support his cherished notion that the Soviet Union had: 'reserv[ed] to itself the right to claim the whole Antarctic continent' [7] (Lukin 2005: 79). The actual wording of the 1958 discussion document, which he cited from the Treaty negotiations, had been vaguer and more cautious: 'The Soviet Union reserves to itself all rights based on discoveries and explorations by Russian navigators and scientists, including the right to advance corresponding territorial claims in the Antarctic' [8] (Anon. 1958). Evidently Lukin was not satisfied with language which had left it unclear whether there were any such rights, and what if any claims might one day be advanced on the basis of them. In a recent version, Lukin added that in 1958 the United States had also based its rights in Antarctica on early exploration (Lukin 2017: 42). The United States did indeed reserve its right to make territorial claims, but it did so on the basis of all its Antarctic exploration over 138 years; it made no suggestion of any right to claim the entire continent; and its successful policy was to suspend the legal status quo, including all national rights, for the duration of the proposed Antarctic Treaty (National Security Council 1958: 483–484).

Koryakin

Apart from Lukin's occasional co-options of the past, recent Russian output on the expedition has been largely confined to reprints of Bellingshausen's book and other primary sources. The first, which included other eyewitness accounts, had some justification for presenting itself as the fourth edition of *Two Seasons* (Bellinsgauzen 2008). The editor was the veteran Antarctic glaciologist Vladislav Sergeevich Koryakin (1933–). Additional primary sources were taken from Sementovskij's 1951 anthology rather than the originals, and some of the notes were also flagged as taken from Soviet commentators. Endnotes and some bibliography for the texts were provided, though more might have been said about how far Bellingshausen's text had been checked against the 1831 original.

Koryakin's introduction gave few citations, and there are mistakes which might not have been made by a professional historian, such as mis-dating Bransfield's sighting of the Peninsula to 7 February 1820, by which date the *Williams* was NEbE of the Elephant and Clarence group (Koryakin 2008: 23). Koryakin gave a vivid picture of the conditions experienced by seamen in the squadron, and compared the accounts left by eyewitnesses. He quoted Bellingshausen's superficially problematic Remark for the first time in 40 years, but explained it away as a response to meeting something totally unforeseen (the same: 19). He might not have proposed that rationale if he had studied the proto-glaciology of Bellingshausen's day and especially the ideas of Buffon and Scoresby about the formation of entirely marine polar ice caps, or if, for that matter, he had seen Simonov's diary entry mentioning that he was looking forward to seeing the phenomenon.

Like V. Lebedev in the 1960s, Koryakin hoped to show, by comparing the primary sources, that Lazarev's account of 28 January 1820 was supported by others and more convincing than the one in Bellingshausen's *Two Seasons*, probably, he suggested, because the latter had been altered by the editor Golenishhev-Kutuzov. In making that familiar case, however, Koryakin mistreated the original texts himself. His version of the account in Lazarev's letter reads as follows:

> On 16th January [O.S.] we reached latitude 69°23′ S, where we met main ice of extraordinary height. It was a fine evening, and looking out from the crosstrees it stretched just as far as our gaze could reach, but we had not long to enjoy that amazing spectacle, because the murk quickly came over again and the usual snow set in. … From there we held our course east, pushing south at every opportunity, but we always met an ice floe main before we reached 70°… The main land in the south had at last been discovered, which they had sought for so long, and whose existence sedentary philosophers had posited—from their cabinets—as being essential for the balance of the terrestrial globe [9] (Koryakin 2008: 17)[10]

---

[10] Koryakin ended the last sentence of his misquotation with a quotation mark, but without either a full point or an ellipsis. Nevertheless readers are bound to understand the sentence, stripped of its opening words 'But considering … concluded that…', as a statement from Lazarev to the effect that the continent had now been discovered—a direct falsification of what he actually said.

For all its seeming authenticity, this was a travesty. Whereas the first ellipsis marked the omission of a short sentence giving the longitude, the second one bridged four manuscript pages of Lazarev's letter, covering an entire year of the voyage. For the last sentence, Koryakin arbitrarily converted a later passage into a stand-alone sentence about an actual discovery of 'main land', and then cobbled it onto the little that Lazarev had said earlier about 28 January 1820. In fact, however, the second passage made no such claim. It was, rather, part of a rebuke to British journalists and others for supposing that the discovery of the South Shetlands amounted to discovering a southern continent. Nothing of the sort, said Lazarev, because we checked it ourselves:

> *But considering that in England and one might say throughout Europe people concluded that* **the main land in the south had at last been discovered, which they had sought for so long,** *and whose existence sedentary philosophers had posited—from their cabinets—as being essential for the balance of the terrestrial globe, so we deemed ourselves bound by our very title of Southern Expedition either to confirm that conclusion or to disprove it completely by passing around that land on the southern side. To that end having arrived in sight of its western termination, which lies in lat. 63°06′ S longitude 62°42′ W, on 23rd January, we passed along its southern side. South Shetlandia is nothing more than a chain of islands of various sizes and of extraordinary altitude, covered in eternal snow and stretching out from NE to SW for about 300 miles [555km], such that its eastern limit lies in latitude 61°10′ S longitude 54°W. There's the* <u>South Continent</u> *for you, and I deliberately note the latitudes and longitudes, because perhaps you might care to glance at a map out of curiosity* [10]. (Lazarev'' 1821: fos 4–4v. Underlined words in English in the original; words hijacked and turned into a separate sentence by Koryakin shown in bold.)

Koryakin cited the Sementovskij anthology (1951: 25–26) as the source for his concoction, but his reference only applies to the hijacked sentence, because the passage about 28 January came two pages before it in that edition. All this, despite the fact that Lazarev's unambiguous satire on what he supposed to be European boasting about the South Shetland Is. was given in Sementovskij just as it is translated here. In short, Koryakin rewrote Lazarev's letter in order to extract a positive claim that the expedition had discovered an Antarctic continent. A preposterous fraud, because as was noted in Chap. 7 it would mean that after mentioning it once to one person Lazarev, of all people, kept silent about a stunning geographical achievement for the next 30 years.

As was pointed out in Chap. 6, a serious difficulty for the 28 January discovery theory in general, and in particular for the claim that Bellingshausen's editor removed anything resembling Lazarev's account of that day from *Two Seasons*, is that Bellingshausen's account of 17 February, both in his reports and in *Two Seasons*, was not unlike what Lazarev said about the earlier date, but that passage was *not* removed by Bellingshausen's editor. Koryakin's solution to this difficulty was as bold as it was illegitimate. He simply ignored what Bellingshausen had said about 17 February 1820.

He did that by first placing Bellingshausen's two descriptions of 28 January, in his report from Sydney and *Two Seasons*, side by side and claiming that they were different (Koryakin 2008: 16–17; see Table 11.1 (a) and (b)). Then, instead of explaining what those supposed differences were, he immediately moved on to the undoubted difference between both of them and the passage in Lazarev's letter. Next, he invited his readers to compare corresponding passages, from the report and the book, about the expedition's probe on 17–18 February 1820, claiming that this was a further example of a coastal sighting having been suppressed by the editor. However, perhaps because the contents of the actual passages about 17 February—(c) and (d) in Table 11.1—are so similar that juxtaposition would have contradicted his interpretation instead of supporting it, he compared (c) with (e) instead, a passage taken from Bellingshausen's description of the *following* day.

But the reason that Bellingshausen made no reference to a barrier-like ice feature under 18 February 1820, (e), so that Koryakin could misrepresent him (or his editor) as underplaying the event, was simply that (e) was a continuation of (d), the long passage in which the phenomenon had been carefully described under the previous day, 17 February, when it was first encountered. (Hence Bellingshausen's cross-reference 'described above' in (e)—an expression which must have baffled any careful reader of Koryakin's analysis of the texts [the same: 18]) That earlier passage (d), however, Koryakin carefully ignored.

However illegitimate, Koryakin needed to resort to some such device if he was to conceal certain aspects of what Bellingshausen actually said about sighting an ice coast, first in his report from Sydney, and then in his book.

(1)   Table 11.1 demonstrates that, like other explorers and doubtless after checking the logbooks etc., Bellingshausen gave a fuller

Table 11.1 Bellingshausen's accounts of 28 January and 17–18 February as quoted, or not, by Koryakin in 2008

	Report from Sydney 20 April 1820, 242v	Two Seasons 1831 1: 171–172, 188–190
28 January 1820	(a) *On the 16th, having reached latitude S 69°25′, longitude 2°10′W, I met continuous ice at its fringes, with one piece jumbled on top of another. Further in to the south, icebergs were visible at various places across it* [11].	(b) … at noon, in latitude 69°21′28″, longitude 2°14′50″, we met ice which, through snow that was falling at the time, looked to us like white clouds. A NE wind was dying down and there was a strong current from NW. The snow meant that we could not see very far; … we made out that continuous [pack] ice extended from east through south to west; our course led straight into that ice field, which was scattered with hillocks [12].
17 February 1820	(c) … *between the 5th and 6th, I reached latitude S 69°7′30″, longitude E 16°15′. There, beyond ice fields comprising small ice and* [ice] *islands, a main of ice was sighted, the edges of which had broken away perpendicularly, and which stretched as far as we could see, rising to the south like land. The flat ice islands that are located close to this main are evidently nothing but detached fragments of this main, since they have edges and upper surfaces which resemble the main* [13].	(d) Eventually, at a quarter past three after noon, we saw a quantity of large, flat, high ice islands, beset with light floes some of which overlay one another in places. The ice formations to SSW join together into hilly, solid standing ice; its edges were perpendicular, forming coves, and its surface rose away to the South, to a distance whose limits we could not make out from the [main] cross-trees. … Seeing that the ice islands had similar surfaces and edges to the large ice formation noted above, which lay before us, we concluded that those huge ice masses and all similar formations get separated from the main coast by reason of their own weight or from other physical causes and, carried by the winds, drift out into the expanse of the Southern Ice Ocean. …[14]
18 February 1820		(*21 lines later in the 1949 edition*) (e) … the ice floes became larger and more frequent, so that it became impossible to push any further S at that point, but half a mile further off in that direction we saw piles of floes, stacked one on top of another. Further [S] still were icebergs similar to those described above, and probably a continuation thereof [15].

account of both dates, 28 January and 17–18 February 1820, in the book than in his report;

(2) In each case the two accounts were broadly similar; when like is compared with like, Bellingshausen's descriptions in the book repeat and amplify what he said in the report;

(3) Bellingshausen's description of 28 January certainly does not resemble his account of 17–18 February or the passage in Lazarev's letter, but the last two, written separately by different people, are similar in some respects, though Lazarev's informal account of 28 January is less detailed than Bellingshausen's account of 17 February. As noted in Chap. 7, there is perhaps a slightly better fit between Lazarev's 'main ice' and Bellingshausen's account of 2(14) February (see Table 7.1). But in neither case is the resemblance strong enough to conclude that the two men were referring to the same event; the difference between the stated positions alone is enough to rule that out.

Point (3), the hypothesis that Lazarev may perhaps have given a January date to something Bellingshausen placed in February, just possibly 2(14) February, must remain a matter for individual judgement. What is certain is that as regards the sighting of an ice coast in 1820 it was not so much Golenishhev-Kutuzov in 1831 as Koryakin in 2008 who suppressed an important part of what Bellingshausen said in his book, namely the passage labelled (d) in Table 11.1, which dates one such sighting to 17 February 1820, 18 days after Bransfield sighted and named Trinity Land at the tip of the Antarctic Peninsula.

As noted in Chap. 4, Novosil'skij echoed Bellingshausen's description of 28 January as an encounter with 'an extensive ice field scattered with small hills' [16] (rather than hillocks), borrowed the latter's phrase about the ship being surrounded on three sides by continuous ice, and explicitly mentioned the risk of embayment: 'it was necessary to make our way out of that bay' [17] (1853c: 29). Koryakin quoted the passage and explained its difference from Lazarev's account of the same day by drawing attention to small differences in the positional coordinates given by the respective officers, which meant, he thought, that Lazarev had described conditions at a slightly later time of day. But his reasoning was ahistorical; positional observations with the technology of the day were nothing like readings from a modern GPS instrument, so that small differences between coordinates given by different observers cannot be converted reliably into time intervals. There may of course have been some time differences, but when an officer gave only one position for a day without qualification that would usually be either the noon position or that reached at an important point, such as obstruction by impassable pack ice. Bellingshausen himself gave different coordinates for noon on 28 January,[11] and then revised those given in his report for the book (see Table 9.1). But, perhaps reluctant to dwell on differences between Bellingshausen and Lazarev over 28 January, Koryakin did not try to resolve those differences as resulting from time intervals, as he did those between Lazarev and Novosil'skij.

Having ignored Bellingshausen's description of 17 February 1820 (Table 11.1), Koryakin made much of Novosil'skij's instead (the same: 35–36), without pausing to consider that the latter probably had his former commander's book open beside him, or that his use of the phrases 'immovable ice', 'ice coast', and 'ice wall' [18] in the early 1850s probably owed much to phrases like 'icy cliff', 'icy barrier', 'permanent barrier', and 'icy coast' which had been published by Dumont d'Urville, Wilkes, and J.C. Ross in the 1840s.

Minor Treatments

Three other reprints of *Two Seasons* did not amount to new editions. A luxurious facsimile, priced at around $3,600 (Bellinsgauzen 2011a), naturally added nothing to the original and has no interest for historians who can download electronic versions of the text volumes, at least, from the Russian State Library in Moscow.

Also in 2011 a lavishly illustrated popular reprint catered for nationalist tastes by replacing Bellingshausen's own title with the question-begging phrase *The Discovery of Antarctica*, and by concocting a cover which portrays the commander in full dress uniform on the ice which he never trod, in front of the Russian tricolour which he may or may not have flown, it having been the ensign of the merchant navy in his day (Bellinsgauzen 2011b).[12] The text was that of the 1949 edition of *Two Seasons* and the

---

[11] The position given in the book is close to that in the track chart (Belov 1963b: Sheet 2), which shows that *Vostok* veered away NWbW after noon but still close to the noon position.

[12] The expedition's artist, Mikhajlov, always showed the ships flying the St Andrew's Cross, the ensign of the Imperial Navy. However it is conceivable that they flew the tricolour of the merchant navy while in the ice fields, because of its much greater visibility. There is no direct evidence for that, but the painting 'Icebergs', completed in 1870 by the artist Ivan Aivazovskij, who had been a friend of Lazarev's, depicts a ship flying the Russian tricolour in front of high icebergs and ice cliffs. The ship is widely believed to be the *Mirnyj*, although

cuts were identified as before. The reprint provided other primary sources alongside Bellingshausen's narrative, although the compilers seem to have been unaware that they were omitting three of the seven letters published by Nikolaj Alekseevich Galkin, *Mirnyj*'s surgeon, about the Pacific phase of the voyage, or that the extracts from Bellingshausen's letter and report to de Traversay, which they reproduced from the versions published in Russia in 1821, had not been exact transcriptions of the commander's dispatches from Sydney a year earlier. From the historiographic perspective, the 2011 reprint took another small step forward by publishing (though not discussing) Bellingshausen's Remark in its original context, those 1821 versions. (Both had appeared on the internet about two years earlier.) A companion volume spliced together various early texts and documents to provide a vivid account of Lazarev's three circumnavigations, but also without breaking new ground (Khoroshevskij 2014). In 2017 yet another reprint of *Two Seasons* offered another replacement title (copyrighted to Bellingshausen)—*To the South Pole on Sloops* Vostok *and* Mirnyj. Readers were provided with Shvede's 1949 introduction, but his notes were amplified by the editor, S. Chertoprud. The '29 islands' claim was reiterated (Bellinsgauzen 2017).

The treatment of the expedition in a recent survey of Antarctic place names suggests that Russian interest in the details of the voyage and its protagonists may be reviving, but still shows little knowledge of the subject (Savatyugin and Preobrazhenskaya 2014: 44–67). Thus the number of expeditioners and the date of Belov's *Karta* (1963a) were given incorrectly, and Bellingshausen's mother, long known to have been Anna Catharina von Folckern and not her sister Elisabeth, was described as an Elizabeth of unknown family (the same: 48). The statements that Bellingshausen lived at his uncle's house in Arensburg (Kuressaare) during his elementary school years, and that he, rather than his far more likely superiors, changed the name of Torson I. to Vysokij,[13] were not supported with evidence (the same: 49, 52). These and other minor flaws make it hard to evaluate these authors' interesting treatment of the supposed discovery, which they dated simply to 'the end of January' despite quoting Lazarev's letter [19]. They also raised the question of whether the

Aivazovskij himself made no such attribution, as he normally did when depicting specific historical subjects, and there is internal evidence to the contrary (Bulkeley 2015b).

[13] Although it is endorsed by SCAR, there are no linguistic grounds for the 'Visokoi' version of this name.

expeditioners understood what they had supposedly achieved, before citing Belov (but not his arguments) as the authority for a positive conclusion (the same: 53–55). With shameless exaggeration they declared that: 'Bellingshausen, Lazarev and *several* members of the crew referred to the Sixth Continent with phrases like "continent of ice", "main of ice", "ice continent", "ice bastion" etc.' [20] (the same: 55—emphasis added).[14]

Turning to the diaspora, the publications of Drs Elena Govor and Aleksandr Massov in Canberra have touched on the expedition from time to time, but have not addressed its main, Antarctic thrust (Govor and Massov 2007). At the University of Tasmania, Dr Irina Gan published a short, bilingual account of Russian exploration and research in Antarctica from 1819 to 2011 (Gan 2011). Apart from summarizing Shvede she highlighted an important passage in Lomonosov, which had already been pointed out by several Soviet authors in the 1960s, but without including the opening context:

> [Near the Strait of Magellan and opposite the Cape of Good Hope] great icebergs pass on their way at about 53°S. That confirms that there are remote islands and main land covered with much unremitting snow, and a great[er] expanse of the earth's surface is burdened with them around the South Pole, than in the North [21]. (Lomonosov" 1763: 256)

Gan slightly mistranslated this by inserting the phrase 'one would find' in her English version (Gan 2011: 3, 59), whereas the passage had simply emphasized that there was probably more, and more glaciated land in the Antarctic than in the Arctic, so that the former would be harder to explore. There are striking parallels with Cook's 'That there may be a … large tract of land, near the Pole, I will not deny; on the contrary I am of opinion there is', and between Cook's 'pestered with ice' and Lomonosov's

---

[14]The true tally is that 'main of ice' was used once by Bellingshausen, and then referred back to three times as a 'main' in the same paragraph; Bellingshausen also once applied 'main' as an adjective to 'enormous' ice formations; 'ice bastion' was used once, in theoretical rather than observational mode, by Simonov; and 'main ice' and 'ice floe main' were used once each by Lazarev. Lazarev also used 'continent' once, in English and without 'ice', when satirizing what he believed were British claims about the South Shetlands (above). The only member of the expedition to apply the word *kontinent* to Antarctic discoveries, once it had entered the Russian language, was Novosil'skij, writing retrospectively in the 1850s after the expeditions of Wilkes and J.C. Ross. Savatyugin and Preobrazhenskaya may well have been misled by Belov's repeated misquotations from Lazarev and Bellingshausen (e.g., Belov 1970a: 204).

'burdened' (Cook 1777 2: 239, 231), but that does not mean we are entitled to ignore, between Lomonosov (and Cook) and Bellingshausen, the contrary influence of Buffon and his followers. Lastly Gan took no stance on the priority question herself, but simply echoed Shvede (1962) with 'This date [28 January] is *counted* in Russia as the day that Antarctica was discovered' [22] (Gan 2011: 10—emphasis added).

In the absence of anything original about the Antarctic achievements of the expedition from the far diaspora, and in the context of Koryakin's mistreatment of important evidence, the self-published historiographic project of Dr Aleksandr Ovlashhenko, at Riga in the near diaspora, takes on added significance (2013, 2014, 2016). Approaching the subject as a jurisprudent and an impassioned enthusiast for Russian rights in Antarctica, Ovlashhenko has focused on the priority question. In the first chapter of his first book he set up Bellingshausen's continental priority as a methodological axiom by claiming that it was established in the nineteenth century (before the continent was known to exist). That relied on confusing the universally accepted Russian priority with regard to land south of the Circle in 1821 with the more debatable matter of who first set eyes on the continent in 1820, Bransfield or Bellingshausen. From that point he set about deploying his priority axiom as a test for the honesty and national loyalty of Soviet historians and other commentators, whose publications are listed and quoted with unrivalled completeness. Regrettably, Ovlashhenko's books are peppered with aggressive denunciations of anyone, living or dead, with whom he disagrees. Despite finding numerous Soviet instances of dissent from the absolute priority position first promulgated in 1949, he refuses to consider that they may have been due to genuine doubts about its validity.

With successive volumes Ovlashhenko has devoted increasing space—beyond his stated Soviet remit—to contemporary scholars who cast doubt on, or who fail to uphold, the dogma of Russian priority, and in particular to the present author. The latter is assailed on the grounds of his nationality and his Twitter account, amongst other things, though not, so far, by analysing his book. (Ovlashhenko declines to state his own nationality.) But whatever Bulkeley and Ovlashhenko may think of each other's work, what matters about their exchanges, historiographically speaking, is that to date not one historian in Russia has had a word to spare for either of them.

In 2016 another ill-supported view of the expedition was aired from the near diaspora. It came from Viktor Rokot, a Belarusian historian who had published a study of Russian America a few years earlier, in an interview with *Radio Mayak* in Moscow. Although he stood by the priority claim, Rokot emphasized that it dated only from 1949. In a somewhat confused exegesis, he declared that Antarctic discovery had not been the main aim of the expedition, but had come about as a by-product of a southerly course that was taken for other, blithely unspecified, reasons; that although Bellingshausen never claimed to have discovered a continent, he and Lazarev had at least supposed there was land somewhere in the region [23]; and that in the 1840s J.C. Ross had declared that Bellingshausen and Lazarev discovered the continent [24]. The statement about Ross was probably taken from Fradkin (1972: 48—above) or a similar source.[15] Nothing was supported by evidence, probably because a radio interview does not lend itself to that. It is regrettable that Rokot has not yet published his views on the expedition in print, where citations could and should be given (Rokot 2016).[16] His remarks were a good example of what nowadays passes for knowledge of the expedition in the Russian-speaking world, which cannot help being based on untrustworthy authorities.

The Naval Archives' Online Exhibition

In about August 2019, when the preparation of this book was already well advanced, the website of the Russian Naval Archives in St Petersburg posted a 'virtual exhibition' to mark the bicentenary of the supposed discovery of Antarctica, which as we have seen has been officially accepted in Russia since 1949 (Kondakova 2019). The compiler makes it clear that this resource is primarily a celebration of the expedition and of the Archives' custody of its records, and puts forward no original insights herself. While the documents selected have considerable interest for

[15] If Rokot based his statement about Ross on Fradkin he should not have done so, because Fradkin merely pointed to Ross's book without a page reference, and there is in fact no such statement in the book, to which Rokot has full access via the internet.

[16] Trusting as he was to memory, Rokot also invented a phrase which did not occur in Lazarev's letter to Shestakov: 'So we probably discovered something after all' [Наверное, мы что-то нашли все-таки]. It makes no sense, because Lazarev knew perfectly well that they made several discoveries, some of which he mentioned to his friend.

historians, it is much to be regretted that other important and still unpublished documents, some of which were envisaged for Kuznetsova's unrealized sourcebook 50 years ago, remain absent from view today.

*   *   *

Perhaps the most important feature of modern Russian perceptions of the Bellingshausen expedition, once interest in the remoter national past revived in the present century, has been their near-universal conformity with the priority claim as formulated in the latter years of Stalin, by contrast with varying levels of dissent in the Soviet period. The author has found only one recent textual example of an 1821 date for the Russian discovery of Antarctica. It came in a thesis about the economic potential of Antarctica published by the Academy of Sciences' Institute for World Economy and Russia's International Relations, where the author dated the discovery of Antarctica, in the shape of Alexander I., to '27 January 1821', doubtless falling into line with Belov's mistake in failing to notice that this was a morning event to which his nautical calendar theory did not apply (Korzun 2009: 14). Interestingly, however, the 'Nature, Science' organization in Kazan held an educational exhibit about the expedition on 10–17 February 2019, thus marking the 199th anniversary of Bellingshausen's second 1820 approach to the coast of Queen Maud Land rather than his first, which remains the generally accepted 'discovery day' elsewhere in Russia.

Assisted by the virtual suspension of critical research, almost all Russians now accept the tale of first discovery as fact. Having been drilled by rote as children in such falsehoods as that the expedition leaders were convinced that they had discovered a new continent (Pleshakov and Others 2018),[17] they may find it hard to exercise their own judgement on the subject. That said, however, recent celebrations of the 28 January 'discovery day' have not been especially conspicuous apart from the usual festivities at Russian Antarctic stations. In 2019 there was a small commemorative exhibition in the library of the Moscow Duma, which also celebrated a forthcoming expedition under sail into the Bellingshausen Sea. According to its leader, Dmitrij Kuznetsov, the eight-person team

---

[17] This is an online answers book for the first year in the geography course which leaves no scope for divergent opinions. Subsequent material contains further doubtful information, such as that Bellingshausen discovered only six islands.

cruised in the vicinity of the Antarctic Peninsula for about a month after setting off from Port Williams in Chile on 1 February 2019.[18] The project was undertaken as an advance celebration of the 2020 bicentenary. Naturally rather more effort was expended in 2020. A new monument to Bellingshausen was set up at the Antarctic research station named after him in 1968 in the South Shetland Is. (Fig. 12.1); Pel′sh's three-part documentary was released (above); a commemorative volume, *The Russian Antarctic*, was published by the Geographical Society—the title's designation of the region, rather than the continent, may or may not be significant (N. Kuznetsov 2020—not seen); and a set of commemorative stamps was issued by the Ministry of Communications. In opting for their virtual exhibition (Kondakova 2019) the Naval Archives were doubtless aware that despite the linguistic challenge such an offering on the internet has the capacity to reach a wider audience worldwide than would have passed the displays of expedition documents at Soviet exhibitions abroad in 1961.[19] On the other hand, by contrast with some of the other commemorations, the virtual exhibition has very little to say about priority beyond assuming it. Meanwhile Bellingshausen's final report and list of discoveries remain unpublished, whereas Rusakov's fulsome exaggerations (1903) have been republished and remain in print.

The contrast between the present situation and the modest variety of views expressed in the Soviet period stands in need of explanation. In the present century the status of the Bellingshausen expedition appears to have risen from that of an asset for a minor element of foreign policy to become a component, still perhaps fairly humble, of national identity—now an extremely important aspect of Russian life and one in which a high degree of conformity is expected from all citizens by the national leadership, if only in public and as a work in progress. If that is what has happened, it seems doubtful whether rational, evidence-based treatments of the matter can interest a significant proportion of the Russian public. The apprehension of the headline achievements of the expedition by those who give it any consideration might best be described, therefore, as one of loyalty to a historical myth which has regrettably acquired the status of fact.

[18] https://duma.mos.ru/ru/0/news/novosti/antarktida-200-let-spustya-ili-30-dney-na-grani-chelovecheskih-vozmojnostey-ekspeditsiya-moskovskih-moreplavateley

[19] With the global pandemic taking hold in Russia from March 2020, both the timing and form of the virtual exhibition and the timing of D. Kuznetsov's cruise were to prove happy decisions.

Original Texts of Quotations

[1] …судовых журналов и штурманских расчетов…

[2] … в записях бортового журнала …

[3] Российская Антарктида живет, говорят ветераны.

[4] Что бы помнили/Крым наш/Арктика наша/Антарктида тоже будет наша

[5] … сезонных полевых базах … Союз, …

[6] … разработкой большого объема нормативно-правовых документов, регулирующих отечественную деятельность в Антарктике.

[7] … резервируя за собой право претензий на весь Антарктический континент…

[8] Советский Союз сохраняет за собой все права, основанные на открытиях и исследованиях русских мореплавателей и ученых, включая право на предъявление соответсвующих территориальных претензий в Антарктике.

[9] *16-го генваря [ст.ст.] достигли мы широты 69°23′S гдѣ встрѣтили матерой ледъ чрезвычайной высоты и въ прекрасный тогда вечеръ, смотря на салeнгу, простирался оный такъ далеко какъ могло только достигать зрѣнiе, но удивительнымъ симъ зрѣлищемъ наслаждались мы недолго, ибо вскорѣ опятъ напасмурило и пошелъ по обыкновенiю снѣгъ. …—отъ Сюда продолжали мы путь свой къ Осту покушаясь при всякойвозможности къ Зюйду, но всѣгда встрѣчали льдяной материкъ не доходя 70° … Открывалась наконец та матерая на Югѣ Земля которую такъ долго искали и Существованiе коей Сидевшiе философы в кабинетахъ своихъ, полагали необходимымъ для равновѣсiя Земного шара.* [купюры Корякина]

[10] *Но такъ какъ в Англiи и можно сказать во всей Европѣ, Заключали что **открывалась наконец та матерая на Югѣ Земля которую такъ долго искали и Существованiе коей Сидевшiе философы в кабинетахъ своихъ**, полагали необходимымъ для равновѣсiя Земного шара, то мы по одному названiю Южной Экспедицiи, обязанностiю почли таковое Заключеннiе или еще болѣе подвердить или вовсе оное опровергнуть тѣмъ, чтобы обойти Землю сiю съ Южной стороны. а потомъ 23iй Генваря придя на видъ западной*

оконечности оной Которая находится въ шир. 63°06′ S долготѣ 62°42′ W пошли по Южную Сторону. Южная Шетландія Есть ничто иное какъ гряда разной величины острововъ чрезмѣрной высоты, покрытыхъ вѣчнымъ снѣгомъ и простирающихся почти на NO и SW около 300 миль, такъ что восточнѣйши[й] край оной находится въ широтѣ 61°10′ S долготѣ 54°′ W. Вотъ тебѣ и South Continent. я нарочно означаю широты и долготы потому что можетъ быть изъ любопытства зохочешь взглянуть на Карту. (**отнятая Корякиномъ фраза**)

[11] *16-го числа, до шедши до Широты S 69°25′ 2°10′ W, въ Стретилъ сплошной ледъ, Украевъ одинъ на другой на бросанный кусками а внутрь к Югу въ разныхъ мѣстахъ по Оному видны ледяные горы.*

[12] … въ полденъ въ широтѣ 69°, 21′, 28″, долготѣ 2°, 14′, 50″, мы встрѣтили льды, которые представились намъ сквозъ шедшій тогда снѣгъ, въ видѣ бѣлыхъ облаковъ. Вѣтръ былъ отъ NO умѣренный, при большой зыби отъ NW, по причинѣ снѣга, зрѣніе наше не далеко простиралось; я привелъ въ бейдевинтъ на SO, и проидя симъ напрвавленіемъ двѣ мили, мы увидели, что сплошные льды простираются отъ Востока чрезъ Югъ на Западъ; путъ нашъ велъ прямо въ сіе льдяное поле, усѣянное буграми.

[13] … съ 5 на 6 число [ст.ст.] дошелъ до широты S. 69°, 7′, 30″, долготы O. 16°, 15′. Здесь за льдеными [так] полями мелкаго льда, и Островами виденъ Материкъ льда. коего края отъ ломаны перпендикулярно, и Которой [так] продолжался по Мерѣ на-шего зренія, возвышаясь къ Югу подобно берегу. Плоскіе льдяные острова, близь сего материка находящіися [так], ясно показываютъ, что онѣ суть отломки сего материка; ибо имѣютъ края и верьхнюю поверьхность подобную материку.

[14] …наконецъ въ четверть четвертаго часа по полудни, увдѣли множество большихъ, плоскихъ, высокихъ льдяныхъ острововъ, затертыхъ плавающими мелкими льдами, и мѣстами одинъ на другомъ лежащими. Льды къ SSW примыкаются къ льду гористому, твердо стоящему; закраины онаго были перпендикулярны и образовали заливы, а поверхность возвышалась отлого къ Югу, на растояніе, предѣловъ котораго мы не могли видѣть съ салинга. …

Видя льдяные острова, поверхностью и краями сходные съ поверхностью и краями большаго вышеупомянутаго льда, предъ нами находящагося, мы заключили, что сіи льдяныя громады и всѣ подобные льды, отъ собственной своей тяжести, или другихъ физическихъ причинъ, отдѣлились отъ матераго берега, вѣтрами отнесенные, плаваютъ по пространству Ледовишаго Южнаго Океана; …

[15] … плавающіе льды становилисъ такъ часты и крупны, что дальнѣйшее въ семъ мѣстѣ покушеніе къ Z, было невозможно, а на полторы мили по сему направленію видны были кучи льдовъ, одна на другую взгроможденныхъ. Далѣе представлялисъ льдяныя горы, подобныя вышеупомянутымъ, и вѣроятно, составляютъ продолженіе оныхъ.

[16] …обширнаго ледянаго поля, усѣяннаго пригорками…

[17 …надо было выйти изъ этой бухты…

[18] неподвижный ледъ/ледянаго берега/ледяная стена

[19] …в конце января 1820…

[20] И Беллинсгаузен, и Лазарев, и *многие* члены экипажа употребляют для обозначения Шестого материка такие слова как «континент льда», «материк льда», «льдяный континент», «льдяный оплот» и другие. [разгрядка наша]

[21] [В близости Магелланского пролива, и против мыса Добрыя надежды] около 53 градусов полуденной ширины великие льды ходят: почему сомневаться не должно, что в большем отдалении островы и матерая земля многими и несходящими снегами покрыты, и что большая обширность земной поверхности около южного полюса, занята оными, нежели в севере.

[22] Эта дата [28 January] в России *считается* днем открытия Антарктиды… [разрядка наша]

[23] …они предположили, что там какая-то земля.

[24] И он [Росс] сказал, что, скорее всего, Беллинсгаузен и Лазарев открыли новый материк.

References[20]

Anon. 1958. Obmen pis'mami mezhdu Gosudarstvennym Departamentom SShA i Posol'stvom SSSR v SShA po voprosu ob Antarktike [Exchange of Letters between the US State Department and the Embassy of the USSR on the Antarctic Question]. *Pravda*, June 4: 5.
———. 2010. Panel Discussion. 190 let nazad byl otkryt shestoj kontinent [190 Years Ago the Sixth Continent was Discovered]. *Rossiya kul'tura*, January 29.
———. 2016. Novosti [Announcements]. *Rossiya I* Website, January. The link to this announcement [https://russia.tv/article/show/article_id/23592] is no longer valid, but the film itself, now owned by VGTRK's *Rossiya 24* Channel, can be viewed in two parts on YouTube: https://www.youtube.com/watch?v=nLCTpawMiVE.
———. 2019. S kreshheniem: èstonskij "Bellinsgauzen" otpravitsya zanogo otkryvat' Antarktidu [A Christening: An Estonian *Bellingshausen* Prepares to Discover Antarctica Again]. *Sputnik eesti*, June 11.
———. n.d.-b. *Osvoenie chelovekom Antarktidy [Human Mastery of Antarctica]*. http://interneturok.ru/geografy/7-klass/materiki-antarktida/osvoenie-chelovekom-antarktidy.
———. n.d.-c. *Razvitie geograficheskikh znanij o Zemle [The Development of Geographical Knowledge about the Earth]*. http://www.nado5.ru/e-book/razvitie-geograficheskikh-znanii-o-zemle
Batchelor, John Calvin. 1983. *The Birth of the People's Republic of Antarctica.* New York: Dial.
Bellinsgauzen, F.F. 2008. *Dvukratnye izyskaniya v yuzhnom ledovitom okeane i plavanie vokrug sveta [Two Seasons of Exploration in the Southern Ice Ocean and a Voyage Around the World]*, ed. V.S. Koryakin. Moscow: Drofa.
———. 2017. *Na shlyupakh "Vostok" i "Mirnyj" k Yuzhnomu polyusu [To the South Pole on Sloops* Vostok *and* Mirnij*]*, ed. S. Chertoprud. Moscow: Algoritm.
Bellinsgauzen", Kapitan". 2011a. *Dvukratnyya izyskaniya v" yuzhnom" ledovitom" okeanye i plavanie vokrug" svyeta v" prodolzhenii 1819, 20 i 21 godov"*, ed. L.I. Golenishhev"-Kutuzov". St Petersburg: Al'faret.
Bellinsgauzen, Faddej Faddeevich. 2011b. *Otkrytie Antarktidy [The Discovery of Antarctica]*, eds. M. Tereshina and Others. Moscow: Èksmo.
Belov, M.I. 1963a. O kartakh pervoj russkoj antarkticheskoj èkspeditsii 1819–1821 gg. [The Maps of the First Russian Antarctic Expedition 1819–1821]. In *Pervaya russkaya antarkticheskaya èkspeditsiya 1819–1821 gg. i eyo otchyotnaya navigatsionnaya karta*, ed. M.I. Belov, 5–56. Leningrad: Morskoj Transport.

[20] Key to archival references: SARN = State Archives of the Russian Navy, F = Fond, S = Series [*Opis'*], P = Piece [*Delo*], fo./fos = folio/s [*list/y*], v = verso [*oborotnoe*]. (The latter is necessary because verso pages were usually unnumbered.) Dates on manuscript documents are given in Universal Time (CE).

————., ed. 1963b. *Pervaya russkaya antarkticheskaya èkspeditsiya 1819–1821 gg. i eyo otchyotnaya navigatsionnaya karta [The First Russian Antarctic Expedition 1819–1821 and its Official Navigational Chart]*. Leningrad: Morskoj Transport.

————. 1970a. Otkrytia ledyanogo kontinenta [The Discovery of the Icy Continent]. *Izvestiya vsesoyuznogo geograficheskogo obshhestva* 102 (3): 201–208.

Berg, L.S. 1951. Bellinsgauzen i Pal′mer [Bellingshausen and Palmer]. *Izvestiya vsesoyuznogo geograficheskogo obshhestva* 83 (1): 25–31.

Bolkhovitinov, N.N., ed. 1997. *Istoriya Russkoj Ameriki [The History of Russian America] (1732–1867)*. Moscow: International Relations.

Bulkeley, Rip. 2015b. Aivazovsky's Icebergs: An Antarctic Mystery. *Polar Record* 51 (6): 644–654.

————. 2021. Bellingshausen in Britain: Supplying the Russian Antarctic expedition, 1819. *The Mariner's Mirror* 107 (1): 40–53.

Campbell, R.J. 2000. *The Discovery of the South Shetland Islands*. London: Hakluyt Society.

Chernousov, A.A. 2011. *Admiral M.P. Lazarev*. St Petersburg: MOO.

Chernykh, V.I. 2015. Nekommercheskij Fond "Russkaya Antarktida 2020" [Not-for-profit "Russian Antarctica 2020"]. M.V. Sachyov website. http://sachev.ru/russkaya_antarktida_2020.htm

Cook, James. 1777. *A Voyage Towards the South Pole, and Round the World; Performed in His Majesty's Ships the Resolution and Adventure, in the Years 1772, 3, 4, and 5*. 2 vols. London: Strahan and Cadell.

Danilov, A.I., and V.L. Mart′yanov. 2017. Sezonnye raboty 62-j Rossijskoj Antarkticheskoj Èkspeditsij [Seasonal Work by the 63rd Russian Antarctic Expedition]. *Rossijskie Polyarnye Issledovanniya* 3 (29): 11–14.

Decree. 1992. Ukaz Prezidenta Rossijskoj Federatsii ot [Decree Issued by the President of the Russian Republic on] 07.08.1992 g. № 824. http://kremlin.ru/acts/bank/1775.

Dmitriev, V.V., ed. 1993–1994. *Morskoj èntsiklopedicheskij slovar′ [Maritime Dictionary]*. St Petersburg: Shipbuilding Magazine.

Favorov, P.A. 1994. *Anglo–russkij voenno-morskoj slovar′ [Anglo–Russian Naval Dictionary]*. Moscow: Military Press.

Fox, Douglas. 2014. Antarctica and the Arctic: A Polar Primer for the New Great Game. *Christian Science Monitor*, January 12. https://www.csmonitor.com/World/Global-Issues/2014/0112/Antarctica-and-the-Arctic-A-polar-primer-for-the-new-great-game

Fradkin, N.G. 1972. *Geograficheskie otkrytiya i nauchnoe poznanie Zemli [Geographical Discoveries and Scientific Knowledge of the Earth]*. Moscow: Mysl′.

Fyodorov, Valerij. 2000. Millenium na Yuzhnom polyuse [The Millenium at the South Pole]. *Moskovskij universitet* 14: 4.

Fyodorovskij, Evgenij. 2001. *Bellinsgauzen*. Moscow: Astrel′.

Gan, Irina. 2011. *Russkie v Antarktide [Russians in Antarctica]*. Homebush, NSW: Russian Historical Society in Australia.

Govor, Elena. 1995. Na mysu Russkikh: iz istorii èkspeditsij 1820 goda [On Russian Point: From the History of Expeditions in 1820]. *Avstraliada* 5: 1–6.

Govor, Elena, and Aleksandr Massov. 2007. *Kogda mir byl shirok: rossijskie moryaki i puteshestvenniki v Avstralii [When the World was Wide: Russian Seamen and Travellers in Australia]*. Canberra: Alcheringa.

Grebenshhikova, G.A. 2010. *Linejnye korabli 1 ranga [Capital Ships of the Line] «Victory» 1765 «Royal Sovereign» 1786*. St Petersburg: Ostrov.

Grejg, Ol'ga. 2008. *Sekretnaya Antarktida, ili Russkaya razvedka na Yuzhnom Polyuse [Secret Antarctica, or Russian Reconnaissance at the South Pole]*. Moscow: Algoritm.

Khoroshevskij, A.Yu., ed. 2014. *M.P. Lazarev—Tri krugosvetnykh puteshestviya [M.P. Lazarev—Three Journeys Around the World]*. Moscow: Èksmo.

Kolesnikova, Maria. 2007. Rediscovering Antarctica: Nearly Two Centuries after Discovering Antarctica's Frigid Frontiers, Russia Is Hoping Its Oil-Forged Riches Can Help It Reclaim Its Position as the Southern Continent's Leading Explorer. *Russian Life*, January/February.

Kondakova, O.N. 2019. *K 200-letiyu otkrytiya Antarktidy Èkspeditsiej F.F. Bellinsgauzena i M.P. Lazareva na shlyupakh "Vostok" i "Mirnyj" [For the Bicentenary of the Discovery of Antarctica by F.F. Bellingshausen and M.P. Lazarev on Sloops Vostok and Mirnyj]*. St Petersburg: Russian Naval Archives. https://rgavmf.ru/virtualnye-vystavki/k-200-letiu-otrritiya-antarktidi.

Kopelev, Dmitrij. 2010. *Na sluzhbe Imperii: nemtsy i Rossijskij flot v pervoj polovine XIX veka [Imperial Service: Germans and the Russian Navy in the First Half of the 19th Century]*. St Petersburg: European University.

Koryakin, V.S. 2008. Kontinent, otkrytyj poslednim [The Last Continent to be Discovered]. In *Dvukratnye izyskaniya v yuzhnom ledovitom okeane i plavanie vokrug sveta*, ed. F.F. Bellinsgauzen, 5–38. Moscow: Drofa.

Korzun, V.A. 2009. *Otsenka vozmozhnostej ispol'zovaniya resursov Antarktiki [An Assessment of the Possibility of Exploiting the Resources of the Antarctic]*. Moscow: Institute for World Economy and Russia's International Relations.

Kotlyakov, V.M. 1994. *Mir snega i l'da [The World of Snow and Ice]*. Moscow: Nauka.

Kozlov, S.A. 2003. *Russkij puteshestvennik èpokhi Prosveshheniia [The Russian Traveller in the Age of Enlightenment]*. Vol. 1. St Petersburg: Istoricheskaya illyustratsiya.

Kudryashov, Konstantin. 2018. Antarktida nasha. Pochemu otkrytie kontinenta Bellingauzenom neosporimo? [Antarctica is Ours. Why is Bellingshausen's Discovery of the Continent Incontestable?], September 20. https://aif.ru/society/history/antarktida_nasha_pochemu_otkrytie_kontinenta_bellinsgauzenom_neosporimo

Kuroedov, V.I., and Others, eds. 2005. *Atlas Okeanov: Antarktika [Atlas of the Oceans: the Antarctic]*. St Petersburg: Ministry of Defence and AANII.

Kuznetsov, Nikita. 2020. *Russkaya Antarktika. 200 let. Istoriya v illyustratsiyakh [The Russian Antarctic: 200 Years of History in Pictures]*. Moscow: Paulsen—not seen.

Lajba, Anatolij. 2015. *Kvadratura polyarnogo kruga [Measuring the Polar Circle]*. Moscow: Poligraf-Tsentr.

Lazarev″, M.P. 1821. Pis′mo Mikhaila Petrovicha Lazareva k″ Aleksyeyu Antonovichu Shestakovu, 24 sentyabrya 1821 [A letter from Mikhail Petrovich Lazarev to Aleksyej Antonovich Shestakov, September 24, 1821 [O.S]]. St Petersburg: SARN—F–315, S–1, P–775, fos 1–6v.

Lazzara, Matthew A., and Others. 2012. Antarctic Automatic Weather Station Program: 30 Years of Polar Observations. *Bulletin of the American Meteorological Society* 93 (10): 1519–1537.

Lomonosov″, M.V. 1763. *Pervyya osnovaniya metallurgii, ili rudnykh″ dyel″ [Elements of Metallurgy, or Mining]*. St Petersburg: Imperial Academy of Sciences.

Lubchenkova, T.Yu. 2001. *Samye znamenitye puteshestvenniki Rossii [The Most Famous Russian Travellers]*. Moscow: Veche.

Lukin, Valerij. 2001. Interview. *Izvestiya*, September 6. http://izvestia.ru/news/251319.

———. 2005. Poisk nevedomogo kontinenta [The Search for the Unknown Continent]. *Vlast′* 10: 75–81.

———. 2006. Interview. Rossiya v Antarktide: vchera, segodnya, zavtra [Russia in Antarctica: Yesterday, Today, Tomorrow]. *Nauka i zhizn′*, February/March. http://www.nkj.ru/interview/2869/.

———. 2011. K 55-letiyu regulyarnykh otechestvennykh issledovanij Antarktiki [55 Years of Regular National Research in the Antarctic]. *Rossijskie Polyarnye Issledovanniya* 1 (3): 41–42.

———. 2013. Interview. Pod antarkticheskij led za novymi formami zhizni [In Search of New Forms of Life Beneath the Antarctic Ice]. *Russian Council on International Affairs Website*, May 16. https://russiancouncil.ru/analytics-and-comments/interview/pod-antarkticheskiy-led-za-novymi-formami-zhizni/?sphrase_id=49506273

———. 2014. Russia's Current Antarctic Policy. *The Polar Journal* 4 (1): 199–222.

———. 2017. Rossijskij vzglyad na budushhee razvitie Sistemy Dogovora ob Antarktike [A Russian Perspective on the Future of the Antarctic Treaty System]. *Rossijskie Polyarnye Issledovanniya* 3 (29): 42–43.

———. 2018. K voprosu o natsional′noj antarkticheskoj strategii [On the Question of a National Antarctic Strategy]. *Mezhdunarodnaya zhizn′* 1: n.p. https://interaffairs.ru/jauthor/material/1963.

————. 2020. Rossijskaya Antarkticheskaya Èkspeditsiya [The Russian Antarctic Expedition]. *Rossijskie Polyarnye Issledovanniya* 1 (39): 62–68.

Lukin, V., N. Kornilov, and N. Dmitriev. 2006. *Sovetskie i rossijskie antarkticheskie èkspeditsiyu v tsifrakh i faktakh (1955–2005gg) [Soviet and Russian Antarctic Expeditions in Figures and Facts, 1955–2005].* St Petersburg: AANII.

Mart'yanov, V.L. 2014. Perspektivy vosstanovleniya stantsii Russkaya kak postoyanno dejstvuyushhej rossijskoj antarkticheskoj stantsii [Prospects for the Reconstruction of Russkaya Station as a Continually Operational Russian Antarctic Station]. *Rossijskie Polyarnye Issledovanniya* 2 (16): 19–21.

Medvedev, Sergej. 2015. Dvuglavyj pingvin: zachem rossijskim politikam nuzhna Antarktida [Double-headed Penguin: Why Antarctica is Necessary for Russian Politicians]. *Forbes website,* January 27. https://www.forbes.ru/mneniya-column/tsennosti/278615-dvuglavyi-pingvin-zachem-rossiiskim-politikam-nuzhna-antarktida.

Moroz, V. 2001. *Antarktida: istoriya otkrytiya [Antarctica: The History of Discovery].* Moscow: Belyj gorod.

National Security Council. 1958. Statement of US Policy on Antarctica, March 8 1958. NSC 5804/1. In *Foreign Relations of the United States 1958–1960.* Vol. 2 (1991). Washington, DC: GPO.

Norchenko, Aleksandr. 2003. *Khronika poluzabytikh plavanij [A Chronicle of Half-Forgotten Voyages].* St Petersburg: Balt.

Novikova, Inna. 2003. Antarktida. Sekretnaya voennaya baza Tret'ego Rejkha khranit strashnye tajny [Antarctica: The Third Reich's Clandestine Military Base Holds Terrible Secrets]. *Pravda* website, January 16. http://www.pravda.ru/science/mysterious/past/16-01-2003/34653-antarctida-0/.

Novosil'skij, P.M. 1853c. *Yuzhnyj polyus″: iz″ zapisok″ byvshago morskago ofitsera [The South Pole: From the Memoirs of a Former Naval Officer].* St Petersburg: Vejmar″.

Order. 2005. Rasporyazhenie Pravitel'stva RF №713-r ot 2 iyunya 2005 g. [Order No. 713-r, Issued by the Government of the Russian Federation on June 2]. http://www.meteorf.ru/documents/9/64/.

————. 2010. Rasporyazhenie Pravitel'stva RF ot 30.10.2010 № 1926-r: O strategii razvitiya deyatel'nosti Rossijskoj Federatsii v Antarktike na period do 2020 goda i na bolee otdalennuyu perspektivu [Order No.1926-r, Issued by the Government of the Russian Federation on October 30 2010: A Strategy for the Development of the Work of the Russian Federation in the Antarctic until 2020 and in the Longer Perspective]. http://science.gov.ru/media/files/file/9odRuhiwbYKHvR2U2zNAClmaxCDUWQgA.pdf.

Ostrovsky, Arkady. 2015. *The Invention of Russia: The Journey from Gorbachev's Freedom to Putin's War.* London: Atlantic.

Ovlashhenko, Aleksandr. 2013. *Materik l'da: pervaya russkaya antarkticheskaya èkspeditsiya i eyo otrazhenie v sovetskoj istoriografii (1920-e–1940-e gody) [The*

Continent of Ice: The First Russian Antarctic Expedition and its Footprint in Soviet Historiography (1920s to 1940s)]. Saarbrücken: Palmarium.

———. 2014. *Antarkticheskij rubikon: tema otkrytiya Antarktidy v sovetskikh istochnikakh nachala 50-kh godov [Antarctic Rubicon: The Discovery of Antarctica as a Theme in Soviet Sources from 1950]*. Saarbrücken: Palmarium.

———. 2016. *Antarkticheskij renessans: provedenie pervykh kompleksnykh antarkticheskikh èkspeditsij i problema otkrytiya Antarktidy [Antarctic Renaissance: The Arrival of the first Combined Antarctic Expeditions and the Problem of the Discovery of Antarctica]*. Saarbrücken: Palmarium.

Pleshakov, A., Vvedenskij, È., and E. Domogatskikh. 2018. *GDZ Geografiya 5 klass* [Geography Homework for Class 5]. https://shkola.center/5-klass/geografiya-5/page,19,32-gdz-geografiya-5-klass-uchebnik-vvedenie-v-geografiyupleshakov-a-vvedenskiy-e-domogackih-e-2018.html

Potapov, R.L. 2012. Popugaj, kotoryj zhil s pingvinami [The Parrot that Lived with Penguins]. *Priroda* 11: 89–96.

Rokot, V. 2016. Transcript of Interview on Radio Mayak, Moscow, May 28. https://radiomayak.ru/shows/episode/id/1303503/.

Rusakov″, V. (pseudonym of S.F. Librovich). 1903. *Russkie Kolumby i Robinzony [Russian Columbuses and Crusoes]*. Moscow: Vol'f.

Savatyugin, L.M., and M.A. Preobrazhenskaya. 1999–2009. *Rossijskie issledovaniya v Antarktike [Russian Exploration in the Antarctic]*. Vol. 4. St Petersburg: Gidromet.

———. 2014. *Karta Antarktidy: imena i sluzhby [The Map of Antarctica: Names and Service Records]*. St Petersburg: Geograf.

Sementovskij, V.N., ed. 1951. *Russkie otkrytiya v Antarktike [Russian Discoveries in the Antarctic]*. Moscow: Geografgiz.

Sergeev, Dmitrij. 2016. Bitva za budushhee: zachem rossijskij flot vozvrashhaetsya v Antarktidu [Conquering the Future: Why the Russian Navy is Returning to Antarctica]. *Teleradiokompaniya Zvezda*, January 25. http://tvzvezda.ru/news/forces/content/201601250744-zkl7.htm.

Sergeeva, Ol'ga Anatol'evna. 2003 (updated 2010). *Antarktida [Antarctica]*. http://festival.1september.ru/articles/582633/.

Shmatkov, Vladimir. 1998. Zagadki, gipotezy, otkrytiya: Pochemu Kuk ne otkryl Antarktidu? [Puzzles, Hypotheses and Discoveries: Why didn't Cook Discover Antarctica?]. *Vokrug sveta* 7: 23–24.

Shvede, E.E. 1949. Pervaya russkaya antarkticheskaya èkspeditsiya 1819–1821 gg. [The First Russian Antarctic Expedition 1819–1821]. In *Dvukratnye izyskaniya v yuzhnom ledovitom okeane i plavanie vokrug sveta v prodolzhenie 1819, 20 i 21 godov*, ed. F.F. Bellinsgauzen and E.E. Shvede, 7–30. Moscow: Geografgiz.

———. 1962. F.F. Bellinsgauzen. In *Lyudi russkoj nauki: geologiya i geografiya*, ed. I.V. Kuznetsov, 419–431. Moscow: Physics and Mathematics Press.

Simakova, Lyudmila. 2015. *Aleksandr Kuchin: Russkij u Amundsena [Alexander Kuchin: A Russian with Amundsen]*. Moscow: Paulsen.

Siryj, S.P. 2010. Krugosvetnoe plavanie kapitana 2 ranga F.F. Bellinsgauzena i lejtenanta M.P. Lazareva na shlyupakh "Vostok" i "Mirnyj" i otkrytie Antarktidy [The Circumnavigation by Junior Captain F.F. Bellingshausen and Lieutenant M.P. Lazarev in the Sloops *Vostok* and *Mirnyj*, and the Discovery of Antarctica]. https://rgavmf.ru/sites/default/files/lib/siry_antarctida.pdf.

Smirnov, V.G. 2005. *Ot kart vetrov i techenij do podvodnykh min [From Wind and Current Charts to Underwater Mines]*. St Petersburg: Gidromet.

———. 2006. *Neizvestnyj Vrangel' [The Unknown Wrangell]*. St Petersburg: Gidromet.

Telitsyn, Vadim. 2013. *Gitler v Antarktike [Hitler in the Antarctic]*. Moscow: Yauza.

Timofeev, P. 2006. Pokorenie nichejnogo materika [The Conquest of No-man's Continent]. *Vash tainyj sovetnik* 6: 22–23.

Tsuker, Viktor. 2015. Vydayushhijsya vklad moryakov rossii [A Notable Contribution from Russian Seamen]. *Vesti morskogo Peterburga* 1: 44–45.

Udintsev, G. 2002. Pervootkryvatelyam Antarktidy [For the First Discoverers of Antarctica]. *Moskovskij zhurnal* January 12: n.p. http://ruskline.ru/monitoring_smi/2002/12/01/pervootkryvatelyam_antarktidy/.

Unkovskij, S.Ya. 2004. *Zapiski moryaka 1803–1819 gg. [Memoirs of a Seaman 1803–1819]*, ed. L. Zakovorotnaya, Moscow: Sabashnikovy.

Usov, Aleksej. 2010. Putin khochet sdelat' Antarktidu russkoj koloniej: $2 milliarda ot Minprirody uplyvut k pingvinam [Putin Wants to Turn Antarctica into a Russian Colony: Minprirody to Waste $2 Billion on Penguins]. *Novyj Region*, October 22. http://vlasti.net/news/107204.

Utusikov, Yu.D. 2012. *Ot Obskikh beregov do mostika «Obi» [From the Shores of the Ob' to the Bridge of the Ob']*. St Petersburg: Morskoe nasledie.

Vaisala. 2010. Automatic Weather Reports from Antarctica. https://www.vaisala.com/sites/default/files/documents/MET%20Antarctica%20success%20story%20B210957EN-A.pdf.

Vdovin, A.I. 2008. Bor'ba s nizkopoklonstvom i kosmopolitizmom v poslevoennye 1940-e gody: prichiny, posledstviya, uroki [The Struggle Against Kowtowing and Cosmopolitanism in the Postwar 1940s: Causes, Results and Lessons]. In *Sluzhenie otechestvu: russkaya traditsiya i sovremennost'*, ed. Evgenij Troitskij, 247–258. Moscow: Granitsa.

Volkov, Dmitrij. 2003. Est' gory kruche Èveresta [There are Mountains Higher than Everest]. *Izvestiya*, February 11. https://iz.ru/news/272868.

Walker, Matt. 1999. Back from the Dead. *New Scientist 2202*, September 4. https://www.newscientist.com/article/mg16322020-200-back-fromthe-dead/.

CHAPTER 12

Conclusions

HISTORIOGRAPHIC SUMMARY

This book has chronicled the changing fortunes of the first Russian Antarctic Expedition in the perceptions and writings of Russian historians and other commentators, and identified the various setbacks that befell Russian knowledge and evaluation of Bellingshausen's achievements over the past 200 years. The first such handicap was the ignorance of the Antarctic that was unavoidable for such a daring early venture deep into the region, but was perhaps made worse by the misleading doctrine of Buffon, to the effect that land was not required for the formation of even the mightiest polar ice cap. No scientist himself, Bellingshausen appears to have been influenced by the latest scientific thought, which was taken on the expedition. Despite the fact that he also had with him Cook's ideas to the contrary (Cook 1777, 2: 240–241), both in English and in translation, the combination of Buffon's influential theory and his own characteristic exactitude may have caused him to interpret the massive, land-like body of ice, sighted on 17 February 1820, as part of a marine polar ice cap with no implications for the presence of land. A second early setback was the failure to preserve the logbooks and journals of the expedition; the author has speculated that the Great Flood which caused extensive damage to St Petersburg in November 1824, wrecking the expedition's ships

© The Author(s), under exclusive license to Springer Nature Switzerland AG 2021
R. Bulkeley, *The Historiography of the First Russian Antarctic Expedition, 1819–21*,
https://doi.org/10.1007/978-3-030-59546-3_12

Vostok and *Mirnyj* along with many others, may also have destroyed some of their documents (Bulkeley 2014a: 19).[1,2]

There were several misapprehensions about the expedition in the nineteenth and early twentieth centuries, both in Russia and abroad, but most were unimportant. Simonov's language should have been clearer in respect of southings, and Novosil'skij contradicted himself by saying that Trinity Land was part of the same mainland as the Alexander I Coast and that Bransfield saw that mainland in 1820, but then also claiming, for a year or two, that the Russians had made the first discovery of the mainland in 1821. Coming from eyewitnesses, those were influential confusions.

By contrast with Novosil'skij, Rabinovich's claim, in 1908, that the discovery of the Alexander I Coast in January 1821 had been the first sighting of the mainland, was justified in the light of contemporary geographical ideas. It remained so until 1936, when it was overtaken by new geographical findings. That did not prevent some commentators from persisting in it (Vvedenskij 1941; V. Vorob'ev 1948). Rabinovich's interpretation remained influential despite radical changes in the relevant geographical picture, and even after the official makeover of the expedition.

In February 1949, at the behest of the Central Committee of the Soviet Communist Party, the expedition's headline achievement was altered from discovery of the first land south of the Antarctic Circle in January 1821 to discovery of the continent of Antarctica in January 1820. In Chaps. 5 and 6 that transformation was placed in the context of new ideas about Antarctica and new knowledge of the Antarctic Peninsula and Alexander I., together with Soviet resistance to the Antarctic policy of the United States. The difficulties that were encountered by the new historical assessment from the outset, including dissent from several Soviet authors, were illustrated in Chap. 8.

Had it been possible to conduct it honestly and rationally, such a reassessment of the expedition would have been an appropriate response to the first roughly complete picture of Antarctica. From the testimony of Soviet historians they were greatly influenced by Debenham's suggestion

---

[1] For the system of romanization used, the rendering of dates, and a few other matters, please consult the Apparatus at the end of the book.

[2] In 1990 Barratt suggested that further documents created by officers in Bellingshausen's expedition might yet be found in Russian archives (Barratt 1990: 97). Some of Bellingshausen's early and unimportant voyage reports, previously unknown, and duplicates of reports already known, have recently been published online by the Naval Archives (Kondakova 2019), but little of major significance.

that, in addition to his valuable 1821 discoveries, Bellingshausen might have seen the land ice which makes up much of the shore of Antarctica in 1820. In an ideal world, Russian experts would have relieved Debenham of the ignorance which caused him to render a standard phrase for 'icebergs' as 'ice-covered mountains' in a key passage (Bellingshausen 1945, 1: 129), and Debenham would have reciprocated by explaining in greater detail why he viewed the approach of 17 February 1820 as a definite sighting, but not that of 28 January, a judgement with which the present author concurs.

In the mid-twentieth century, however, the cultural practice and international relations of the Soviet Union were far from ideal, a situation which was by no means entirely of Soviet making. The historians and other commentators who upgraded Debenham's tentative and quickly retracted suggestion into the absolute, Bransfield-beating priority of 28 January 1820 did not enjoy such luxuries as freedom of thought and expression. Soviet understanding of the expedition was seriously compromised by the government-sponsored pronouncement from the Geographical Society, in February 1949, in favour of that poorly supported reading of events, and by its entrenchment as dogma from 1950 onwards, as the supposed priority became enshrined in Soviet Antarctic diplomacy. From that point all citizens of a state founded on Marxism-Leninism were expected to conform to the view endorsed by the Central Committee of the Communist Party. Indeed, there being no 'class basis' for dissent, as Georgij Maksimilianovich Malenkov (1901–1988), explained to the 19th Party Congress (Malenkov 1952), the official opinion on such matters *was* everyone's opinion, whether or not they personally agreed with it.

Examples were found of Soviet authors who either stood by the pre-1949 assessment, that Bellingshausen's first discoveries south of the Circle came in 1821, or who at least treated the glorification of 28 January 1820 with conspicuous reserve. But open scholarly discussion, about whether the case for absolute priority was or was not conclusive, was out of the question. The condemnation of Andreev's work on Peter the Great demonstrates the tendency of the Soviet authorities to value a historical investigation less for its soundness than for the utility of its conclusions and the chauvinistic style of its presentation.

In the early 1960s Belov and Lebedev began, very cautiously, to open the subject up by mentioning that some primary sources remained unexamined and by publishing Bellingshausen's superficially negative but in reality innocuous Remark, about seeing no signs of land in 1820, for the first time in almost 140 years. Their separate failures to place the 1820 discovery claim on an unassailable basis of evidence and interpretation were explained as only partly due to unsound arguments; another factor, the paucity of surviving eyewitness accounts of the voyage, had spurred them on to push linguistic interpretation and factual reconstruction beyond reasonable bounds. As Brezhnev's project to restore Soviet faith in the official version of Communism took hold, even the small spark of intellectual freedom regarding the expedition kindled by those two scholars appears to have been extinguished. The present author doubts very much, however, whether he would have found the courage to take the first small steps away from orthodoxy that they attempted, if he had been in their shoes. For whatever reason, and with a handful of honourable exceptions, original research or commentary on the expedition became rare in the last 25 years of the Soviet Union, but some variety of views could still be found. Reversals of the expedition's claimed discovery of Antarctica, from 1820 back to the older, 1821 version, were published both quite early and quite late in that period.

On the face of it, post-Soviet Russian commentators on the expedition have enjoyed greater freedom, and with it less cause for irrationality, than their Soviet predecessors.[3] But any attempt at forming a general opinion of modern Russian treatments of the subject is frustrated by a single fact. Since 1990 no professional historian inside Russia has published anything about the main historical questions which surround the expedition. Over the past seven years none has responded, for example, to the author's presentation, in Russian and in the Academy's leading journal of the history of science, of the case for doubting the reliability of Lazarev's letter, the single account of a single day, 28 January 1820, on which the entire case for Russian priority depends (Balkli 2013).[4]

---

[3] It might however be argued that since 1991 loyalty to the Motherland (*Rodina*), already an important element in 'correct thinking' during the Soviet period, has replaced loyalty to the Communist Party in Russian culture. Malenkov's dictum could thus be reworded to say that there is currently no 'patriotic basis' for Russians to query their country's supposed priority in Antarctica. That was certainly the impression given by one Russian publishing house, when they refused to have anything to do with this book, unread, simply because it raises such doubts.

[4] It is regrettable that the journal chose not to publish a rebuttal submitted by Dr Ovlashhenko. However, if that text was anything like the critique of (Balkli 2013) which he published shortly afterwards (Ovlashhenko 2014: 447–480), it may have exhibited a reluctance to address the main issue, namely the factual reliability of Lazarev's letter (Chap. 7), as

In the present century, and despite the near-total absence of any restate-ment of its intellectual grounds, the 1949 Soviet reassessment has been revived strongly, both on the internet and in the still scanty published lit-erature, which mainly consists of republications of Bellingshausen's book. There is in practice less variety of opinion about the expedition today than under the Soviet Union. Lastly, a possible connection was found between that development and the, as yet unofficial, demand that Russia should receive a disproportional share in the economic resources of the Antarctic, as climate change may render them more accessible.

A LITANY OF SHORTCOMINGS

The chief defects in Russian, and especially in Soviet, treatments of the expedition over the past 70 years come down to these:

- *Misuse of sources.* Cuts were made in primary sources, sometimes without signalling that this had been done (Andreev; Samarov). Quoted texts were selected or rearranged to hide what was not required, or to force them to say what was—see for example Vvedenskij with Cook and Koryakin with Lazarev. Some primary sources, such as Bellingshausen's final report, Simonov's main scien-tific report, or the extracts from his diary, were largely or completely disregarded. After the early 1950s few scholars apart from Belov and Kuznetsova did much archival research. That reflected a general lack of interest in the details of the voyage, from which important docu-ments have yet to be published and analysed in Russia. In some ways the most neglected source of all was Lazarev's letter, since neither its authenticity nor its reliability was ever examined.
- *Neglect of the cultural background.* The universal Russian neglect of proto-glaciological European thought, beyond Lomonosov, might be overlooked as petty parochialism if it had not led to a more seri-ous error, the failure to understand that there was more than one theory about the origins of polar ice caps for Bellingshausen to choose from, and that he opted for the latest, which happened to be mistaken.

---

well as other idiosyncratic features, such as a tendency to regard sympathetic commentators as primary sources (the fallacious argument from authority).

- *Indifference to foreign texts.* Over the past 200 years, very few Russian commentators on the expedition have been familiar with relevant foreign texts, such as accounts of the Bransfield expedition down to and including the latest compilation of original sources (Campbell 2000). Some, including Novosil'skij, Shokal'skij, V. Lebedev, Tryoshnikov and Belov, did better than others, but especially during the Soviet period that was often a case of doing somewhat better, with much still wanting.

- *Revisions of fact.* This heading covers the elevation of Lazarev to co-commander; doubling the number of defaulting German naturalists from one to two; Belov's revisions of the voyages of Bransfield and Palmer; turning Lazarev's informal letter, written after the voyage, into a shipboard journal (Tsuker, Krechetnikov); Lukin's displacement of the Bellingshausen–Palmer meeting; Fradkin's false but thereafter endlessly repeated claim about J.C. Ross's views on Bellingshausen; the headlining of '29 islands' rather than the more accurate tally of 18, etc.. Some of the textual cuts suppressed unwanted facts such as the extent of British assistance at Sydney. The repeated upgrading of supposed proximity into sighting and discovery could fall under this or other headings.

- *Resort to fiction.* The insertion of imagined speech and action into purportedly factual accounts amounted to a special form of factual revision.

- *Cosmetics.* Awkward facts, such as Bellingshausen's Remark or Simonov's failure to publish most of his data, were frequently suppressed or glossed over.

- *Fallacious reasoning.* Despite the widespread provision of courses in logic within the Soviet educational system, many commentators from the expedition's Soviet heyday appear to have been unaware that the arguments from authority, from popularity, and from personal characteristics of the opponent are fallacious modes of reasoning. The commonest fallacy was the red herring (*ignoratio elenchi*), in which evidence for uncontested facts, such as the sighting of 17 February 1820 or Bellingshausen's admirable skill and courage as a navigator, was presented as if it supported a sighting on 28 January 1820, which it did not. In the matter of the South Shetlands even a competent scholar like Alejner was obliged to resort to the fallacy of false comparison. But in the end, rhetoric is no substitute for reason.

- *Conformity with political demands.* As already said, although this is a blemish in Soviet treatments of the expedition in particular, the blame lay not with the historians, but with their rulers. The Soviet 'command research' model was a regrettable attitude for a modern state to adopt towards historical research. It was also pointless, because it invalidated the process and its outcomes. As for prestige, that too was sacrificed, as the Vatican learned in the case of Galileo, a supposedly favourite example for Soviet ideologues.

It is to the great credit of Soviet historians that they achieved so much in their authoritarian circumstances. Nevertheless those who worked with and misused the original sources for the voyage, the historians, bear some of the responsibility for the sorry state of mythopoeic confusion to which the achievements of the Bellingshausen expedition have now been reduced. It is small wonder that tertiary commentators, such as Krechetnikov, Lukin, and Tsuker, doubtless assuming that so much boastful certainty was founded on solid evidence, should continue mistakenly to upgrade what are frequently unclear or unreliable texts into the sort of material, such as logbooks, that might rationally be expected to have underpinned the findings of Soviet historians.

CONCLUSIONS

The final verdict on the Soviet claim that the Bellingshausen expedition was the first to sight the Antarctic mainland can only be a qualified one. On the one hand, it was concocted and subsequently maintained by the regrettable, sometimes even deplorable, methods which have been examined here. On the other hand, until someone invents a time machine the paucity of surviving records from the expedition leaves open the very improbable possibility that Bellingshausen, Novosil'skij, and Simonov irresponsibly ignored on 28 January 1820 the spectacular phenomenon of a massive, land-like body of ice which the first two of those witnesses recorded for 17 February 1820. If and only if that were the case, against all reasonable odds, there would be a very small chance that Lazarev's account of the day, in an informal and sometimes inaccurate letter, was more reliable than those of his comrades. In short the Russian discovery of Antarctica is a demonstrable fake, even if it came about, at least in part, through some sort of muddle. (One can almost hear the experts, back in 1948, wishing they had put things more clearly.) But while that can eliminate the priority claim from the sum total of what rational people should believe about the past, it can never place it completely beyond the limits of possibility.

Between 1949 and 1970, the brief heyday of research on the expedition, those Soviet historians and commentators who tackled the subject may well have felt that they were being required by their government to manufacture the bricks of priority without the needful straw of evidence, or at least without enough straw. Their lack of success points up the difficulty they faced. At the same time, they could only be as honest, as open, and as rational as the foregone patriotic conclusion to their work permitted. Although the constraints are no longer the same, a modern Russian author setting out to breach the lockstep of national glory, by finding against the 70-year-old claim that their countrymen discovered Antarctica, would probably incur widespread opprobrium. While the overall decline of scholarly interest in the expedition over the past half-century is disappointing, therefore, not least because it leaves the way clear for irrational belief in Russian priority to flourish unchecked, it is hardly surprising that scholars have gravitated towards topics in maritime history which are free from such pressures.

Perhaps the most interesting finding yielded by this survey has been the broad correlation, in Russian treatments of the Bellingshausen expedition over the past 70 years, between what is sometimes referred to as 'firm government' and acceptance of the revised assessment. At first glance it seems that, during the final years of Stalin, the early years of Brezhnev, and the current Putin era, the claim that the expedition discovered Antarctica on 28 January 1820 has been more firmly lodged both in public opinion and with historians and other commentators than during the intervals between those periods.

That the claim has relied on such external, non-rational support should count against it in the scales of truth. But the correlation is not a simple one, and any counter-evidence must be weighed carefully. The first objection arises from the dissent that was either voiced or tacitly signalled during the first period of firm government, under Stalin. Several points can be made in reply. First, the seemingly reluctant protagonists of the Bellingshausen makeover would have been very familiar with the long-accepted previous assessment, that the expedition's most important achievements were its Antarctic discoveries in January 1821, whether or not they should be deemed the discovery of Antarctica as Rabinovich had urged. The figurehead for the new assessment, Berg, was himself on record to that effect 20 years earlier. Next, the revised version of the expedition was asserted on the basis of very little evidence for about two years, before someone noticed Lazarev's letter. And lastly, doubtless constrained by its official character, none of the dissenters rejected the revision outright. The combination of lip-service to the new line with veiled dissent confirms, rather than disproves, the role of government in establishing the 1820 discovery claim.

Another possible counter-example to the firm government hypothesis is the ardent commitment to the 1820 discovery story of two of its most productive proponents, V. Lebedev and Belov, working during a period of slightly less firm government, the Khrushhev years. But first, the hypothesis does not require that commentators should only have taken the official line in firm government periods. It is more a question of whether awkward parts of the evidence, such as Bellingshausen's Remark, could be aired in public. Lebedev certainly did that, though Belov only referred to it indirectly. And second, two such historians, one amateur and the other professional, were unlikely to put as much effort as they did into their work on the expedition without feeling that it was important to do so.

It must be admitted, however, that the clearest confirmation of the firm government hypothesis lies in the contrast between the wavering of the late Soviet period and the certainties of the Putin era, disregarding the years between 1990 and 2000 because nothing new was published on the subject then. Today, both state control of education and the information media, and the length of time—two generations—which has by now elapsed since the original view, that the expedition achieved its most important discoveries in January 1821, last prevailed—may contribute to a greater conformity with the 1949 revision than was detected in the Soviet period.

On the other hand, this book should not close without acknowledging recent signs that a move away from the nationalist dogma of absolute priority may be under way, at least amongst experts. The neutrality of the Naval Archives' online exhibition has already been remarked (Kondakova 2019). The expedition's page in the Russian edition of Wikipedia, entirely written by Russian or Russian-speaking editors, draws extensively on the work of non-Russian scholars such as Barratt, Tammiksaar, and the present author. On the question of names in the South Shetlands, the RAE does not endorse those applied to the main group by Bellingshausen in 1831, and is even the nominating party to the SCAR Gazetteer for the English name of one of the largest, Livingston. They do still stand by the

Fig. 12.1 Plaques on the monument at Russia's Bellingshausen Antarctic research station on King George Island, South Shetlands, dedicated in January 2020. (Photograph by Timo Palo and published with his permission)

Russian names for the Elephant and Clarence group, however, which is unfortunate because as we have seen that claim is valid only for Rozhnov/ Gibbs. Lastly, the official plaque on the new monument to Bellingshausen at the eponymous and much-visited Russian Antarctic research station (Fig. 12.1) describes him as 'one of the first to see Antarctica', rather than the very first, just as Sergej Markov did 68 years earlier (Chap. 8). Anyone who knows anything about Antarctic history would agree with that statement without hesitation. A divergence may be slowly opening, therefore, between the informed opinion of some (but not all) Russian experts and the maintenance by some of its custodians, such as the Yeltsin Presidential Library or the *Russia 1* television channel, of the Soviet claim to a discovery on 28 January 1820.

* * *

The hypothesis has been advanced, in Chaps. 10 and 11, that the underlying cause for the general conformity of modern Russian opinion with the 1820 discovery claim is that it matters even more today than it did in the Soviet period—not that it receives much careful attention as a result. Having finally abandoned the Soviet project Russia, like many other countries including Britain, finds few objective grounds to justify its existence beyond the fact itself. In those circumstances, and also like other countries, positive stories about the national past are deployed to create and uphold what in the Russian case is known as the *Rodina* (Motherland).[5] The heavy economic and personal costs of forsaking the Soviet framework have combined with the brief and troubled history of the new nation to inject great urgency into forging a stable national identity from the available material. Suitable components include the two 'patriotic' (because largely defensive) wars against Napoleon and Hitler, and selected cultural achievements from the many to which Russia can lay claim. Amongst the latter, however, the tale of the discovery of Antarctica is perhaps easier to imprint in the minds of young children than the work of, say, Pushkin or Lomonosov. Furthermore one has only to think of comparable Soviet achievements in the early years of space exploration to see that the Bellingshausen expedition's status as a pre-Soviet, simple, but also broadly *peaceful* achievement makes it fairly special, even though, unlike the reigns of Ivan the Terrible, Peter the Great, and Alexander III, and several other topics in the syllabus, it contributed little to building or maintaining the old, pre-Soviet nation.

This 'Motherland' or 'crown jewels' interpretation of the altered role of the supposed discovery of Antarctica in the present century might be thought to imply the corollary that, if its former foreign policy role were to be revived in advance of the 2048 review of the Antarctic Treaty's Protocol on Environmental Protection, then Russian perceptions of the expedition would begin to vary, as they did in the Soviet period. But that does not really follow, because individual memes can play more than one role. Thus for example the decisive victory achieved, at an immense sacrifice, by the Soviet people in World War II forms both a major component in the modern national identity and an asset to be deployed in Russia's European policy. If that is possible for such a central crown jewel, how much more so for a relatively minor gem like the Bellingshausen expedition, located as it is somewhere towards the back of that metaphorical headpiece? Such a duality was indeed detected in Russia's current Antarctic

---

[5] The 'Third Rome' perspective on Russia's destiny espoused by some intellectuals in 'Holy Russia' (which extends beyond the boundaries of the modern state) is not so much an alternative to this approach as a special case of it (Saulkin 2018).

policy (Chap. 11), but the dominant theme in popular treatments, including school syllabuses, remains that of national glory. However both the recent displacement of the discovery of Antarctica meme, away from foreign policy and towards domestic ideology, and any future reversal of that change are only adjustments in the balance between those two aspects, both of which have been present since 1949. All that is being claimed here is that any such shift may affect the frequency of dissent from the meme in Russia, not only because of official constraints but also because the level of public tolerance for nonconformity may fall when the needs of the Motherland take priority.[6]

The last point to be made about this proposed explanation for the modified status of the supposed discovery of Antarctica in modern Russia is that it is not tied to the formal ending of the Soviet Union in 1991 or any other date marker. As historians have long understood, changes in human society, including cultural changes, often get under way before they become visible on the surface of events. In the case of the Bellingshausen expedition an increased emphasis on its Motherland role, which was always there anyway, may well date back to the Brezhnev era, when critical analysis began to dwindle although expressions of dissent still occurred.

* * *

The bottom line to the long sequence of Russian commentary that has been set out in this book is that Russian priority in discovering Antarctica has never been proved, because the available evidence is insufficient to prove it. Lazarev's letter to Shestakov is certainly an important document, but its credentials are weakened by the presence of errors of fact, and by the absence of any confirmation from other sources that the expedition sighted an ice coast of the Antarctic mainland on 28 January 1820, the date required for absolute priority. The hypothesis that Bellingshausen's editor allowed such a sighting on 17 February 1820 to remain in *Two Seasons*, but arbitrarily removed a similar description of 28 January, is too

---

[6]An alternative explanation, or a refinement of this one, might be that the two requirements from the supposed discovery of Antarctica meet the needs of different groups, with its foreign policy role valued by the political class and its Motherland role valued both by that group and by such members of the public (probably a minority) as recollect the tale of discovery from their schooldays.

contrived to be plausible. That being the case, the four other surviving descriptions of 28 January 1820, two of them written on board *Vostok* and a third, *Two Seasons*, compiled from shipboard records, which together describe that encounter as a threatened embayment by solid pack ice, have the stronger claim on our acceptance. As stated above, it all comes down to a balance of probabilities. It is more likely that Lazarev made a minor error, from his point of view at the time, than that Bellingshausen gave such a detailed and emphatic description of the land-like ice cliffs seen on 17 February 1820, but stayed silent about a similar phenomenon encountered 20 days earlier.

The Russians did see the ice coast of Queen Maud Land on the later date, well south of the Antarctic Circle, whereas the rocky peaks at the tip of the Antarctic Peninsula, seen by Bransfield and Smith and their shipmates on 30 January, lie to its north. Overall, the aims and achievements of the two expeditions were quite different. Bransfield and Smith made a thorough survey of the South Shetland Is. in little more than two months, during a voyage lasting just under four months. Bellingshausen spent about two months in or close to the ice fields in his first Antarctic season alone, and his voyage lasted just over two years. The British expedition (commanded by an Irishman) achieved regional significance by adding completeness and precision to Smith's breakthrough discovery of the islands in 1819. The Russian expedition (commanded by a Baltic German) attained global significance by being the first to sight the ice coast of Queen Maud Land in February 1820, by discovering the first land south of the Circle in January 1821, and by boldly enduring and recording the ice fields for long periods of time. Those achievements are undeniable, regardless of the distracting pursuit of continental priority. This book, however, has been less concerned with the history of the expedition than with the extraordinary measures adopted by Russian historians and other commentators in order to support the claim to first discovery that was launched by the Soviet government in 1949.

The academic silence in modern Russia, with respect to the expedition, resembles one of those awkward football matches in which one team leads by a goal which was only awarded because of a lamentable error by the referee, an error which every fair-minded spectator would prefer had never happened. The mistake cannot be remedied, and the lucky (or some might say, unlucky) team may win the match. But the rules of knowledge are not like those of football. If one part of an otherwise competent body of work was deficient, it is just as necessary and just as possible to re-examine it as it would be if new data had been discovered or a seminal new hypothesis proposed. And there is no final whistle.

As for the future, the Bellingshausen expedition has become something of a miner's canary for Russian culture. Much work would be needed for Russian scholars to re-evaluate their country's Cold War claim to priority in discovering Antarctica. But only time will tell whether an open-minded, democratic discussion will develop, in which all the surviving sources are re-examined, or whether reasons of state and deference to authority, including the tyranny of internet-fuelled public opinion, will prolong the present 'post-truth' situation indefinitely. In principle, the dodgy methods sometimes used to uphold the priority claim (Belov; Koryakin) should undermine themselves by exhibiting their authors' own lack of faith in the available evidence, and so provide the impetus for re-opening the question. There are now some welcome signs that such a transformation has begun, although the concern remains that for some voices rationality is still not an important consideration.

Even if no widespread change develops, it should be stressed that with very few exceptions living Russians are more victims in this matter than offenders. Perhaps they should be, or become, more critical of what everyone gets taught about the expedition at school, but how many people anywhere live lives of constant self-examination? After all, Pierre Bezukhov is a character in fiction. All the same, with the possible exception of Ballinacurra, County Cork, the village where Bransfield was born, Russia today is the only part of the world where the small differences in date between equally imperialist British and Russian sightings of the Antarctic mainland, 200 years ago, matter in the slightest to anyone—all thanks to a minor episode in the Cold War. Russia is also so unfortunate as to be the only place where some people cherish the illusion that Antarctica belongs exclusively to them. The sooner such unnecessary miseries are ended the better, but they can be ended only by the sufferers themselves.

The question of which expedition first sighted the ice-bound coast of Antarctica is a trivial one, but the question of whether Russian historians have examined it honestly and rationally is not. By dint of insistence and repetition, and using often faulty arguments to which most non-Russians do not have access, the 70-year-old Russian bluff which asserts that priority has, for many people outside Russia, overborne the scepticism with which it was formerly received (Armstrong 1951, 1971). It is high time for that bluff to be called, before it reaches its centenary and is presented, perhaps, during the revision process for the Protocol on Environmental

Protection in Antarctica. To borrow a phrase from Cook, the author is under no illusion that this book will put a final end to deep-rooted Russian superstitions about the Bellingshausen expedition, least of all the cherished myth of Russian priority in discovering Antarctica. But hopefully a few readers, both in Russia and elsewhere, will welcome any light that has been shed here on a matter that has for too long confused and corrupted the early history of Antarctica.

REFERENCES[7]

Armstrong, Terence. 1951. Four Eye-Witness Accounts of Bellingshausen's Antarctic Voyage of 1819–21. *Polar Record* 6 (41): 85–87.

———. 1971. Bellingshausen and the Discovery of Antarctica. *Polar Record* 15 (99): 887–889.

Balkli, Rip. 2013. Pervye nablyudeniya materikovoj chasti Antarktidy: popytka kriticheskogo analiza [First Sightings of the Mainland of Antarctica: Towards a Critical Analysis]. *Voprosy istorii estestvoznaniya i tekhniki* 4: 41–56.

Barratt, Glynn. 1990. *Melanesia and the Western Polynesian Fringe*. Vancouver: University of British Columbia Press.

Bellingshausen, Captain. 1945. *The Voyage of Captain Bellingshausen to the Antarctic Seas, 1819–1821*. ed. Frank Debenham, 2 vols. London: Hakluyt Society.

Bulkeley, Rip. 2014a. *Bellingshausen and the Russian Antarctic Expedition, 1819–21*. Basingstoke: Palgrave Macmillan.

Campbell, R.J. 2000. *The Discovery of the South Shetland Islands*. London: Hakluyt Society.

Cook, James. 1777. *A Voyage Towards the South Pole, and Round the World; Performed in His Majesty's Ships the Resolution and Adventure, in the Years 1772, 3, 4, and 5*. 2 vols. London: Strahan and Cadell.

Kondakova, O.N. 2019. *K 200-letiyu otkrytiya Antarktidy Èkspeditsiej F.F. Bellinsgauzena i M.P. Lazareva na shlyupakh "Vostok" i "Mirnyj"* [For the Bicentenary of the Discovery of Antarctica by F.F. Bellingshausen and M.P. Lazarev on Sloops Vostok and Mirnyj]. St Petersburg: Russian Naval Archives. https://rgavmf.ru/virtualnye-vystavki/k-200-letiu-otrritiya-antarktidi.

---

[7] Key to archival references: SARN = State Archives of the Russian Navy, F = Fond, S = Series [*Opis'*], P = Piece [*Delo*], fo./fos = folio/s [*list/y*], v = verso [*oborotnoe*]. (The latter is necessary because verso pages were usually unnumbered.) Dates on manuscript documents are given in Universal Time (CE).

Malenkov, G.M. 1952. Otchyotnyj doklad XIX s″ezdu VKP(b) 5 oktyabrya 1952—vyderzhka [Report to the 19th Congress of the Soviet Communist Party, October 5 1952—extract]. In *Stalin i Kosmopolity*, A.A. Zhdanov and G.M. Malenkov, 2012. Moscow: Algoritm.

Ovlashhenko, Aleksandr. 2014. *Antarkticheskij rubikon: tema otkrytiya Antarktidy v sovetskikh istochnikakh nachala 50-kh godov [Antarctic Rubicon: The Discovery of Antarctica as a Theme in Soviet Sources from 1950]*. Saarbrücken: Palmarium.

Saulkin, V.A. 2018. Tretij Rim protiv novogo Karfagena [The Third Rome Versus the New Carthage], March 29. https://nic-pnb.ru/analytics/tretij-rim-protiv-novogo-karfagena/.

Vorob'ev, V.I. 1948. Obshhij ocherk [Overview]. In *Materialy po lotsii Antarktiki*, ed. I.F. Novoselov, 1–38. Moscow: Hydrographic Department, Soviet Navy.

Vvedenskij, N. 1941. K voprosu o russkikh otkrytiyakh v Antarktike v 1819–1821 godakh, v svete novejshikh geograficheskikh issledovanij [On the Question of Russian Discoveries in the Antarctic in 1819–1821, in the Light of Recent Geographical Research]. *Izvestiya vsesoyuznogo geograficheskogo obshhestva* 73 (1): 118–122.

Russian Versions of Tables

Таблица 7.1   Лазарев о 28 января и Беллинсгаузен о 2 Февраля 1820

28 января 1820 в письме Лазарева	2 февраля 1820 в книге Беллинсгаузена
(a) *16-го генваря [ст.ст.] достигли мы широты 69°23′ S гдѣ встретили матерой ледъ чрезвычайной высоты и въ прекрасный тогда вечеръ, смотря на саленгу, простирался оный такъ далеко какъ могло только достигать зрѣніе, но удивительнымъ симъ зрѣлищемъ наслаждались мы недолго, ибо вскорѣ опять напасмурило и пошелъ по обыкновенію снѣгъ.* (л. 2)	(b) Мы продолжали путь на Югъ, при тихомъ вѣтрѣ отъ SOTO и ясной погодѣ. ... на Югѣ становилось часъ отъ часу свѣтлѣе. ... далѣе къ Югу представилось до пятидесяти льдяныхъ разнообразныхъ громадъ, заключающихся въ срединѣ льдянаго поля. Обозрѣвая пространство сего поля ..., мы не могли видѣть предѣловъ онаго; конечно было продлженіемъ того, которое видѣли въ пасмурную погоду 16го Генваря, но по причинѣ мрачности и снѣга хорошенько разсмотрѣть не могли. ... ... Недолго наслаждались мы ясною погодою, въ 6 часовъ небо покрылось облоками, а въ 8 нашла пасмурность съ снѣгомъ и градомъ, ... (Беллинсгаузен 1831 1: 177–178)

R. Bulkeley, *The Historiography of the First Russian Antarctic Expedition, 1819–21*, https://doi.org/10.1007/978-3-030-59546-3

Таблица 9.1 Беллинсгаузен о 17 февраля 1820

Рапорт от Сиднея, апреля 1820	Двукратныя изыслания, 1831
… съ *5 на 6 число дошелъ до широты S. 69°, 7′, 30″, долготы О. 16°, 15′. Здесь за льдеными* [так] *полями мелкаго льда, и Островами виденъ Материкъ льда. коего края отъ ломаны перпендикулярно, и Которой* [так] *продолжался по Мерѣ на-шего зрѣнія, возвышаясь къ Югу подобно берегу. Плоскіе льдяные острова, близь сего материка находящіися* [так], *ясно показываютъ, что онѣ суть отломки сего материка; ибо имѣютъ края и верьхную повѣрьхность подобную материку.* (Беллинсгаузенъ 1820а: 242об)	… увдѣли множество большихъ, плоскихъ, высокихъ льдяныхъ островов, затертыхъ плавающими мелкими льдами, и мѣстами одинъ на другомъ лежащими. Льды къ SSW примыкаются къ льду гористому, твердо стоящему; закраины онаго были перпендикулярны и образовали заливы, а поверхность возвышалась отлого къ Югу, на растояніе, предѣловъ котораго мы не могли видѣть съ салинга. … Видя льдяные острова, поверхностью и краями сходные съ поверхностью и краями большаго вышеупомянутаго льда, предъ нами находящагося, мы заключили, что сіи льдяныя громады и всѣ подобные льды, отъ собственной своей тяжести, или другихъ физическихъ причинъ, отдѣлились отъ **матераго берега**, вѣтрами отнесенные, плаваютъ по пространству Ледовишаго Южнаго Океана; … (Беллинсгаузенъ 1831, I: 188–189 – разгрядка наша)

Таблица 11.1 Донесения Беллинсгаузена о 28 января и 17–18 февраля 1820 так, как изложил или не изложил Корякин (2008)

	Рапорт от Сиднея 9 апреля 1820	Двукратные изыскания 1831 т. 1
28 января 1820	(а) *16 числа, дошедши до широты S 69°25′, долготы 2°10′ W, въ стретилъ сплошной ледъ Украевъ одинъ на другой набросанный кусками а внутрь къ Югу въ разныхъ мѣстахъ по оному видны ледяные горы.* (л. 8)	(b) …, въ полдень въ широтѣ 69°21′28″, долготѣ 2°14′50″, мы всшрѣтили льды, которые представились намъ сквозь шедшій тогда снѣгъ, въ видѣ бѣлыхъ облаковъ. Бѣтръ былъ отъ NO умѣренный, при большой зыби отъ NW, по причинѣ снѣга, зрѣніе наше не далеко простиралось; … мы увидѣли, что сплошные льды простираются отъ Востока чрезъ Югъ на Западъ; путь нашъ велъ прямо въ сіе льдяное поле, усѣянное буграми. (ст. 171–172)

(*continued*)

Таблица 11.1   (continued)

17 февраля 1820	(c) ... *съ 5 на 6е число дошел до широты S 69° 7′ 30″, долглоты E 16° 15′.*	(d) ... увидѣли множество большихъ, плоскихъ, высокихъ льдяныхъ острововъ, затертыхъ плавающими мелкими льдами, и мѣстами одинъ на другомъ лежащими. Льды къ SSW примыкаются къ льду гористому, твердо стоящему; закраины онаго были перпендикулярны и образовали заливы, а поверхность возвышалась отлого къ Югу, на разстоянiе, предѣловъ котораго мы не могли видѣть съ салинга.

Здесь за льдеными [так] *полями мелкаго льда, и Островами виденъ Материкъ льда. коего края отъ ломаны перпендикулярно, и Которой* [так] *продолжалс[я] по Мѣрѣ на-шего зрѣнiя, возвышаясь къ Югу подобно берегу. плоскiе льдяные острова, близь сего материка находящiися, ясно показываютъ, что онѣ суть отломки Сего Материка; ибо имѣютъ края и верьхнюю поверьхность подобную материку.* (л. 8)

...

Видя льдяные острова, поверхностью и краями сходные съ поверхностью и краями большаго вышеупомянутаго льда, предъ нами находящагося, мы заключили, что сiи льдяныя громады и всѣ подобные льды, отъ собственной своей тяжести, или другихъ физическихъ причинъ, отдѣлились отъ матераго берега, вѣтрами отнесенные, плавають по пространству Ледовитаго Южнаго Океана; ... (ст. 188–189)

[*ниже 30 строками*]

(e) ... плавающiе льды становились такъ часты и крупны, что далнѣйшее въ семъ мѣстѣ покушенiе къ Z, было невозможно, а на полторы мили по сему направленiю видны были кучи льдовъ, одна на другую взгромозденныхъ. Далѣе представлялись льдяныя горы, подобныя вышеупомянутымъ, и вѣроятно, составляуть продолженiе оныхъ. (ст. 190)

18 февраля 1820

Apparatus

Footnotes are indicated with superscript numbers and the original texts of translated quotations are indicated with numbers in square brackets. Those texts are provided at the end of each chapter *in the original spelling*, even where a modern Russian edition converts a pre-revolutionary text into modern spelling. All passages from manuscript documents are shown in italics. However the older, 'three-legged' form of the letter T, which looked like an inverted Ш, is not available in computer fonts except for the lower-case italic (cursive) form, so that normal, upright passages from older texts cannot be shown exactly as published. Russian versions of Tables 1, 2, and 3 are provided in the endmatter.

The original texts of the epigraphs cannot be shown in the chapter notes. They were as follows: Renan—'L'oubli, et je dirai même l'erreur historique, sont un facteur essentiel de la création d'une nation, et c'est ainsi que le progrès des études historiques est souvent pour la nationalité un danger' (1882: 4); Polish saying—'Przeszłość zawsze się zmienia, tylko przyszłość pozostaje taka sama.'

Russian words have been romanized according to GOST 2002(B), one of the two codes accepted in Russia today. GOST 2002(B) is more user-friendly than other codes because it dispenses with diacritics, thus freeing up the choice of font, and provides English-speakers with some phonetic guidance. In line with that aim, however, four letters have not been rendered according to GOST 2002(B). 'Ы' and 'Э' (the open vowel at the

R. Bulkeley, *The Historiography of the First Russian Antarctic Expedition, 1819–21*,
https://doi.org/10.1007/978-3-030-59546-3

start of words like 'expedition') are shown as 'Y' and 'È', the GOST 2002(A) alternatives, and 'X' is shown as 'KH', following the old Soviet GOST 1971(2) code. For the sake of English-speakers 'Ц' has been romanized as 'TS' rather than 'C'. It should be noted that GOST 2002(B) transcribes the pre-revolutionary 'Ѣ', or 'yat', as 'YE', but the Russian letter 'E' is often iotized to varying degrees, depending on context. Although the capital letters 'KH', 'TS', 'YA', and 'YU' might strictly be so transcribed, as the likely majority non-Russian readers have been spared such pedantry.

Familiar Russian personal names and place names, some of which were cyrillizations of German originals, take their usual translated or German forms, such as 'Bellingshausen', in the main text. For bibliographic purposes, however, all names published in Russian are transcribed from either their pre-revolutionary (Bellinsgauzen″) or their modern (Bellinsgauzen) spelling, as appropriate, in both citations and references. The honorifics, forenames, and initials of authors follow their title pages. An author known only by their initials is listed in the References by the *first* initial rather than that of the unknown surname. In citations, the volume number is given before the pagination only if more than one volume is referenced in the book.

Where a non-English title is repeated in successive editions, the translation is not repeated. Translated book titles are capitalized as they would be if published in English, but their transliterated titles follow the originals, which often had only initial capitals. In conveying the identity of Russian publishers, some licence has been exercised as between translation, romanization, or the use of standard abbreviations. Nothing, unfortunately, can be done about the tendency of nineteenth-century Russian periodicals to number the pages of their sections separately.

To avoid the awkward differences between the rank of Midshipman in Britain and *Michman* in Russia, two occurrences of the latter have been translated simply as 'officer', whereas the rank of *Gardemarine* has been translated as 'Midshipman', to which it closely corresponds.

Dates in the author's text are New Style (Common Era) unless designated (O.S.) for Old Style. Dates translated in quoted Russian passages are shown as they were written. Before 1918 almost all dates followed the Old Style (Julian calendar), which in this context meant that they were usually 12 days earlier than the equivalent Common Era dates. However between re-entering the western hemisphere in December 1820 and mid-February 1821, when Bellingshausen adjusted his calendar, expedition dates were only 11 days behind Common Era rather than 12. Soviet and post-Soviet authors have usually retained the Old Style dates in their treatments of the expedition, but apart from Belov they have seldom noticed the ship's dates in primary sources for early 1821. Where clarification seemed advisable the author has interpolated [O.S.] for Old Style and sometimes [s.d.] for ship's date. The question of whether Bellingshausen or other protagonists rendered some dates according to nautical, rather than civil, usage is discussed when surveying the work of Mikhail Belov in Chap. 9.

References[1]

A.N., and M.G., trans. 1823. Izvestie ob″ otkrytiyakh″ kap. Villiyama Skorresbi na vostochnom″ beregu zapadnoj Grenlandii [A Report on Captain William Scoresby's Discoveries on the West Coast of Greenland]. *Severnyj Arkhiv″* 7 (13): 50–59, 7 (14): 189–200, 7 (15): 240–258—not seen.

A.R.H. 1939. On Some Misrepresentations of Antarctic History. *Geographical Journal* 94: 309–330.

Aagaard, Bjarne. 1940. *Who Discovered Antarctica?* Proceedings of the Sixth Pacific Science Congress of the Pacific Science Association, held at the University of California, Berkeley, Stanford University, and San Francisco, July 24th—August 12th 1939, vol. 2, 675–707. Berkeley, CA: University of California Press.

Adamov, Arkadij. 1949. Antarktida—russkoe otkrytie [Antarctica—A Russian Discovery]. *Znanie–sila* 7: 26–29.

———. 1951. Bessmertnyj podvig russkikh moryakov [An Immortal Victory by Russian Seamen]. *Nauka i zhizn'* 1: 41–43.

Adams, C.C. 1904. "Antarctica" The New Continent Larger than Europe or the United States. *The New York Times*, March 13: SM2.

---

[1] **Key to archival citations**: SARN = State Archives of the Russian Navy, F = Fond, S = Series [*Opis'*], P+= Piece [*Delo*], fo./fos = folio/s [*list/y*], v = verso [*oborotnoe*] (because verso pages were unnumbered). Dates on manuscript documents are given in Universal Time (CE).

© The Author(s), under exclusive license to Springer Nature Switzerland AG 2021
R. Bulkeley, *The Historiography of the First Russian Antarctic Expedition, 1819–21*,
https://doi.org/10.1007/978-3-030-59546-3

Afanas'ev, Yu.N., and Others. 1996. *Sovetskaya istoriografiya [Soviet Historiography]*. Moscow: Russian State University for Humanities.

Afonin, M. 1948. Bor'ba vokrug Antarktiki [The Dispute About the Antarctic]. *Izvestiya*, March 19: 3.

Alejner, A.Z. 1949. Geograficheskie predstavleniya ob Antarktike s drevnejshikh vremen do pervoj russkoj antarkticheskoj èkspeditsii i ikh otrazhenie na kartakh [Geographical Representations of the Antarctic from Ancient Times to the First Russian Antarctic Expedition, and their Reflection in Maps]. *Izvestiya vsesoyuznogo geograficheskogo obshhestva* 81 (5): 484–496.

———. 1958a. Osnovnye ètapy geograficheskogo issledovaniya Antarktiki [Basic Stages in Antarctic Exploration]. In *Antarktika: materialy po istorii issledovaniya i po fizicheskoj geografii*, ed. E.N. Pavlovskij and S.V. Kalesnik, 55–66. Moscow: Geografgiz.

———. 1958b. Geograficheskie naimenovaniya v Antarktike [Geographical Nomenclature in the Antarctic]. In *Antarktika: materialy po istorii issledovaniya i po fizicheskoj geografii*, ed. E.N. Pavlovskij and S.V. Kalesnik, 407–443. Moscow: Geografgiz.

Aleksandrov, A.I. 1949. Antarktika [The Antarctic]. *Priroda* 8: 25–31.

Alexandrow, A. (pseud.). 1904. *A Complete Russian-English Dictionary*. 3rd ed. London: Nutt.

Anderson, Alun. 2009. *After the Ice: Life, Death and Politics in the New Arctic*. London: Virgin.

Andreev, A.I. 1947a. *Pyotr Velikij—sbornik statej [Peter the Great—An Anthology]*. Moscow–Leningrad: Academy of Sciences.

———. 1947b. Pyotr I v Anglii v 1698 g. [Peter I in England in 1698]. In *Pyotr Velikij—sbornik statej*, 63–103. Moscow–Leningrad: Academy of Sciences.

———. 1947c. Rabota S.M. Solov'eva nad «Istoriej Rossij» [S.M. Solov'ev's Work on *The History of Russia*]. *Trudy Istoriko-arkhivnogo instituta* 3: 3–16.

———. 1949a. Russkie v Antarktike v 1819–1821 gg. [Russians in the Antarctic 1819–1821]. *Izvestiya Akademii Nauk: seriya istorii i filosofii* 6 (1): 77–78.

———. 1949b. Èkspeditsiya F.F. Bellinsgauzena—M.P. Lazareva v Yuzhnyj Ledovityj okean v 1819–1821 gg. i otkrytie russkimi moryakami Antarktidy [The Expedition of F.F. Bellingshausen and M.P. Lazarev to the Southern Ice Ocean in 1819–1821, and the Discovery of Antarctica by Russian Seamen]. In *Plavanie shlyupov «Vostok» i «Mirnyj» v Antarktiku v 1819, 1820 i 1821 godakh*, 5–16. Moscow: Geografgiz.

———. 1949c. *Plavanie shlyupov «Vostok» i «Mirnyj» v Antarktiku v 1819, 1820 i 1821 godakh [The Antarctic Voyage of the Sloops Vostok and Mirnyj in 1819, 1820 and 1821]*. Moscow: Geografgiz.

———. 2012. Pis'mo A.I. Andreeva O.M. Medushevske, iyunya 1949 [A Letter from A.I. Andreev to O.M. Medushevskaya, June 1949]. In Pis'ma

O.M. Medushevskoj A.I. Andreevu, ed. V.G. Anan'ev. *Vestnik Rossijskogo Gosudarstvennogo Gumanitarnogo Universiteta* 21: 11–23.

Anon. 1793. *Captain Cook's Third and Last Voyage to the Pacific Ocean*. London: Fielding and Stockdale.

———. 1804a. *Slovar' geograficheskij Rossijskago Gosudartsva [Geographical Dictionary of the Russian Empire]*. Vol. 3. (K–M). Moscow: University Press.

———. 1804b. Fortsetzung des Aufsatzes: Das nordliche Amerika [North America—Continued]. *Skizzen in Kupfern geografisch-istorisch-artistisch-ökonomischen Inhalts* 36, Vienna: 284–288.

———. 1813. *Nouveau dictionnaire Russe-Français-Allemand [A New Russian, French, and German Dictionary]*. Vol. 1. St Petersburg: Glasounov.

———. 1818. Foreign Articles. *Niles Weekly Register* 15 (381). Baltimore.

———. 1819. n.t. SARN—F–166, S–1, P–660a, fos 347–348.

———. 1821/1822. Schedule of Rations Consumed on *Vostok*. n.d. but after August 27. SARN—F–132, S–1, P–813, fos 45–53v.

———. 1821a. Dal'nyejshiya svyedeniya o plavanii otryada, sostoyashhago iz" shlyupov" Vostoka i Mirnago, pod" komandoyu Kapitana Bellinsgauzena [Further Information about the Voyage of the Squadron Comprising Sloops Vostok and Mirnyj, Commanded by Captain Bellingshausen]. *Russkij Invalid* 196: 786–788.

———. 1821b. New Shetland. *Literary Gazette* 5: 691–692, 712–713, 746–747.

———. 1821c. New Shetland. *The Philosophical Magazine and Journal* 58 (280): 144–145.

———. 1824. Return of the Russian Antarctic Expedition. *Philosophical Magazine* 43: 462.

———. 1829. Pis'mo k * [Letter to *]. *Kazanskij vyestnik"* 35 (1): 46–56.

———. 1830. Südpolarländer [Antarctic Lands]. *Allgemeine deutsche Real–Encyklopädie für die gebildeten Stände*. Vol. 10. Leipzig: Brockhaus.

———. 1840. Discovery of the Antarctic Continent. *Sydney Herald*, March 13: 2.

———. 1841. Geograficheskiya otkrytiya v" pervyya tridtsat' let" tekushhago stolyetiya [Geographical Discoveries in the First 30 Years of the Present Century]. *Zhurnal" Ministerstva Narodnago Prosvyeshheniya* 32 (5): 1–24.

———. 1842. Otkrytie novago materika v Yuzhnom" polusharii [Discovery of a New Mainland in the Southern Hemisphere]. *Syn" Otechestva* 1: 24–32.

———. 1868. Expeditions to the North Pole. *Hours at Home* 7: 138–146.

———. 1922. Ledyanyya gory [Icebergs]. *Segodniya* 33: 3. Riga.

———. 1939. Razdel Antarktiki [Partition of Antarctica]. *Problemy Arktiki* 3 (5): 107.

———. 1940. Otdelenie istorii geograficheskikh znanii [Section for the History of Geographical Knowledge]. *Izvestiya vsesoyuznogo geograficheskogo obshhestva* 72 (2): 845–846.

————. 1946a. O zhurnalakh «Zvezda» i «Leningrad» [On the Magazines *Zvezda* and *Leningrad*]. *Pravda*, August 21: 1.

————. 1946b. Edward Bransfield's Antarctic Voyage, 1819–1820, and the Discovery of the Antarctic Continent. *Polar Record* 4: 385–393.

————. 1948a. O rabote Instituta istorii Akademii nauk SSSR [The Work of the Institute of History, Soviet Academy of Sciences]. *Voprosy istorii* 11: 144–149.

————. 1948b. Protiv ob"ektivizma v istoricheskoj nauki [Against Objectivism in Historiography]. *Voprosy istorii* 12: 3–12.

————. 1949. Russkie otkrytiya v Antarktike [Russian Discoveries in the Antarctic]. *Izvestiya*, February 11: 3.

————. 1950. Memorandum Sovetskogo Pravitel'stva po voprosu o rezhime Antarktiki [The Soviet Government's Memorandum on the Question of an Antarctic Regime]. *Izvestiya*, June 10: 2.

————. 1958. Obmen pis'mami mezhdu Gosudarstvennym Departamentom SShA i Posol'stvom SSSR v SShA po voprosu ob Antarktike [Exchange of Letters between the US State Department and the Embassy of the USSR on the Antarctic Question]. *Pravda*, June 4: 5.

————. 1960. Russia "Discovered Antarctica". *The Times*, June 16: 10.

————. 1962. Trail Blazers of Antarctica. *UNESCO Courier* 1: 17.

————. 2010. Panel Discussion. 190 let nazad byl otkryt shestoj kontinent [190 Years Ago the Sixth Continent was Discovered]. *Rossiya kul'tura*, January 29.

————. 2016. Novosti [Announcements]. *Rossiya I* Website, January. The link to this announcement [https://russia.tv/article/show/article_id/23592] is no longer valid, but the film itself, now owned by VGTRK's *Rossiya 24* channel, can be viewed in two parts on YouTube: https://www.youtube.com/watch?v=nLCTpawMiVE.

————. 2019. S kreshheniem: èstonskij "Bellinsgauzen" otpravitsya zanogo otkry- vat' Antarktidu [A Christening: An Estonian *Bellingshausen* Prepares to Discover Antarctica Again]. *Sputnik eesti*, June 11.

————. n.d.-a. *Bellinsgauzen Faddej Faddeevich [Bellingshausen, Faddej Faddeevich]*. Morskaya Èntsklopediya Website. http://95.31.135.131/card/view/24.

————, ed. n.d.-b. *Osvoenie chelovekom Antarktidy [Human Mastery of Antarctica]*. http://interneturok.ru/geografy/7-klass/materiki-antarktida/osvoenie-chelovekom-antarktidy.

————. n.d.-c. *Razvitie geograficheskikh znanij o Zemle [The Development of Geographical Knowledge about the Earth]*. http://www.nado5.ru/e-book/razvitie-geograficheskikh-znanii-o-zemle.

Anpilogov, G. 1948. «*Pyotr Velikij»—sbornik statej* pod redaktsiej d-ra istoricheskikh nauk A.I. Andreeva [*Peter the Great—An Anthology*, ed. Dr of History A.I. Andreev]. *Voprosy istorii* 4: 120–123.

Aristov, V.V. 1990. Zhizn' i deyatelnost' I.M. Simonova [The Life and Work of I.M. Simonov]. In *Dva plavaniya vokrug Antarktidy*, ed. T.Ya. Sharipova, 8–14. Kazan: Kazan University Press.

Armstrong, Terence. 1950. Recent Soviet Interest in Bellingshausen's Antarctic Voyage of 1819–21. *Polar Record* 5 (39): 475–478.

———. 1951. Four Eye-Witness Accounts of Bellingshausen's Antarctic Voyage of 1819–21. *Polar Record* 6 (41): 85–87.

———. 1971. Bellingshausen and the Discovery of Antarctica. *Polar Record* 15 (99): 887–889.

Arsen'ev, V.A., and V.A. Zemskij. 1951. *V strane kitov i pingvinov [In the Land of Whales and Penguins]*. Moscow: Moscow Society of Naturalists.

———. 1954. *V strane kitov i pingvinov*. 2nd ed. Moscow: Moscow University Press.

Auburn, F.M. 1982. *Antarctic Law and Politics*. London: Hurst.

Aver'yanov, V.G., and Ya.P. Koblents. 1970. 150-letie otkrytiya Antarktidy i sovetskie issledovaniya v Antarktike [150th Anniversary of the Discovery of Antarctica, and Soviet Antarctic Research]. *Byulleten' Sovetskoj Antarkticheskoj Èkspeditsii* 77: 5–18.

Avsyuk, G.A., and S.N. Kartashov. 1971. 150 let otkrytiya Antarktidy russkoj èkspeditsiej [150 Years Since the Discovery of Antarctica by a Russian Expedition]. *Antarktika—doklady kommisii* 1968: 5–12.

Aycke, J.C. 1836. Bemerkungen über Grundeis [Remarks on Grundeis]. *Annalen der Physik und Chemie* 2 (9): 122–129.

Baird, G.H. 1821. Latitude and Longitude of Places in New South Britain. *The Edinburgh Philosophical Journal* 5 (9): 233.

Bakaev, V.G., and Others. 1966. *Atlas Antarktiki [Atlas of the Antarctic]*. Vol. 1. Moscow–Leningrad: Geodesic and Cartographic Department.

Balch, Edwin Swift. 1909. Stonington Antarctic Explorers. *Bulletin of the American Geographical Society* 41 (8): 473–492.

———. 1925. The First Sighting of Western Antarctica. *Geographical Review* 15 (4): 650–653.

Balkli, Rip. 2013. Pervye nablyudeniya materikovoj chasti Antarktidy: popytka kriticheskogo analiza [First Sightings of the Mainland of Antarctica: Towards a Critical Analysis]. *Voprosy istorii estestvoznaniya i tekhniki* 4: 41–56.

Baltijskij, N. (pseudonym of O.V. Kuusinen). 1945. O patriotizme [On Patriotism]. *Novoe vremya* 1: 3–10.

Bardin, Vladimir. 1989. *V gorakh i na lednikakh Antarktidy [On the Mountains and Glaciers of Antarctica]*. Moscow: Znanie.

Barratt, Glynn. 1988. *The Russians and Australia*. Vancouver: University of British Columbia Press.

———. 1990. *Melanesia and the Western Polynesian Fringe*. Vancouver: University of British Columbia Press.

Batchelor, John Calvin. 1983. *The Birth of the People's Republic of Antarctica*. New York: Dial.

Baughman, T.H. 1994. *Before the Heroes Came: Antarctica in the 1990s.* Lincoln, NE: University of Nebraska Press.

Behrens, C.F. 1738. *Der wohlversuchte Südländer: Reise um die Welt [The Much Sought After Southerner: A Circumnavigation].* Leipzig: Self.

Bellingshausen, Captain. 1945. *The Voyage of Captain Bellingshausen to the Antarctic Seas, 1819–1821,* ed. Frank Debenham, 2 vols. London: Hakluyt Society.

Bellinsgauzen, F.F. 1949. *Dvukratnye izyskaniya v yuzhnom ledovitom Okeene i plavanie vokrug sveta v prodolzhenie 1819, 1820, i 1821 godov,* ed. E.E. Shvede. Moscow: Geografgiz.

———. 1960. *Dvukratnye izyskaniya v yuzhnom ledovitom okeane i plavanie vokrug sveta v prodolzhenie 1819, 20 i 21 gg.,* ed. E.E. Shvede. Moscow: Geografgiz.

———. 2008. *Dvukratnye izyskaniya v yuzhnom ledovitom okeane i plavanie vokrug sveta [Two Seasons of Exploration in the Southern Ice Ocean and a Voyage Around the World],* ed. V.S. Koryakin. Moscow: Drofa.

Bellinsgauzen, Faddej Faddeevich. 2011. *Otkrytie Antarktidy [The Discovery of Antarctica],* eds. M. Tereshina and Others. Moscow: Èksmo.

———. 2017. *Na shlyupakh "Vostok" i "Mirnyj" k Yuzhnomu polyusu [To the South Pole on Sloops* Vostok *and* Mirnij*],* ed. S. Chertoprud. Moscow: Algoritm.

Bellinsgauzen", F.F. 1820a. Report to the Marquis de Traversay from Sydney, April 20 1820. St Petersburg: SARN—F–166, S–1, P–660b, fos 239–245v.

———. 1820b. Personal letter to the Marquis de Traversay from Sydney, April 20 1820. St Petersburg: SARN—F–166, S–1, P–660b, fos 246–249v.

———. 1820c. Report to the Marquis de Traversay from Sydney, November 2 1820. St Petersburg: SARN—F–166, S–1, P–660b, fos 354–359.

Bellinsgauzen", F.F. Kap. 1821a. Report to the Marquis de Traversay from Rio de Janeiro, March 17 1821. St Petersburg: SARN—F–166, S–1, P–660b, fos 352–353v.

Bellinsgauzen", Kapitan". 1821b. Vypiska iz" doneseniya Kapitana 2 ranga Bellinsgauzena k" Morskomu Ministru ot" 8 Aprelya 1820 goda [O.S.] iz" Porta Zhaksona [Extract from the Report of Junior Captain Bellingshausen from Port Jackson to the Minister of Marine, April 8 1820 [O.S.]]. *Syn" Otechestva,* April 23 1821 (O.S.), 69: 133–135.

———. 1821c. Vypiska iz" pis'ma Kapitana Bellinsgauzena k" Morskomu Ministru ot" 8 Aprelya 1820 goda [O.S.] iz" Porta Zhaksona [Extract from Captain Bellingshausen's letter to the Minister of Marine from Port Jackson, April 8 1820 [O.S.]]. *Syn" Otechestva,* April 23 1821 (O.S.), 69: 135–137.

———. 1821d. Karta Plavaniya Shlyupov" Vostoka i Mirnago vokrug" Yuzhnago polyusa v" 1819, 1820 i 1821 godakh" pod" Nachal'stvom" Kapitana Billensgauzena [Chart of the Voyage of Sloops *Vostok* and *Mirnyj* around the South Pole in 1819, 1820 and 1821 under the Command of Captain Bellingshausen]. St Petersburg: SARN—F–1331, S–4, P–536, fos 5–19.

Bellinsgauzen″, F.F. 1821e. Report to the Marquis de Traversay from Kronstadt, August 5 1821. St Petersburg: SARN—F–203, S–1, P–826, fos 1–18v.

Bellinsgauzen″, Kap. 1823. Donesenie Kapitana 2 ranga Bellingauzena iz″ Porta Zhaksona, o svoem″ plavanii [The Report of Junior Captain Bellingshausen from Port Jackson, About His Voyage]. *Zapiski izdavaemyya Gosudarstvennym″ Admiraltejskim″ Departamentom″* 5: 201–219.

Bellinsgauzen″, Kapitan″. 1831. *Dvukratnyya izyskaniya v″ yuzhnom″ ledovitom″ okeanye i plavanie vokrug″ svyeta v″ prodolzhenii 1819, 20 i 21 godov″ [Two Seasons of Exploration in the Southern Ice Ocean and a Voyage around the World, During the Years 1819, 1820 and 1821]*, ed. L.I. Golenishhev″-Kutuzov″, 2 vols. plus *Atlas*. St Petersburg: Glazunovs.

———. 2011. *Dvukratnyya izyskaniya v″ yuzhnom″ ledovitom″ okeanye i plavanie vokrug″ svyeta v″ prodolzhenii 1819, 20 i 21 godov″*, ed. L.I. Golenishhev″-Kutuzov″. St Petersburg: Al'faret.

Belov, M.I. 1948. *Semyon Dezhnev 1648–1948*. Moscow: GlavSevMorPut.

———. 1954. Vvedenie [Introduction]. In *Geograficheskij sbornik III*, 5–12. Moscow: Academy of Sciences.

———. 1956–1969. *Istoriya otkrytiya i osvoeniya severnogo morskogo puti [History of the Discovery and Conquest of the Northern Sea Route]*. 4 vols. Moscow: Morskoj Transport.

———. 1961a. Zabytaya karta [A Forgotten Chart]. *Vodnyj Transport* 21 (138): 3.

———. 1961b. Otchyotnaya karta pervoj russkoj antarkticheskoj èkspeditsii [The Official Chart of the First Russian Antarctic Expedition]. *Byulleten' Sovetskoj Antarkticheskoj Èkspeditsii* 31: 5–14.

———. 1962. Shestaya chast' sveta otkryta russkimi moryakami (Novye materialy…) [The Sixth Continent Discovered by Russian Seamen—New Sources…]. *Izvestiya vsesoyuznogo geograficheskogo obshhestva* 94 (2): 105–114.

———. 1963a. O kartakh pervoj russkoj antarkticheskoj èkspeditsii 1819–1821 gg. [The Maps of the First Russian Antarctic Expedition 1819–1821]. In *Pervaya russkaya antarkticheskaya èkspeditsiya 1819–1821 gg. i eyo otchyotnaya navigatsionnaya karta*, ed. M.I. Belov, 5–56. Leningrad: Morskoj Transport.

———. 1963b. *Pervaya russkaya antarkticheskaya èkspeditsiya 1819–1821 gg. i eyo otchyotnaya navigatsionnaya karta [The First Russian Antarctic Expedition 1819–1821 and its Official Navigational Chart]*. Leningrad: Morskoj Transport.

———. 1966a. Proekt pervoj sovetskoj èkspeditsii v Antarktidu [A Proposal for the First Soviet Expedition to Antarctica]. *Byulleten' Sovetskoj Antarkticheskoj Èkspeditsii* 58: 64–67.

———. 1966b. Slavu pervogo tochnogo vychisleniya mestopolozheniya yuzhnogo magnitnogo polyusa anglichanin Ross dolzhen razdelit' s Bellinsgauzenom [The Englishman Ross should Share the Glory of the First Accurate Determination of the South Magnetic Pole with Bellingshausen]. *Nauka i Zhizn'* 8: 21–23.

————. 1969. Istoriya otkrytiya i issledovaniya Antarktiki [History of the Discovery and Exploration of Antarctica]. In *Atlas Antarktiki [Atlas of the Antarctic]*, ed. E.I. Tolstikov, vol. 2, 35–97. Moscow–Leningrad: Geodesic and Cartographic Department.

————. 1970a. Otkrytia ledyanogo kontinenta [The Discovery of the Icy Continent]. *Izvestiya vsesoyuznogo geograficheskogo obshhestva* 102 (3): 201–208.

————. 1970b. The First Maritime Magnetic Survey around Antarctica. In *Problems of Polar Geography*, 244–252. (Trans. of *Trudy Arkticheskogo i Antarkticheskogo nauchnogo issledovatel'skogo instituta*, 185). Jerusalem: Israel Program for Scientific Translations.

————. 1971. Comment by Professor M.I. Belov. *Polar Record* 15 (99): 890–891.

Belov, M.I., and V.V. Kuznetsova. 1974. Pervonachal'nyj proekt russkoj ekspeditsii v Yuzhnyj i Severnyj ledovityj okeany [The Initial Proposal for a Russian Expedition to the Southern and Northern Ice Oceans]. *Izvestiya vsesoyuznogo geograficheskogo obshhestva* 106 (6): 491–497.

Berg, L.S. 1929. *Ocherk istorii russkoj geograficheskoj nauki (vplot' do 1923 g.) [Towards a History of Russian Geography Down to 1923]*. Leningrad: Academy of Sciences.

————. 1949a. Russkie otkrytiya v Antarktike i sovremennyj interes k nej [Russian Discoveries in the Antarctic and Modern Interest in the Region]. *Vestnik Akademii Nauk: seriya geograficheskaya i geofizicheskaya* 3: 39–46.

————. 1949b. *Ocherki po istorii russkikh geograficheskikh otkrytii [Towards a History of Russian Geographical Discoveries]*. 2nd ed. of (Berg 1929). Moscow: Academy of Sciences.

————. 1949c. Sovremennyj interes k Antarktike [Modern Interest in the Antarctic]. *Zvezda* 2: 91–96.

————. 1949d. Russkie otkrytiya v Antarktike i sovremennyj interes k nej [Russian Discoveries in the Antarctic and Modern Interest in the Region]. *Vestnik Akademii Nauk* 3: 39–46.

————. 1949e. Russkie otkrytiya v Antarktike [Russian Discoveries in the Antarctic]. *Nauka i zhizn'* 3: 22–25.

————. 1950. *Velikie russkie puteshestvenniki [Great Russian Travellers]*. Moscow–Leningrad: Detgiz.

————. 1951. Bellinsgauzen i Pal'mer [Bellingshausen and Palmer]. *Izvestiya vsesoyuznogo geograficheskogo obshhestva* 83 (1): 25–31.

[Berkh″, V.N.], ed. 1823. *Khronologicheskaya istoriya vsyekh″ puteshestvii v syevernyya polyarnyya strany [Chronological History of All Journeys Into Arctic Regions]*. 2nd ed. St Petersburg: Imperial Academy of Sciences.

Berkh″, V.N. 1828. Dopolnenie k zhizneopisaniyu M.V. Lomonosova [Supplement to the Biography of M. V. Lomonosov]. *Moskovskij Telegraf* 21 (11): 289–314.

Berlyant, A.M., T.F. Vasil'ev, and I.A. Suetova. 1981. Prostranstvennye soötnosheniya lednikovogo i korennogo rel'efa Antarktidy [Spatial Relationships
between the Glacial and Basic Relief of Antarctica]. *Antarktika—doklady kommisii 1980*, n.p.

Bird, Richard Evelyn. 1930. *Little America*. New York: Putnam's.

———. 1935. *Discovery*. New York: Putnam's.

Bogoyavlenskij, G.P. 1955. *Russkie geografy i puteshestvenniki [Russian Geographers
and Travellers]*. Moscow: State Library of the USSR.

———. 1963. *Fizicheskaya geografiya: bibliograficheskoe posobie poso uchitelej
[Physical Geography: A Bibliographic Guide for Teachers]*. Moscow:
Textbook Press.

Bolkhovitinov, N.N., ed. 1997. *Istoriya Russkoj Ameriki [The History of Russian
America] (1732–1867)*. Moscow: International Relations.

Bolkhovitinov, V., and Others. 1950. *Rasskazy o russkom pervenstve [Tales of
Russian Priority]*. Moscow: Young Guard.

Bolotnikov, N.Ya. 1953. Faddej Faddeevich Bellinsgauzen i Mikhail Petrovich
Lazarev [Faddej Faddeevich Bellingshausen and Mikhail Petrovich Lazarev]. In
Russkie moreplavateli, ed. V.S. Lupach, 183–209. Moscow: Military Press.

Brandenberger, David. 2010. Stalin's Populism and the Accidental Creation of
Russian National Identity. *Nationalities Affairs* 38 (5): 723–739.

de Brosses, C. 1756. *Histoire des Navigations aux Terres Australes [A History of
Voyages to Southern Lands]*. Vol. 1. Paris: Durand.

Bruce, William S. 1894. The Story of the Antarctic. *Scottish Geographical Magazine*
10: 57–62.

Buache, Philippe. 1740. *Carte des Terres Australes [Map of Southern Lands]*. Paris:
Académie Royale des Sciences. https://www.davidrumsey.com/luna/servlet/
detail/RUMSEY~8~1~299835~90070885:Carte-des-Lieux-ou-les-
Differentes-.

Buache, [P.]. 1763. Geographical and Physical Observations, Including a Theory
of the Antarctic Regions, and the Frozen Sea which They are Supposed to
Contain. Includes: Chart of the Antarctic Polar Circle, Opp. p.32. *Gentleman's
Magazine* 33: 32–36.

de Buffon, M. le Comte. 1769. *Histoire naturelle générale et particulière [General
and Specific Natural History]*. Vol. 1. Paris: Imprimerie Royale.

———. 1778. *Histoire naturelle générale et particulière. Supplément*. Vol. 5. Paris:
Imprimerie Royale.

———. 1860. *Correspondance inédite de Buffon [Unpublished Correspondance]*.
Vol. 2. Paris: Hachette.

de Buffon, M. le Comte, and L.-J.-M. Daubenton. 1749. *Histoire naturelle
générale et particulière*. Vol. 1. Paris: Imprimerie Royale.

Bulkeley, Rip. 2008. Aspects of the Soviet IGY. *Russian Journal of Earth
Sciences* 10: 1–17.

————. 2011. Cold War Whaling: Bellingshausen and the *Slava* Flotilla. *Polar Record* 47 (2): 135–155.

————. 2014a. *Bellingshausen and the Russian Antarctic Expedition, 1819–21*. Basingstoke: Palgrave.

————. 2014b. Bellingshausen on Cook—'Glorious' or What? *Cook's Log* 37 (1): 20–23.

————. 2015a. The Bellingshausen-Palmer Meeting. *Polar Record* 51 (2): 212–222.

————. 2015b. Aivazovsky's *Icebergs*: An Antarctic Mystery. *Polar Record* 51 (6): 644–654.

————. 2016. Naming Antarctica. *Polar Record* 52 (1): 2–15.

————. 2019. Bellingshausen's 'Mountains': The 1820 Russian Sighting of Antarctica and Bellingshausen's Theory of the South Polar Ice Cap. *Polar Record* 55 (6): 392–401.

————. 2021. Bellingshausen in Britain: Supplying the Russian Antarctic expedition, 1819. *The Mariner's Mirror* 107 (1): 40–53.

Bush, W.M. 1982. *Antarctica and International Law*. Vol. 3. Dobbs Ferry, NY: Oceana.

Butler, Raymond A. About. 1958. *The Age of the Sealer. Map 2. Antarctica*. Washington, DC: US Antarctic Programs.

Campbell, R.J. 2000. *The Discovery of the South Shetland Islands*. London: Hakluyt Society.

Charcot, J.-B. 1906. *Le "Français" au Pôle Sud [The Français at the South Pole]*. Paris: Flammarion.

Cherevichnyj, I.I. 1963. *V nebe Antarktidy [In the Fog of Antarctica]*. Moscow: Morskoj Transport.

Chernous'ko, L.D. 1949. Russkie—pervoötkryvateli Antarktidy [Russians—The First Discoverers of Antarctica]. *Morskoj sbornik* 4: 95–99.

Chernousov, A.A. 2011. *Admiral M.P. Lazarev*. St Petersburg: MOO.

Chernykh, V.I. 2015. Nekommercheskij Fond "Russkaya Antarktida 2020" [Not-for-profit "Russian Antarctica 2020"]. M.V. Sachyov website. http://sachev.ru/russkaya_antarktida_2020.htm.

Cook, James. 1777. *A Voyage Towards the South Pole, and Round the World; Performed in His Majesty's Ships the Resolution and Adventure, in the Years 1772, 3, 4, and 5*. 2 vols. London: Strahan and Cadell.

————. 1784. *A Voyage to the Pacific Ocean: Undertaken, by the Command of His Majesty, for Making Discoveries in the Northern Hemisphere: … in the Years 1776, 1777, 1778, 1779, and 1780*. Vol. 2. Dublin: Chamberlain and Others.

Cook, Jacques, et Autres. 1778. *Voyage au Pôle Austral et autour du monde [A Voyage Towards the South Pole, and Round the World]*. Vol. 5. Paris: Merigot.

Cook, James, and Others. 1814. *Captain Cook's Three Voyages to the Pacific Ocean*. Vol. 1. New York: Duyckinck.

Crantz, David. 1767. *The History of Greenland*. Vol. 1. London: Society for the Furtherance of the Gospel.

Croome, Angela. 1960. Geophysics and Space Research. *Discovery* 21 (9): 401–402.

Dal', V.I. 1863. *Tolkovyj slovar' zhivago velikorusskago yazyka [Reference Dictionary of the Greater Russian Language]*. Vol. 1. Moscow: Semyon.

———. 1865. *Tolkovyj slovar' zhivago velikorusskago yazyka [Reference Dictionary of the Greater Russian Language]*. Vol. 2. St Petersburg–Moscow: Ris'.

Dal', Vladimir. 1881. *Tolkovyj slovar' zhivago velikorusskago yazyka [Reference dictionary of the Greater Russian language]*. Vol. 2. St Petersburg–Moscow: Vol'f.

Daly, J.C.K. 1991. *Russian Seapower and 'the Eastern Question' 1827–41*. Basingstoke: Macmillan.

Danilov, A.I., and V.L. Mart'yanov. 2017. Sezonnye raboty 62-j Rossijskoj Antarkticheskoj Èkspeditsij [Seasonal Work by the 63rd Russian Antarctic Expedition]. *Rossijskie Polyarnye Issledovanniya* 3 (29): 11–14.

Davis, J. 1595. *The Worlds Hydrographicall Discription*. London: Self.

Debenham, F. 1937. Letter to H.R. Mill, 13 February 1937. Cambridge: Archives of the Scott Polar Research Institute: MS/100/23/48.

———. 1939. Foreword. *Polar Record* 3 (17): 1–2.

Debenham, F. Frank. 1959. *Antarctica*. London: Herbert Jenkins.

Decree. 1992. Ukaz Prezidenta Rossijskoj Federatsii ot [Decree Issued by the President of the Russian Republic on] 07.08.1992 g. № 824. http://kremlin.ru/acts/bank/1775.

Desmarest, [Nicolas]. 1803. *Encyclopédie Méthodique: Géographie-Physique [Systematic Encyclopedia: Physical Geography]*. Vol. 2. Paris: Agasse.

Dmitriev, V.V., ed. 1993–1994. *Morskoj èntsiklopedicheskij slovar' [Maritime Dictionary]*. St Petersburg: Shipbuilding Magazine.

Dubrovin, L.I. 1976. *Chelovek na ledyanom kontinente [Man in the Icy Continent]*. Leningrad: Gidromet.

Dubrovin, L.I., and M.A. Preobrazhenskaya. 1987. *O chem govorit karta Antarktiki? [What does the Map of the Antarctic Tell Us?]*. Leningrad: Gidromet.

Duperrey, L.I. 1841. Notice sur la position des pôles magnétiques de la terre [The Position of the Earth's Magnetic Poles]. *Bulletin de la Société de Géographie* 16: 314–324.

Durdenevskij, V.N. 1950. Problema pravovogo rezhima pripolyarnykh oblastej (Antarktika i Arktika) [The Problem of the Legal Regime in Polar Regions (the Antarctic and Arctic)]. *Vestnik Moskovskogo Universiteta* 7: 111–114.

E.A. 1949. O prioritete russkikh v otkrytii Antarktiki [On Russian Priority in Discovering Antarctica]. *Morskoj flot* 4: 46–47.

Esakov, V.A. 1964. Otkrytie Antarktidy i issledovaniya èkspeditsii F.F. Bellinsgauzena i M.P. Lazareva [The Discovery and Exploration of Antarctica by the Bellingshausen–Lazarev Expedition]. In *Russkie okeanicheskie i morskie issledovaniya v XIX—nachale XX v*, 53–69, eds. V.A. Esakov, A.F. Plakhotnik, and

A.I. Alekseev. Moscow: Nauka. https://flot.com/publications/books/shelf/explorations/7.htm.

Evans, John. 1824. *Revision and Explanation of the Geographical and Hydrographical Terms.* Bristol: Rose.

Favorov, P.A. 1994. *Anglo–russkij voenno-morskoj slovar' [Anglo–Russian Naval Dictionary].* Moscow: Military Press.

Fedoseev, I.A. 1979. F.F. Bellinsgauzen. *Voprosy istorii estestvoznaniya i tekhniki* 67–68: 122–123.

Firsov, Ivan. 1983. *I Antarktida, i Navarin [Both Antarctica and Navarino].* Yaroslavl': Verkhnevol'sk Press.

———. 1988. *Polveka pod parusami [Half a Century Under Sail].* Moscow: Mysl'.

de Fonvielle, Wilfrid. 1889. *Le Pôle Sud [The South Pole].* Paris: Hachette.

Forman, Paul. 1971. Weimar Culture, Causality, and Quantum Theory, 1918–1927: Adaptation by German Physicists and Mathematicians to a Hostile Intellectual Environment. *Historical Studies in the Physical Sciences* 3: 1–115.

Forster, John R. 1778. *Observations Made During a Voyage Round the World.* London: Robinson.

Fox, Douglas. 2014. Antarctica and the Arctic: A Polar Primer for the New Great Game. *Christian Science Monitor*, January 12. https://www.csmonitor.com/World/Global-Issues/2014/0112/Antarctica-and-the-Arctic-A-polar-primer-for-the-new-great-game.

Fradkin, N.G. 1972. *Geograficheskie otkrytiya i nauchnoe poznanie Zemli [Geographical Discoveries and Scientific Knowledge of the Earth].* Moscow: Mysl'.

Fricker, Karl. 1898. *Antarktis [The Antarctic].* Berlin: Schaff & Grund.

Fyodorov, Valerij. 2000. Millenium na Yuzhnom polyuse [The Millenium at the South Pole]. *Moskovskij universitet* 14: 4.

Fyodorovskij, E. 1976. Svezhij veter okeana [Fresh Ocean Breeze]. *Iskatel'* 3: 1–74.

Fyodorovskij, E. Evgenij. 2001. *Bellinsgauzen.* Astrel': Moscow.

G[auss]. 1840. Abweichungen der Magnetnadel, beobachtet vom Capitaine Bellingshausen in den Jahren 1819–1821 [Magnetic Declinations Observed by Captain Bellingshausen in 1819–1821]. *Beobachtungen des magnetischen Vereins im Jahre* 1839: 117–119.

G[ershau], P. 1892. Admiral" Fhaddej Fhaddeevich" Bellinsgauzen" [Admiral Faddej Faddeevich Bellingshausen]. *Russkaya starina* 75: 373–395.

Gamalyeya, P.Ya. 1806–1808. *Teoriya i praktika korablevozhdeniya [Theory and Practice of Navigation].* St Petersburg: Naval Press.

Gan, Irina. 2011. *Russkie v Antarktide [Russians in Antarctica].* Homebush, NSW: Russian Historical Society in Australia.

de Gerebtzoff, Nicholas. 1858. *Essai sur l'histoire de la civilisation en Russie [On the History and Civilization of Russia].* Vol. 2. Paris: Amyot.

Gidromettsentr Rossii. 2015. 195 let nazad 28 yanvarya Russkaya èkspeditsiya otkryla Antarktidu [On January 28 195 Years Ago a Russian

Expedition Discovered Antarctica]. https://meteoinfo.ru/news/1-2009-10-01-09-03-06/10474-28012015-195-28-.

Golant, V. 1947. Polyarnaya likhoradka v Amerike [Pole Fever in America]. *Zvezda* 11: 173–179.

Golubev, D. 1949. *Russkie v Antarktike [Russians in the Antarctic]*. Moscow: Goskult'prosvetizdat.

Goncharova, N.N. 1973. Khudozhnik krugosvetnoj ėkspeditsii 1819–1822 godov E. Korneev [E. Korneev, Artist with the Circumnavigating Expedition of 1819–1822]. *Izvestiya vsesoyuznogo geograficheskogo obshhestva* 105 (1): 67–72.

Gorskij, N.N., and V.I. Gorskaya. 1957. *English-Russian Oceanographic Dictionary*. Moscow: State Publishing House for Technical and Theoretical Literature.

Gould, Laurence M. 1960. Statement of Dr Laurence M. Gould. *Hearings Before the Committee on Foreign Relations: United States Senate, Eighty-Sixth Congress Second Session*, June 14: 74–77.

Gould, R.T. 1925. The First Sighting of the Antarctic Continent. *Geographical Journal* 65: 220–225.

———. 1941. The Charting of the South Shetlands, 1819–28. *The Mariner's Mirror* 27: 206–242.

Govor, Elena. 1995. Na mysu Russkikh: iz istorii ėkspeditsij 1820 goda [On Russian Point: From the History of Expeditions in 1820]. *Avstraliada* 5: 1–6.

Govor, Elena, and Aleksandr Massov. 2007. *Kogda mir byl shirok: rossijskie moryaki i puteshestvenniki v Avstralii [When the World was Wide: Russian Seamen and Travellers in Australia]*. Canberra: Alcheringa.

Grebenshhikova, G.A. 2010. *Linejnye korabli 1 ranga [Capital Ships of the Line] «Victory» 1765 «Royal Sovereign» 1786*. St Petersburg: Ostrov.

Grejg, Ol'ga. 2008. *Sekretnaya Antarktida, ili Russkaya razvedka na Yuzhnom Polyuse [Secret Antarctica, or Russian Reconnaissance at the South Pole]*. Moscow: Algoritm.

Grekov, B.D., and Others. 1947. *Istoriya SSSR. Vol. 1. S drevnejshikh vremen do kontsa XVIII veka [History of the USSR. Vol. 1. From Ancient Times to the End of the 18th Century]*. Moscow: Politizdat.

Grigor'ev, A. 1949. Russkie otkrytiya v Antarktike [Russian Discoveries in the Antarctic]. *Smena* 6: 10–11.

Grigor'ev, A.A., and D.M. Lebedev. 1949. Otkrytie antarkticheskogo materika russkoj ėkspeditsiej Bellinsgauzena–Lazareva 1819–1821 gg. [The Discovery of the Antarctic Continent by the Russian Expedition of Bellingshausen and Lazarev 1819–1821]. *Izvestiya Akademii Nauk SSSR: seriya geograficheskaya i geofizicheskaya* 13 (3): 185–193.

———. 1950. *Prioritet russkikh otkrytii v Antarktike [The Priority of Russian Discoveries in the Antarctic]*. Moscow: Pravda.

Grigor'ev″, S.G. 1906. *Vokrug″ yuzhnago polyusa [Around the South Pole]*. Moscow: Ryabushinskie.

———. 1937. *Vokrug yuzhnogo polyusa*. 3rd ed. Moscow: Textbook Press.

Grushinskij, N.P., and A.G. Dralkin. 1988. *Antarktida [Antarctica]*. Moscow: Nedra.

Guretskij, V.O. 1954. Russkie geograficheskie nazvaniya v Antarktike [Russian Geographical Names in the Antarctic]. *Izvestiya vsesoyuznogo geograficheskogo obshhestva* 86 (5): 457–465.

Gusev, A.M. 1972. *Ot Èl'brusa do Antarktidy [From Elbrus to Antarctica]*. Moscow: Soviet Russia.

Gvozdetskij, N.A. 1947. Pervoe morskoe puteshestvie rossiyan vokrug sveta [The First Russian Circumnavigation]. *Priroda* 1: 85–89.

Hattersley-Smith, G. 1991. *The History of Place-Names in the British Antarctic Territory*. Cambridge: British Antarctic Survey.

Henry, David. 1774. Journal of a Voyage to Discover the North East Passage; Under the Command of the Hon. Commodore Phipps, and Capt. Skiffington Lutwych, ... In *An Historical Account of all the Voyages Round the World, Performed by English Navigators...* Vol. 4. Supplement: 29–118. London: Newbery.

Hinks, Arthur R. 1940. The Log of the *Hero*. *Geographical Journal* 96 (6): 419–430.

Hobbs, William Herbert. 1939a. The Discoveries of Antarctica within the American Sector, as Revealed by Maps and Documents. *Transactions of the American Philosophical Society* 31 (1): 1–71.

———. 1939b. The Discovery of Antarctica: A Reply to Professor R.N. Rudmose Brown. *Science* 89: 580–582.

Horensma, Pier. 1991. *The Soviet Arctic*. Milton Park, Abingdon: Routledge.

Houben, H.H. 1934. *Sturm auf den Südpol [Assault on the South Pole]*. Berlin: Ullstein.

Hunter, Douglas. 2009. *Half Moon: Henry Hudson and the Voyage that Redrew the Map of the New World*. London: Bloomsbury Press.

Hydrographic Department. 1930. *The Antarctic Pilot*. London: HMSO.

———. 1948. *The Antarctic Pilot*. London: HMSO.

Hydrographic Office. 1952. *A Functional Glossary of Ice Terminology*. Washington, DC: U.S. Navy Hydrographic Office.

Imperial Chancellery. 1830. *Polnoe sobranie zakonov" Rossijskoj Imperii [Complete Collection of the Laws of the Russian Empire]*. Vol. 18, 1767–1769. St Petersburg: Imperial Chancellery.

Imperial Russian Academy. 1814. *Slovar' Akademii Rossijskoj [Dictionary of the Russian Academy]*. Pt 3 K–N. St Petersburg: Imperial Academy of Sciences.

———. 1822. *Slovar' Akademii Rossijskoj [Dictionary of the Russian Academy]*. Pt 6 S–end. St Petersburg: Imperial Russian Academy.

Isachsen, Gunnar. 1931. Norske undersøkelser ved Sydpollandet 1929–1931 [Norwegian Explorers in the Antarctic 1929–1931]. *Norsk geografisk tidsskrift* 3 (5–8): 345–351.

Isakov, I.S., ed. 1950. *Morskoj atlas [Maritime Atlas]*. Vol. 1. Naval General Staff: Moscow.

Ivashintsov", N. 1872. *Russkiya krugosvyetnyya puteshestviya, c" 1803 po 1849 god" [Russian Circumnavigations from 1803 to 1849]*. St Petersburg: Ministry of Marine.

Jones, A.G.E. 1982. *Antarctica Observed*. Whitby: Caedmon.

Kabo, V.P., and N.M. Bondareva. 1974. Okeaniskaya kollektsiya I.M. Simonova [I.M. Simonov's Oceania Collection]. *Sbornik muzeya antropologii i Ėtnografii* 30: 101–111.

Kalenikin, Sergej. 1989. Pyl' v Antarktide [Dust in Antarctica]. *Smena* 17: 12–15.

Kalesnik, S.V. 1949a. K 130-letiyu russkoj ėkspeditsii v Antarktidu [The 130th Anniversary of the Russian Antarctic Expedition]. *Priroda* 8: 80–82.

———. 1949b. Russkie otkrytie v Antarktike [Russian Discoveries in the Antarctic]. *Slavyane* 4: 19–22.

Kaminer, L.V., and Others. 1963. *Istoriya estestvoznaniya 1951–1956: literatura, opublikovannaya v SSSR [Bibliography of the History of Science in the USSR 1951–1956]*. Moscow: Academy of Sciences.

Kant, I. 1801. *Physische Geographie [Physical Geography]. Part 1*. Mainz: Vollmer.

Kardin, E.V. 1966. Legendy i Fakty [Legends and Facts]. *Novyj Mir* 2: 237–250.

Kasimenko, V.A. 1948. *Kak lyudi otkryvali zemlyu [How People Discovered the Earth]*. Moscow: Lenin Library.

Kemtts", Dr. 1848. Ob" uspyekhakh" zemlevyedyeniya s" pervoj poloviny XVIII stolyetiya [Results of Exploration since the First Half of the 18th Century]. In *Karmannaya knizhka dlya lyubitelej zemlevyedyeniya, izdavaemaya ot" Russkago Geograficheskago Obshhestva*, ed. Anon, 7–142. St Petersburg: Imperial Chancellery.

Khlebnikov, K.T. 1979. *Russkaya Amerika v neopublikovannykh zapiskakh [Russian America in Unpublished Memoirs]*. Leningrad: Nauka.

Khoroshevskij, A.Yu., ed. 2014. *M.P. Lazarev—Tri krugosvetnykh puteshestviya [M.P. Lazarev—Three Journeys Around the World]*. Moscow: Ėksmo.

Khvat, L. 1956. *Zagadochnyj materik [Mysterious Continent]*. Moscow: Geografgiz.

Khvostov", D.I. 1825. *Russkie morekhodtsy na ledovitom" okeane [Russian Seafarers in the Ice Ocean]*. St Petersburg: Ministry of Education.

Kippis, Andrew. 1788. *The Life of James Cook*. Vol. 2. Basle: Tourneisen.

Kisilev", Egor". 1819–1821. *Pamit'nik" prinadlezhit" matrozu 1j stat'ej Egoru Kisilevu [The Notebook of Seaman 1st Class Egor Kisilev]*. Manuscripts Division, Russian State Library, Moscow: Fond 178, MS 10897.8.

Koblents, Ya.P. 1970. Schislenie vremeni na russkom flote i khronologiya sobitij pervoj russkoj antarkticheskoj ėkspeditsii [The Reckoning of Time in the

Russian Navy and the Chronology of Events in the First Russian Antarctic Expedition]. *Byulleten' Sovetskoj Antarkticheskoj Èkspeditsii* 80: 5–23.

Kolesnikova, Maria. 2007. Rediscovering Antarctica: Nearly Two Centuries after Discovering Antarctica's Frigid Frontiers, Russia Is Hoping Its Oil-Forged Riches Can Help It Reclaim Its Position as the Southern Continent's Leading Explorer. *Russian Life*, January/February.

Kolgushkin, V.V., and P.R. Maksimov. 1958. *Opisanie starinnykh atlasov, kart i planov XVI, XVII, XVIII vekov i pervoj poloviny XIX veka [A Description of Old Atlases, Maps and Plans from the 16th to the First Half of the 19th Century].* Leningrad: Hydrographic Department, Soviet Navy—not seen.

Kondakova, O.N. 2019. *K 200-letiyu otkrytiya Antarktidy Èkspeditsiej F.F. Bellinsgauzena i M.P. Lazareva na shlyupakh "Vostok" i "Mirnyj" [For the Bicentenary of the Discovery of Antarctica by F.F. Bellingshausen and M.P. Lazarev on Sloops* Vostok *and* Mirnyj*]*. St Petersburg: Russian Naval Archives. https:// rgavmf.ru/virtualnye-vystavki/k-200-letiu-otrritiya-antarktidi.

Kondratov, A.M. 1988. *Atlantidy pyati okeanov [Atlantises of Five Oceans].* Leningrad: Gidromet.

Konetskij, Viktor. 1983. *Tretij lishnij [Three's a Crowd].* Leningrad: Soviet Author.

Kopelev, Dmitrij. 2010. *Na sluzhbe Imperii: nemtsy i Rossijskij flot v pervoj polovine XIX veka [Imperial Service: Germans and the Russian Navy in the First Half of the 19th Century].* St Petersburg: European University.

Kornilovich″, A. 1825. Izvyestie ob″ èkspeditsiyakh″ v″ Syeverovostochnuyu Sibir′ flota Lejtenantov″ Barona Vrangelya i Anzhu v″ 1821, 1822, i 1823 godakh″ [News of the Expeditions of Lieutenants Baron Wrangell and Anjou to North Eastern Siberia in 1821–23]. *Severnyj Arkhiv* 13 (4): 334–378.

Koryakin, V.S. 2008. Kontinent, otkrytyj poslednim [The Last Continent to be Discovered]. In *Dvukratnye izyskaniya v yuzhnom ledovitom okeane i plavanie vokrug sveta*, ed. F.F. Bellinsgauzen, 5–38. Moscow: Drofa.

Korzun, V.A. 2009. *Otsenka vozmozhnostej ispol'zovaniya resursov Antarktiki [An Assessment of the Possibility of Exploiting the Resources of the Antarctic].* Moscow: Institute for World Economy and Russia's International Relations.

Kostritsyn, B.V. 1951. K voprosu o rezhime Antarktiki [On the Question of an Antarctic Regime]. *Sovetskoe gosudarstvo i pravo* 3: 38–43.

Kotlyakov, V.M. 1994. *Mir snega i l'da [The World of Snow and Ice].* Moscow: Nauka.

Kotukhov, M.P. 1951. *Velikij podvig: otkrytie Antarktidy [A Great Victory: The Discovery of Antarctica].* Moscow: Ministry of the Navy—not seen.

———. 1955. *Velikij podvig: otkrytie Antarktidy.* 2nd ed. Moscow: State Geographical Press.

Kotzebue, O. 1821. *Entdeckungsreise in die Süd-See und nach der Berings-Strasse, zur Erfahrung einer nordöstichen Durchfahrt [Voyage of Discovery to the Pacific Ocean and through the Bering Strait, to find out a North-East Passage].* Vol. 1. Weimar: Hoffman.

von Kotzebue, O., and J.F. Eschscholtz. 1830. *A New Voyage Round the World in the Years 1823, 24, 25, and 26*. London: Colburn and Bentley.

Kozlov, S.A. 2003. *Russkij puteshestvennik èpokhi Prosveshheniia [The Russian Traveller in the Age of Enlightenment]*. Vol. 1. St Petersburg: Istoricheskaya illyustratsiya.

Kozlovskij, A.M. 1988. *Vokrug tol'ko lyod [Icebound]*. Leningrad: Gidromet.

Kozlyakov, V.N., and Others. 2016. Zayavlenie o lishenii Vladimira Rostislavovicha Medinskogo uchenoj stepeni doktora istoricheskikh nauk [Statement About Withdrawing the Degree of Doctor of History from Vladimir Rostislavovich Medinskij]. Ministry of Education, Russian Federation. April 25. http://wiki. dissernet.org/tools/vsyakosyak/MedinskyVR_ZoLUS.pdf.

Krajner, N.P. 1962. P.A. Kropotkin o proiskhozhdenii valunov [P.A. Kropotkin on the Origin of Boulders]. *Trudy Instituta Istorii Estestvoznaniya i tekhniki* 42 (3): 195–211.

Krasheninnikov″, S.P. 1853. Bellinsgauzena Èkspeditsiya v Yuzhnyj Ledovityj okean″ [Bellingshausen's Expedition to the Southern Ice Ocean]. In *Voennyj Èntsiklopedicheskij Leksikon″*, vol. 2, 241–242. St Petersburg: Institute for Military Education.

Krechetnikov, A. 2020. Kak rossiyane i britantsy odnovremenno otkryli Antarktidu [How Russians and Britons Simultaneously Discovered Antarctica]. BBC Russian Service, January 28. https://www.bbc.com/russian/features-51264899.

Kruchinin, Yu.A. 1965. *Shel'fovye ledniki Zemli Korolevy Mod [Ice Tongues of Queen Maud Land]*. Leningrad: Gidromet.

von Krusenstern, Vice-Admiral. 1833. Über die Entdeckung des südlichen Continents [On the Discovery of the Southern Continent]. *Annalen der Erd-, Völker- und Staatenkunde* 8 (4): 95–96.

Kruzenshtern″, Vice-Admiral. 1836. *Dopolnenie k″ izdannym″ v″ 1826 i 1827 ob″yasneniyam″ osnovanij … Atlasa Yuzhnago Morya [Supplement to the Explanatory Principles of the Atlas of the Pacific Ocean, 1826 and 1827]*. St Petersburg: Scientific Committee, Ministry of the Navy.

Kublitskij, Georgij. 1949a. Kolumby Antarktiki [Columbuses of the Antarctic]. *Vokrug sveta* 4: 4–9.

———. 1949b. *Otkryvateli Antarktidy [The Discoverers of Antarctica]*. Moscow– Leningrad: Children's Press, Ministry of Education.

———. 1957. *Po materikam i okeanam [Across Continents and Oceans]*. Moscow: Children's Press, Ministry of Education.

Kucherov, I.P. 1963. Navigatsionnye karty Antarktiki, sostavlennye v èkspeditsii Bellinsgauzena–Lazareva v 1819–1821 gg. [Navigational Charts of the Antarctic, Created During the Bellingshausen–Lazarev Expedition 1819–1821]. *Antarktika—doklady kommisii 1962* (3): 153–165.

Kucherov, I.P., and K.A. Bogdanov. 1962. Svidetel' nauchnogo podviga i geroizma [Testimony to a Heroic Scientific Achievement]. *Priroda* 5: 89–91.

Kudryashov, Konstantin. 2018. Antarktida nasha. Pochemu otkrytie kontinenta Bellinsgauzenom neosporimo? [Antarctica is Ours. Why is Bellingshausen's Discovery of the Continent Incontestable?], September 20. https://aif.ru/society/history/antarktida_nasha_pochemu_otkrytie_kontinenta_bellinsgauzenom_neosporimo.

Kuk, Dzhems. 1948. *Puteshestvie k yuzhnomu polyusu i vokrug sveta [A Journey Towards the South Pole and Around the World]*, ed. I. Magidovich. Moscow: Geografgiz.

Kuroedov, V.I., and Others, ed. 2005. *Atlas Okeanov: Antarktika [Atlas of the Oceans: the Antarctic]*. St Petersburg: Ministry of Defence and AANII.

Kuznetsov, Nikita. 2020. *Russkaya Antarktika. 200 let. Istoriya v illyustratsiyakh [The Russian Antarctic: 200 Years of History in Pictures]*. Moscow: Paulsen—not seen.

Kuznetsova, V.V. 1967. Novye dokumenty pervoj russkoj antarkticheskoj ekspeditsii [New Documents from the First Russian Antarctic Expedition]. *Byulleten' Sovetskoj Antarkticheskoj Ekspeditsii* 66: 5–11.

———. 1968. Novye dokumenty o russkoj ekspeditsii k severnomu polyusu [New Documents on the Russian Expedition to the North Pole]. *Izvestiya vsesoyuznogo geograficheskogo obshhestva* 100 (3): 237–245.

Lajba, Anatolij. 2015. *Kvadratura polyarnogo kruga [Measuring the Polar Circle]*. Moscow: Poligraf-Tsentr.

Larionov, A.L. 1963. Korabli Pervoj russkoj antarkticheskoj ekspeditsii—shlyupi «Vostok» i «Mirnyj» [The Ships of the First Russian Antarctic Expedition—Sloops *Vostok* and *Mirnyj*]. In *Pervaya russkaya antarkticheskaya ekspeditsiya 1819–1821 gg. i eyo otchyotnaya navigatsionnaya karta*, ed. M.I. Belov, 128–142. Leningrad: Morskoj Transport.

Laurie, James. 1842. *System of Universal Geography: founded on the works of Malte-Brun and Balbi*. Edinburgh: Black.

Lazarev″, M.P. 1821. Pis'mo Mikhaila Petrovicha Lazareva k″ Aleksyeyu Antonovichu Shestakovu, 24 sentyabrya 1821 [A letter from Mikhail Petrovich Lazarev to Aleksyej Antonovich Shestakov, 24 September 1821 [O.S]]. St Petersburg: SARN—F–315, S–1, P–775, fos 1–6v.

———. 1918. Letter 1 in: Pis'ma Mikhaila Petrovicha Lazareva k″ Aleksyeyu Antonovichu Shestakovu v″ g. Krasnyj Smolenskoj gubernii [Letters from Mikhail Petrovich Lazarev to Aleksyej Antonovich Shestakov at Krasnyj in the Smolensk Gubernorate]. *Morskoj sbornik″* 403 (1): 51–66.

Lazzara, Matthew A., and Others. 2012. Antarctic Automatic Weather Station Program: 30 Years of Polar Observations. *Bulletin of the American Meteorological Society* 93 (10): 1519–1537.

Lebedev, D.M. 1949. Obsuzhdenie dokladov [A Comment on the Papers]. In *Voprosy istorii otechestvennoj nauki*, ed. S.I. Vavilov, 819–831. Moscow: Academy of Sciences.

Lebedev, D.M., and V.A. Esakov. 1971. *Russkie geograficheskie otkrytiya i issledo-vaniya [Russian Geographical Discoveries and Exploration]*. Moscow: Mysl'.

Lebedev, V.L. 1957. *Antarktika [The Antarctic]*. Moscow: Geografgiz.

Lebedev, V. 1960. Who Discovered the Antarctic. *Soviet Union* 127: 52.

Lebedev, V.L. 1961. Geograficheskie nablyudenya v Antarktike èkspeditsij Kuka 1772–1775 gg. i Bellinsgauzena–Lazareva 1819–1821 gg. [Geographical Observations in the Antarctic by the Cook, 1772–1775, and Bellingshausen–Lazarev, 1819–1821, Expeditions]. *Antarktika—doklady kommisii 1960* (1): 7–24.

———. 1962. Kto otkryl Antarktidu? (otvet za pis'mo g-na Ternera) [Who Discovered Antarctica? A reply to Mr Turner's Letter]. *Antarktika—doklady kommisii 1961* (2): 153–166.

———. 1963. Reshenie spornykh voprosov Antarkticheskoj istorii na novoj osnove [A Solution for Contentious Issues in Antarctic History on a New Basis]. *Antarktika—doklady kommisii 1962* (3): 176–179.

———. 1964. O raznom tolkovanii nekotorykh mest iz dokumentov èkspeditsii Bellinsgauzena i Lazareva [On Alternative Interpretations of Some Passages in Documents from the Bellingshausen–Lazarev Expedition]. *Antarktika—doklady kommisii 1963* (4): 170–174.

———. 1975. Ot kuda v more led? [Where does Sea Ice Come From]. *Khimiya i zhizn'* 2: 35–38.

Lemeshhuk, G.P. 1984. *Iz goroda na Neve: moreplavateli i puteshestvenniki [From a City on the Neva: Navigators and Travellers]*. Leningrad: Lenizdat.

Leskov", A.S. 1823. Letter to Admiral Moller, April 2 1823. St Petersburg: SARN—F–116, S–1, P–2596, fo. 3. (Text in Belov, 1963a: 8.)

Litke, Fyodor. 1828. *Chetyrekratnoe puteshestvie v" Syevernyj Ledovityj Okean" [Four Seasons of Exploration in the Northern Ice Ocean]*. St Petersburg: Naval Press.

Lloyd, C., ed. 1949. *The Voyages of Captain James Cook Round the World*. London: Chanticleer.

Lomonosov, M.V. 1949. *O sloyakh zemnykh i drugie raboty po geologii* [On Strata and Other Geological Works], ed. G.G. Lemmlejn. Moscow–Leningrad: Gostekhizdat.

———. 1953. Mysli o proiskhozhdenii ledyanykh gor v severnykh moryakh [Thoughts on the Origin of Icebergs in Northern Seas] in *Polnoe sobranie sochinenii* [Complete Collected Works], vol. 3: 447–459. Moscow and Leningrad: Soviet Academy of Sciences.

Lomonosov", M.V. 1763. *Pervyya osnovaniya metallurgii, ili rudnykh" dyel" [Elements of Metallurgy, or Mining]*. St Petersburg: Imperial Academy of Sciences.

———. 1847. Kratkoe opisanie raznykh" puteshestvij po syevernym" moryam", i pokazanie vozmozhnago prokhodu Sibirskim" okeanom" v" Vostochnuyu

Indiyu [A Short Description of Various Journeys Through Northern Seas and An Exposition of a Possible Route to the East Indies Via the Siberian Ocean]. In *Sochineniya Lomonosova*, ed. Aleksandr" Smirdin". St Petersburg: Imperial Chancellery—not seen.

———. 1854. Kratkoe opisanie raznykh" puteshestvij po syevernym" moryam", i pokazanie vozmozhnago prokhodu Sibirskim" okeanom" v" Vostochnuyu Indiyu [A Short Description of Various Journeys Through Northern Seas and An Exposition of a Possible Route to the East Indies Via the Siberian Ocean]. In *Proekt" Lomonosova i èkspeditsiya Chichagova*, ed. A. Sokolov", 3–141. St Petersburg: Hydrographic Department.

Lomonosow, Michael. 1763. Tankar, om Is-bergens ursprung uti de Nordiska Hafven [Thoughts on the Origin of Icebergs in Northern Seas]. *Kongl. Vetenskaps Academiens Handlingar* 24: 34–40.

Lowe, F. 1842. Bellingshausens Reise nach der Südsee und Entdeckungen im südlichen Eismeer [Bellingshausen's Voyage to the Pacific and Discoveries in the Southern Ice Ocean]. *Archiv für wissenschaftliche Kunde von Russland* 2: 125–174.

Lubchenkova, T.Yu. 2001. *Samye znamenitye puteshestvenniki Rossii [The Most Famous Russian Travellers]*. Moscow: Veche.

Luchininov, S.T. 1973a. *Shlyup Vostok [Sloop Vostok]*. Moscow: DOSAAF.

———. 1973b. *Shlyup Mirnyj [Sloop Mirnyj]*. Moscow: DOSAAF.

Lüdecke, Cornelia, and Colin Summerhayes. 2012. *The Third Reich in Antarctica*. Norwich: Erskine.

Lukin, Valerij. 2001. Interview. *Izvestiya*, September 6. http://izvestia.ru/news/251319.

———. 2005. Poisk nevedomogo kontinenta [The Search for the Unknown Continent]. *Vlast'* 10: 75–81.

———. 2006. Interview. Rossiya v Antarktide: vchera, segodnya, zavtra [Russia in Antarctica: Yesterday, Today, Tomorrow]. *Nauka i zhizn'*, February/March. http://www.nkj.ru/interview/2869/.

Lukin, Valerij V.V. 2011. K 55-letiyu regulyarnykh otechestvennykh issledovanij Antarktiki [55 Years of Regular National Research in the Antarctic]. *Rossijskie Polyarnye Issledovanniya* 1 (3): 41–42.

Lukin, Valerij. 2013. Interview. Pod antarktticheskij led za novymi formami zhizni [In Search of New Forms of Life Beneath the Antarctic Ice]. *Russian Council on International Affairs Website*, May 16. https://russiancouncil.ru/analytics-and-comments/interview/pod-antarkticheskiy-led-za-novymi-for-mami-zhizni/.

Lukin, Valerij V.V. 2014. Russia's Current Antarctic Policy. *The Polar Journal* 4 (1): 199–222.

———. 2017. Rossijskij vzglyad na budushhee razvitie Sistemy Dogovora ob Antarktike [A Russian Perspective on the Future of the Antarctic Treaty System]. *Rossijskie Polyarnye Issledovanniya* 3 (29): 42–43.

Lukin, Valerij. 2018. K voprosu o natsional'noj antarkticheskoj strategii [On the Question of a National Antarctic Strategy]. *Mezhdunarodnaya zhizn'* 1: n.p. https://interaffairs.ru/jauthor/material/1963.

Lukin, Valerij, and V.V. 2020. Rossijskaya Antarkticheskaya Èkspeditsiya [The Russian Antarctic Expedition]. *Rossijskie Polyarnye Issledovanniya* 1 (39): 62–68.

Lukin, V., N. Kornilov, and N. Dmitriev. 2006. *Sovetskie i rossijskie antarkticheskie èkspeditsiyu v tsifrakh i faktakh (1955–2005gg) [Soviet and Russian Antarctic Expeditions in Figures and Facts, 1955–2005]*. St Petersburg: AANII.

Lyalin, A.Ya., and Others. 2007. *Morskoj Muzej Rossii [The Russian Naval Museum]*. St Petersburg: Central Naval Museum.

Lyalina, M.A. 1898. *Russkie moreplavateli arkticheskie i krugosvyetnye [Russian Arctic Voyages and Circumnavigations]*. St Petersburg: Devrien.

Magidovich, I. 1948. Dzhems Kuk, ego dejstvitel'nye i mnimye otkrytiya [James Cook, his Real and So-called Discoveries]. In *Puteshestvie k yuzhnomu polyusu i vokrug sveta*, ed. Dzhems Kuk and I. Magidovich, 3–34. Moscow: Geografgiz.

Magidovich, I.P. 1949. *Ocherki po istorii geograficheskikh otkrytii [Towards a History of Geographical Discoveries]*. Vol. 1. Moscow: Textbook Press.

———. 1953. Entries for Bellingshausen and Lazarev in: *Russkie moreplavateli*, ed. V.S. Lupach, 473–578. Moscow: Military Press.

Maksimovich'', L.M. 1788. *Novyj i polnyj geograficheskij slovar' Rossijskago gosudarstva [New Complete Geographical Dictionary of the Russian Empire]*. Vol. 3. Moscow: University Press and Novikov.

Malenkov, G.M. 1952. Otchyotnyj doklad XIX s''ezdu VKP(b) 5 oktyabrya 1952—vyderzhka [Report to the 19th Congress of the Soviet Communist Party, October 5 1952—extract]. In *Stalin i Kosmopolity*, A.A. Zhdanov and G.M. Malenkov, 2012. Moscow: Algoritm.

Markov, S.N. 1944. Klady «Kolumbov Rossijskikh»: dokumenty o russkoj morskoj slave [Treasures of the 'Russian Columbuses': Documents on Russian Naval Glory]. *Morskoj sbornik* 8–9: 76–81; 10: 81–88.

Markov, K.K. 1950. Oshibki Akademika A.A. Grigor'eva [The Errors of Academician A.A. Grigor'ev]. *Izvestiya vsesoyuznogo geograficheskogo obshhestva* 82 (5): 453–471.

Markov, S.N. Sergej. 1973. *Vechnye sledy [Eternal Footsteps]*. Moscow: Young Guard.

———. 1982. *Vechnye sledy*. Moscow: Sovremennik.

Markov, K.K., and Others. 1968. *Geografiya Antarktidy [Geography of Antarctica]*. Moscow: Mysl'.

Markov, S.N., and Sergej Nikolaevich. 1952. Otkryvatel' Antarktidy [Discoverer of Antarctica]. *Vokrug sveta* 1: 65.

Markova, G. 1985. Tikhoökanskaya kartoteka Sergeya Markova [Sergej Markov's Pacific Card Index]. *Al'manach bibliofila* 19: 107–124.

Marks″, A.F. 1916. *Ustrojstvo poverkhnosti i rastitel′nyj pokrov″ Zemnogo shara [The Distribution of Land Surfaces and Vegetation across the Earth]*. Map. St Petersburg: Marks″. Source: https://www.davidrumsey.com/luna/servlet/detail/RUMSEY~8~1~255643~5519943.

Mart′yanov, V.L. 2014. Perspektivy vosstanovleniya stantsii Russkaya kak postoyanno dejstvuyushhej rossijskoj antarkticheskoj stantsii [Prospects for the Reconstruction of Russkaya Station as a Continually Operational Russian Antarctic Station]. *Rossijskie Polyarnye Issledovanniya* 2 (16): 19–21.

Martens, Friderich. 1675. *Spitzbergische oder Groenlandische Reisebeschreibung gethan im Jahr 1671 [Account of a Voyage to Spitsbergen and Greenland in 1671]*. Hamburg: Schultzen.

Martin, L. 1938. An American Discovered Antarctica. In *Comptes rendus du Congrés international de géographie, Amsterdam, 1938*, pt 2 (4): 215–218. Leiden: International Geographical Union.

Martin, Lawrence. 1940. Antarctica Discovered by a Connecticut Yankee, Captain Nathaniel Brown Palmer. *Geographical Review* 30 (4): 529–552.

Martin, Stephen. 1996. *A History of Antarctica*. Sydney: State Library of New South Wales Press.

McCannon, John. 1998. *Red Arctic*. Oxford: Oxford University Press.

Medvedev, Sergej. 2015. Dvuglavyj pingvin: zachem rossijskim politikam nuzhna Antarktida [Double-headed Penguin: Why Antarctica is Necessary for Russian Politicians]. *Forbes* website, January 27. https://www.forbes.ru/mneniya-column/tsennosti/278615-dvuglavyi-pingvin-zachem-rossiiskim-politikam-nuzhna-antarktida.

Men′shikov″, M. 1891. Bellingsgauzen″ [thus], Fhaddej (Fabian″) Fhaddeevich″ [Bellingshausen, Faddej (Fabian) Faddeevich]. In *Kritiko-biograficheskij slovar′ russkikh″ pisatelej i uchonykh″*, ed. S.A. Vengerov″, vol. 2, 388–392. St Petersburg: Efron.

Miers, J. 1820. Account of the Discovery of New South Shetland. *Edinburgh Philosophical Journal* 6: 367–380.

Mill, Hugh Robert. 1903. Bellingshausen's Antarctic Voyage. *Geographical Journal* 31: 150–159.

———. 1905. *The Siege of the South Pole*. London: Alston Rivers.

Miller, D.H. 1928. Political Rights in Polar Regions. In *Problems of Polar Research*, ed. W.L.G. Joerg. New York: American Geographical Society.

Mitin, L.I. 1990. Sovremennaya otsenka nauchnoj deyatel′nosti I.M. Simonova v ėkspeditsii 1819–1821 gg. [A Modern Evaluation of the Scientific Work of I.M. Simonov on the Expedition]. In *Dva plavaniya vokrug Antarktidy* T.Ya. Sharipova, 280–312. Kazan: Kazan University Press.

Mitin, Rear Admiral L., and Lt. Cmdr S. Dorogokupets. 1983. Novaya vstrecha s «Terra australis» [A New Encounter with *Terra australis*]. *Vokrug sveta* 6: 28–32, 7: 39–41.

————. 1984. Marshrutom pervoötkryvatelej [Along the Track of the First Discoverers]. *Nauka i zhizn'* 4: 144–151.

Mogil'nitskij, Valerij. 2006. Vechnye sledy Markova [Markov's Eternal Footsteps]. *Vesti Saryarki*, March 14. 9: 6. http://old-site.karlib.kz/semenov/vechnye_sledy_markova.pdf.

Moller, A.V. 1821a. Raport [Report]. August 5. SARN—F-203, S-1, P-826, fos 23–26.

————. 1821b. Raport [Report]. August 27. SARN—F-203, S-1, P-826, fo. 33.

Molodtsov, S.V. 1954. *Sovremennoe mezhdunarodno-pravovoe polozhenie Antarktiki [The Situation of the Antarctic in Modern International Law]*. Moscow: Gosyurizdat.

Moroz, V. 2001. *Antarktida: istoriya otkrytiya [Antarctica: The History of Discovery]*. Moscow: Belyj gorod.

Morozov, P.F., and K.I. Nikul'chenkov. 1949. Èkspeditsiya F.F. Bellinsgauzena—M.P. Lazareva v Yuzhnyj Ledovityj okean i otkrytie Antarktidy [The Expedition of F.F. Bellingshausen and M.P. Lazarev to the Southern Ice Ocean and the Discovery of Antarctica]. *Morskoj sbornik* 9: 51–62.

Muldashev, Ernst R. 2013. *Propavshee zoloto Levanevskogo [The Lost Gold of Levanevskij]*. Moscow: Molma.

Myasnikov, V.F. 1986. *Puteshestvie v stranu belogo sfinksa [A Journey to the Land of the White Sphinx]*. Simferopol': Tavriya.

National Security Council. 1958. Statement of US Policy on Antarctica, March 8 1958. NSC 5804/1. In *Foreign Relations of the United States 1958–1960*. Vol. 2 (1991). Washington, DC: GPO.

Norchenko, Aleksandr. 2003. *Khronika poluzabytikh plavanij [A Chronicle of Half-Forgotten Voyages]*. St Petersburg: Balt.

Nordenskjöld, N.O.G., and G. Anderson. 1905. *Antarctica, or Two Years Amongst the Ice of the South Pole*. London: Hurst and Blackett.

Novikova, Inna. 2003. Antarktida. Sekretnaya voennaya baza Tret'ego Rejkha khranit strashnye tajny [Antarctica: The Third Reich's Clandestine Military Base Holds Terrible Secrets]. *Pravda* website, January 16. http://www.pravda.ru/science/mysterious/past/16-01-2003/34653-antarctida-0/.

Novosil'skij, P.M. 1853a. Yuzhnyj polyus": iz" zapisok" byvshago morskago ofitsera [The South Pole: From the Memoirs of a Former Naval Officer]. *Panteon"* 11 (9): 31–80; (10): 19–62.

————. 1853b. Shestoj kontinent" [The Sixth Continent]. *Panteon"* 12 (11): 99–116.

————. 1853c. *Yuzhnyj polyus": iz" zapisok" byvshago morskago ofitsera [The South Pole: From the Memoirs of a Former Naval Officer]*. St Petersburg: Vejmar".

————. 1854a. *Shestoj kontinent" [The Sixth Continent]*. 1st ed. St Petersburg: Vejmar".

————. 1854b. *Shestoj kontinent"*. 2nd ed. St Petersburg: Vejmar".

————. 1854c. *Shestoj kontinent"*. 3rd ed. St Petersburg: Imperial Academy of Sciences.

————. 1855. O geograficheskikh" otkrytiyakh" v" Yuzhnom" Polyarnom" morye, v" istekshej polovinye XIX stolyetiya [On Geographical Discoveries in the South Polar Sea in the First Half of the 19th Century]. *Zhurnal" Ministerstva Narodnago Prosvyeshheniya* 87: 16–31.

Nudel'man, A.V. 1959. *Sovetskie ekspeditsii v Antarktiku 1955–1959 gg. [Soviet expeditions to the Antarctic 1955–1959]*. Moscow: Academy of Sciences.

————. 1960. *Sovetskie ekspeditsii v Antarktiku 1958–1960 gg*. Moscow: Academy of Sciences.

————. 1962. *Sovetskie ekspeditsii v Antarktiku 1959–1961 gg*. Moscow: Academy of Sciences.

————. 1965. *Sovetskie ekspeditsii v Antarktiku 1961–1963 gg*. Moscow: Nauka.

O'Reilly, Bernard. 1818. *Greenland, the Adjacent Seas, and the North-West Passage to the Pacific Ocean*. London: Baldwin, Cradock, and Joy.

Odhams. 1935. *The New Pictorial Atlas of the World*. London: Odhams.

Okhuizen, Edwin. 2005. Dutch Pre-Barentsz Maps and the Pomor Thesis About the Discovery of Spitsbergen. *Acta Borealia* 22 (1): 21–41.

Order. 2005. Rasporyazhenie Pravitel'stva RF №713-r ot 2 iyunya 2005 g. [Order No. 713-r, Issued by the Government of the Russian Federation on June 2]. http://www.meteorf.ru/documents/9/64/.

————. 2010. Rasporyazhenie Pravitel'stva RF ot 30.10.2010 № 1926-r: O strategii razvitiya deyatel'nosti Rossijskoj Federatsii v Antarktike na period do 2020 goda i na bolee otdalennuyu perspektivu [Order No.1926-r, Issued by the Government of the Russian Federation on October 30 2010: A Strategy for the Development of the Work of the Russian Federation in the Antarctic until 2020 and in the Longer Perspective]. http://science.gov.ru/media/files/file/9odRuhiwbYKHvR2U2zNAClmaxCDUWQgA.pdf.

Orlov, B. 1949. Russian Antarctic Discoveries of 1821 are Basis of Soviet Claim. *USSR Information Bulletin* 9: 296–297.

Osipov, K. 1950. *Kak russkie lyudi otkryli Antarktidu [How Russian People Discovered Antarctica]*. Moscow: Geografgiz.

Osokin, Cap S. 1961. Nauchnyj podvig russkikh moryakov [A Scientific Victory by Russian Seamen]. *Krasnaya zvezda* 227: 5.

Ostrovskij, B.G. 1949a. Novoe ob istoricheskom pokhode Bellinsgauzena–Lazareva v Antarktiku [A New Source for Bellingshausen and Lazarev's Historic Voyage to the Antarctic]. *Zvezda* 2: 96–99.

————. 1949b. O pozabytykh istochnikakh i uchastnikakh antarkticheskoj ekspeditsii Bellinsgauzena–Lazareva [On Forgotten Sources for and Members of the Bellingshausen–Lazarev Antarctic Expedition]. *Izvestiya vsesoyuznogo geograficheskogo obshhestva* 81 (2): 239–249.

Ostrovskij, B.G.B. 1966. *Lazarev*. Moscow: Young Guard.

Ostrovsky, Arkady. 2015. *The Invention of Russia: The Journey from Gorbachev's Freedom to Putin's War*. London: Atlantic.

Ovlashhenko, Aleksandr. 2013. *Materik l'da: pervaya russkaya antarkticheskaya èkspeditsiya i eyo otrazhenie v sovetskoj istoriografii (1920-e–1940-e gody) [The Continent of Ice: The First Russian Antarctic Expedition and its Footprint in Soviet Historiography (1920s to 1940s)]*. Saarbrücken: Palmarium.

———. 2014. *Antarkticheskij rubikon: tema otkrytiya Antarktidy v sovetskikh istochnikakh nachala 50-kh godov [Antarctic Rubicon: The Discovery of Antarctica as a Theme in Soviet Sources from 1950]*. Saarbrücken: Palmarium.

———. 2016. *Antarkticheskij renessans: provedenie pervykh kompleksnykh antarkticheskikh èkspeditsij i problema otkrytiya Antarktidy [Antarctic Renaissance: The Arrival of the first Combined Antarctic Expeditions and the Problem of the Discovery of Antarctica]*. Saarbrücken: Palmarium.

de Pagès, [P.M.F.]. 1783. Tagebuch einer Seefahrt gegen den Nordpol [Journal of a voyage towards the North Pole]. Extract translated from the French original. *Der Teutsche Merkur*. 3rd Viertelj. 193–242.

Pasetskij, V.M. 1981. Russkie geograficheskie otkrytiya i issledovaniya pervoj poloviny XIX veka [Russian Geographical Discoveries and Exploration in the First Half of the 19th Century]. *Voprosy istorii* 12: 98–108.

Pavlovskij, E.N., and S.V. Kalesnik, eds. 1958. *Antarktika: materialy po istorii issledovaniya i po fizicheskoj geografii [Antarctica: Towards a History of its Exploration and its Physical Geography]*. Moscow: Geografgiz.

Pavlovskij, I.Ya. 1879. *Russko–nyemetskij slovar' [Russian–German Dictionary]*. 2nd ed. Riga: Kummel.

Penck, A. 1904. Antarktika. *Deutsche geographische Blätter* 37: 1–9.

Petermann, A. 1863. Neue Karte der Süd-Polar-Regionen [A New Map of the Antarctic]. *Petermann's Geographische Mittheilungen* 9: 407–428.

———. 1865–1867. Die Erforschung der arktischen Central-Region durch eine Deutsche Nordfahrt [Exploration of the Central Arctic by a German Expedition]. *Petermann's Geographische Mittheilungen Ergänzungsband IV*: 1–14.

Petrov, N., and O. Èdel'man. 1997. Novoe o sovetskikh geroyakh [A New Perspective on Soviet Heroes]. *Novyj Mir* 6: 140–151.

Petrova, Evgenia, and Others, ed. 2012. *Pavel Mikhailov 1786–1840: Voyages to the South Pole*. St Petersburg: Palace.

Phipps, C.J. 1774. *A Voyage towards the North Pole*. London: Nourse.

Pimenova, È.K. 1925. *Geroi Yuzhnogo Polyusa (Lejtenant Shekl'ton i Kapitan Skott) [Heroes of the South Pole (Lieutenant Shackleton and Captain Scott)]*. Leningrad–Moscow: Kniga.

Pinegin, Nikolaj. 1934. *Sem' desyat dnej bor'by za zhizn' [Seventy Days in a Fight for Life]*. Arkhangel'sk: Northern Regions Press.

Pleshakov, A., Vvedenskij, È., and E. Domogatskikh. 2018. *GDZ Geografiya 5 klass* [Geography Homework for Class 5]. https://shkola.center/5-klass/geografiya-5/page,19,32-gdz-geografiya-5-klass-uchebnik-vvedenie-vgeo-grafiyupleshakov-a-vvedenskiy-e-domogackih-e-2018.html.

Polevoj, N.A. 1833. Retsenziya na *Poyezdku k Ledovytomu moryu Fr. Belyanskogo* i *Poyezdku v Yakutsk″* [Review of *A Journey to the Arctic Ocean by Fr. Belyanskij* and *Journey to Yakutsk*]. *Moskovskij Telegraf″* 52 (14): 216–252.

Polovtsov″, A.A., ed. 1900. *Russkij biograficheskij slovar′ [Russian Biographical Dictionary]*. Vol. 2. St Petersburg: Imperial Estate.

Popov, L.A. 1979. *God v Antarktike [A Year in the Antarctic]*. Moscow: Nauka.

Potapov, R.L. 2012. Popugaj, kotoryj zhil s pingvinami [The Parrot that Lived with Penguins]. *Priroda* 11: 89–96.

Powell, G. 1822. *Chart of South Shetland Including Coronation Island, &c. From the Exploration of the Sloop* Dove *in the Years 1821 and 1822 by George Powell Commander of the Same*. London: Laurie. https://collections.rmg.co.uk/collections/objects/540915.html.

Prévost d'Exiles, Antoine-François. 1759. *Histoire générale des voyages [A General History of Voyages]*. Vol. 15. Paris: Didot.

Purdy, John. 1822. *Memoir, descriptive and explanatory, to Accompany the New Chart of the Ethiopic or South Atlantic Ocean*. London: Laurie.

———. 1824. *A Chart of the World on Mercator's Projection*. London: Laurie. https://texashistory.unt.edu/ark:/67531/metapth193446/.

Rabinovich″, I.O. 1908. *Shestaya chast′ svyeta [The Sixth Continent]*. St Petersburg: Stepanova.

Rajkhenberg, M. 1941. Otkrytie pervoj zemli v Antarktike [The Discovery of the First Land in the Antarctic]. *Sovetskaya Arktika* 2: 61–69.

Ravenstein, L. 1867. *Meyer's Hand-Atlas der neuesten Erdbeschreibung in 100 Karten [Meyer's Portable Atlas]*. Hildburghausen: Bibliographisches Institut—not seen.

Renan, Ernest. 1882. *Qu'est-ce qu'une nation? [What Constitutes a Nation?]*. Paris: Calmann Lévy.

Robertson, William. 1780. *The History of America*. Vol. 2. 3rd ed. London: Strahan and Cadell.

Robin, G. de Q. 1962. Discovery, Exploration, Adventure & Courage. *UNESCO Courier* 1: 14–20.

Rodomanov, B.B. 1959. O ponyatiakh «materik», «kontinent» i «chast′ sveta» [On the Meanings of *materik, kontinent* and *chast′ sveta*]. *Izvestiya vsesoyuznogo geograficheskogo obshhestva* 91 (2): 159–160.

Rokot, V. 2016. Transcript of Interview on Radio Mayak, Moscow, May 28. https://radiomayak.ru/shows/episode/id/1303503/.

Ross, Captain John. 1819. *A Voyage of Discovery, Made Under the Orders of the Admiralty, in His Majesty's Ships Isabella and Alexander*. Vol. 1. London: Longman and Others.

Ross, Captain Sir James Clark. 1847. *A Voyage of Discovery and Research in the Southern and Antarctic Regions During the Years 1839–43.* 2 vols. London: John Murray.

Rusakov", V. (pseudonym of S.F. Librovich). 1903. *Russkie Kolumby i Robinzony* [Russian Columbuses and Crusoes]. Moscow: Vol'f.

Russwurm, Carl. 1870. *Nachrichten über die adeliche und freiherrliche Familie von Bellingshausen [The Noble and Landed Bellingshausen Family].* Reval: Lindfors' Erben.

Ryabchikov, E. 1955. K zemle tajn! [To the Land of Secrets!]. *Ogonyok* 49: 18–19.

Rybakov, S.N. 1976. *Zhivaya Antarktika [The Living Antarctic].* Leningrad: Gidromet.

Rymill, J.R. 1936. An Antarctic Illusion: The Coasts of Grahamland. *The Times*, December 12: 13.

———. 1938a. British Graham Land Expedition, 1934–37. *Geographical Journal* 91 (4): 297–312; 424–438.

———. 1938b. *Southern Lights: the Official Account of the British Graham Land Expedition, 1934–1937.* London: Chatto and Windus.

Sabine, Edward. 1850–1853. *Observations Made at the Magnetical and Meteorological Observatory at Hobarton, in Van Diemen island, and by the Antarctic Naval Expedition.* 3 vols. London: Hobart Observatory.

de Saint-Pierre, Bernardin. 1796. *Etudes de la Nature [Studies of Nature].* Vol. 1. rev. ed. London: Spilsbury.

Sale, Richard, and Eugene Potapov. 2010. *The Scramble for the Arctic: Ownership, Exploitation and Conflict in the Far North.* London: Lincoln.

Samarov, A.A., ed. 1952. *M.P. Lazarev—dokumenty [M.P. Lazarev—Documents].* Vol. 1. Moscow: Ministry of the Navy.

Saulkin, V.A. 2018. Tretij Rim protiv novogo Karfagena [The Third Rome Versus the New Carthage], March 29. https://nic-pnb.ru/analytics/tretij-rim-protiv-novogo-karfagena/.

Savatyugin, L.M., and M.A. Preobrazhenskaya. 1999–2009. *Rossijskie issledovaniya v Antarktike [Russian Exploration in the Antarctic].* Vol. 4. St Petersburg: Gidromet.

———. 2014. *Karta Antarktidy: imena i sluzhby [The Map of Antarctica: Names and Service Records].* St Petersburg: Geograf.

Schwab, Jakob Friedrich, ed. 1804. *III Naturlehre. Das Meer. [Natural History Part 3—the Oceans].* Oestereichischer Toleranz-Bote. Vienna: Nehm.

Scoresby, William, Jr. 1818. On the Greenland or Polar Ice. *Memoirs of the Wernerian Society* 2: 261–388.

———. 1820. *An Account of the Arctic Regions.* Vol. 1. Edinburgh: Constable.

Seaborn, Captain Adam. (pseudonym of J.C. Symmes). 1818. *Symzonia: A Voyage of Discovery.* New York: Seymour.

Sementovskij, V.N. 1950. Ideo-politicheskaya napravlennost' kursa fizicheskoj geografii SSSR [The Politico-ideological Orientation of a Course in the Physical Geography of the USSR]. *Voprosy geografii* 18: 21–33.

————., ed. 1951. *Russkie otkrytiya v Antarktike [Russian Discoveries in the Antarctic]*. Moscow: Geografgiz.

Sementovskij, V.N., and N.N. Vorob'ev. 1940. *Fiziko-geograficheskie èkskursii v okrestnostyakh g.* In *Kazani [Excursions in Physical Geography around Kazan]*. Kazan: Tatgosizdat.

Senex, Iohn. 1725. *A Map of the World*. London: Self. https://nla.gov.au/nla. obj-230683748/view.

Sergeev, Dmitrij. 2016. Bitva za budushhee: zachem rossijskij flot vozvrashhaetsya v Antarktidu [Conquering the Future: Why the Russian Navy is Returning to Antarctica]. *Teleradiokompaniya Zvezda*, January 25. http://tvzvezda.ru/news/forces/content/201601250744-zkl7.htm.

Sergeeva, Ol'ga Anatol'evna. 2003 (updated 2010). *Antarktida [Antarctica]*. http://festival.1september.ru/articles/582633/.

Sharipova, T.Ya., ed. 1990. *Dva plavaniya vokrug Antarktidy [Two Voyages around Antarctica]*. Kazan: Kazan University Press.

Shepilov, D. 1947. Sovetskij patriotizm [Soviet Patriotism]. *Pravda*, 11 August 1947: 2–3.

Shhedrovskij, I.S. 1844. *Ego vysokoprevoskhoditel'stvu Admiralu Faddeyu Faddeyevichu Bellinsgauzenu [For His Excellency Admiral Faddej Faddeevich Bellingshausen]*. Kronstadt: Self—not seen.

Shherbakov, D.I. 1955. Zagadki Antarktidy [Mysteries of Antarctica]. *Ogonyok* 49: 17–18.

————. 1956. Nauchnye rezul'taty antarkticheskikh èkspeditsij [Scientific Results from Antarctic Expeditions]. In *Zagadochnyj materik*, ed. L. Khvat, 259–285. Moscow: Geografgiz.

Shherbakov, D. 1976. «Vostok» i «Mirnyj» otkryvayut Antarktidu [*Vostok* and *Mirnyj* Discover Antarctica]. In *Korabli-geroi*, ed. Adm V.N. Alekseev. Moscow: DOSAAF.

Shmatkov, Vladimir. 1998. Zagadki, gipotezy, otkrytiya: pochemu Kuk ne otkryl Antarktidu? [Puzzles, Hypotheses and Discoveries: Why didn't Cook Discover Antarctica?]. *Vokrug sveta* 7: 23–24.

Shokal'skij, Yu.M. 1898. Polyarnyya strany Yuzhnago polushariya [Polar Countries of the Southern Hemisphere]. In *Èntsiklopedicheskij slovar'*, ed. K.K. Arsen'ev″ and F.F. Petrushevskij, vol. 24, 489–495. St Petersburg: Brokgauz and Efron.

————. 1928. Stoletie so vremeni otpravleniya Russkoj antarkticheskoj èkspeditsij pod komandoyu F. Bellinsgauzena i M. Lazareva 4 iyulya 1819 g. iz Kronshtadta [Centenary of the Departure from Kronstadt of the Russian Antarctic Expedition, Commanded by F. Bellingshausen and M. Lazarev, on July 4 1819 [O.S.]]. *Izvestiya gosudarstvennogo russkogo geograficheskogo obshhestva* 60 (2): 176–212.

————. 1937. Novosti ob Antarktide [News about Antarctica]. *Izvestiya gosudarstvennogo geograficheskogo obshhestva* 69 (4): 666–667.

————. 1939. Ostrov Petra I [Peter I Island]. *Izvestiya gosudarstvennogo geogra-ficheskogo obshhestva* 71 (9): 1393–1396.

Shteppa, Konstantin F. 1962. *Russian Historians and the Soviet State*. New Brunswick, NJ: Rutgers University Press.

Shur, L.A. 1971. *K beregam Novogo Sveta [To the Shores of the New World]*. Moscow: Nauka.

Shvede, E.E. 1947. Puteshestvie kapitana Bellinsgauzena v antarkticheskie morya 1819–1821 [The Voyage of Captain Bellingshausen to the Antarctic Seas 1819–1821]. *Izvestiya vsesoyuznogo geograficheskogo obshhestva* 79 (3): 357–358.

————. 1949. Pervaya russkaya antarkticheskaya ekspeditsiya 1819–1821 gg. [The First Russian Antarctic Expedition 1819–1821]. In *Dvukratnye izyskaniya v yuzhnom ledovitom okeane i plavanie vokrug sveta v prodolzhenie 1819, 20 i 21 godov*, ed. F.F. Bellinsgauzen and E.E. Shvede, 7–30. Moscow: Geografgiz.

————. 1952. *Otkrytie Antarktidy russkimi moryakami [The Discovery of Antarctica by Russian Seamen]*. Moscow: Znanie.

————. 1960. Otkrytie Antarktidy russkimi moreplavatelyami v 1819–1821 gg. [The Discovery of Antarctica by Russian Navigators in 1819–1821]. In *Dvukratnye izyskaniya v yuzhnom ledovitom okeane i plavanie vokrug sveta v prodolzhenie 1819, 20 i 21 gg*, ed. E.E. Shvede, 9–52. Moscow: Geografgiz.

————. 1962. F.F. Bellinsgauzen. In *Lyudi russkoj nauki: geologiya i geografiya*, ed. I.V. Kuznetsov, 419–431. Moscow: Physics and Mathematics Press.

Simakova, Lyudmila. 2015. *Aleksandr Kuchin: Russkij u Amundsena [Alexander Kuchin: A Russian with Amundsen]*. Moscow: Paulsen.

Simonov, I.M. 1951. Shlyupy «Vostok» i «Mirnyj» ili plavanie rossiyan v Yuzhnom Ledovitom okeane i okolo sveta [Sloops *Vostok* and *Mirnyj*, or the Voyage by Russians in the Southern Ice Ocean and around the World]. In *Russkie otkryt-iya v Antarktike*, ed. V.N. Sementovskij, 51–175. Moscow: Geografgiz.

————. 1955. Avtobiografiya I.M. Simonova (1848 g.) [Autobiography of I.M. Simonov (1848)]. *Istoriko-astronomicheskie issledovaniya* 1: 268–277.

————. 1990a. Slovo ob uspekhakh plavaniya shlyupov «Vostok» i «Mirnyj» okolo sveta i osobenno v Yuzhnom Ledovitom more, v 1819, 1820 i 1821 godakh [An Address about the Results from the Voyage of the Sloops *Vostok* and *Mirnyj* around the World and Especially in the Southern Ice Ocean, in 1819, 1820 and 1821]. In *Dva plavaniya vokrug Antarktidy*, ed. T.Ya. Sharipova, 18–40. Kazan: Kazan University Press.

————. 1990b. «Vostok» i «Mirnyj» [*Vostok* and *Mirnyj*]. In *Dva plavaniya vokrug Antarktidy*, ed. T.Ya. Sharipova, 46–248. Kazan: Kazan University Press.

Simonov″, Prof. i Kav. 1822a. Plavanie shlyupa *Vostoka* v″ Yuzhnom″ Ledovitom″ Morye [The Voyage of the Sloop *Vostok* in the Southern Ice Ocean]. *Kazanskij vyestnik″* 4 (3): 156–165, 4 (4): 211–216, 5 (5): 38–42, 5 (7): 174–181, 6 (10): 107–116, 6 (12): 226–232.

————. 1822b. *Slovo o uspyekhakh'' plavaniya shlyupov'' Vostoka i Mirnago okolo svyeta i osobenno v'' Yuzhnom'' Ledovitom'' morye, v'' 1819, 1820 i 1821 godakh''* [*An Address about the Results from the Voyage of the Sloops Vostok and Mirnyj around the World and Especially in the Southern Ice Ocean, in 1819, 1820 and 1821*]. Kazan: Kazan University Press.

Simonov'', Prof. 1822c. Kratkij otchyot'' [A Brief Report]. *Kazanskij vyestnik''* 3 (10): 98–107.

————. 1825. O raznosti temperatury v'' Yuzhnom'' i Syevernom'' polushariyakh'' [On the Difference in Temperature between the Southern and Northern Hemispheres]. *Kazanskij vyestnik''* 14: 99–119.

Simonov'', I. 1828. *Opredyelenie geograficheskogo polozheniya myest'' yakornago stoyaniya shlyupov'' VOSTOKA i MIRNAGO u.m.∂.* [*A Determination of the Geographical Location of the Anchorages of Sloops Vostok and Mirnyj etc.*]. St Petersburg: Department of Education.

Siryj, S.P. 2010. Krugosvetnoe plavanie kapitana 2 ranga F.F. Bellinsgauzena i lejtenanta M.P. Lazareva na shlyupakh "Vostok" i "Mirnyj" i otkrytie Antarktidy [The Circumnavigation by Junior Captain F.F. Bellingshausen and Lieutenant M.P. Lazarev in the Sloops *Vostok* and *Mirnyj*, and the Discovery of Antarctica]. https://rgavmf.ru/sites/default/files/lib/siry_antarctida.pdf.

Skorezbi, M.L. Villiam''. 1825. *Podennyya zapiski o plavanii na syevernyj kitovyj promysl''* [*Journal of a Voyage to the Northern Whale Fishery*]. St Petersburg: Naval Press.

Smirnov'', N.M. 1853. *Sobranie russkikh'' voennykh'' razskazov''* [*A Collection of Russian War Stories*]. St Petersburg: Military Education Department.

Smirnov, V.G. 2005. *Ot kart vetrov i techenij do podvodnykh min* [*From Wind and Current Charts to Underwater Mines*]. St Petersburg: Gidromet.

————. 2006. *Neizvestnyj Vrangel'* [*The Unknown Wrangell*]. St Petersburg: Gidromet.

Sokolov, A.V. 1950a. Obzor deyatel'nosti otdeleniya istorii geograficheskikh znanij i istoricheskoj geografii za period s oktyabrya 1947 g. po dekabr' 1949 g. [An Overview of the Work of the Historical Section from October 1947 to December 1949]. *Voprosy geografii* 17: 241–254.

————. 1950b. Obzor deyatel'nosti otdeleniya istorii geograficheskikh znanij i istoricheskoj geografii s dekabrya 1949 g. po aprel' 1950 g. [An Overview of the Work of the Historical Section from December 1949 to April 1950]. *Voprosy geografii* 20: 333–337.

Sokolov, A.I. 1951. Predislovie [Foreword]. In *Russkie otkrytiya v Antarktike*, ed. V.N. Sementovskij, 3–6. Moscow: Geografgiz.

Sokolov, A.V. 1953. Obzor deyatel'nosti otdeleniya istorii geograficheskikh znanij i istoricheskoj geografii s maya 1950 g. po yanvar' 1953 g. [An Overview of the Work of the Historical Section from May 1950 to January 1953]. *Voprosy geografii* 31: 274–285.

———. 1960. Obzor deyatel'nosti otdeleniya istorii geograficheskikh znanij i istoricheskoj geografii s fevralya 1953 g. po maj 1958 g. [An Overview of the Work of the Historical Section from February 1953 to May 1958]. *Voprosy geografii* 50: 238–252.

———. 1969. V moskovskom filiale geograficheskogo obshhestva SSSR [At the Moscow Branch of the Soviet Geographical Society]. *Voprosy istorii estestvoznaniya i tekhniki* 26 (1): 87–89.

———. 1970. Obzor deyatel'nosti otdeleniya istorii geograficheskikh znanij i istoricheskoj geografii s oktyabrya 1958 po dekabr' 1963 g. [An Overview of the Work of the Historical Section from October 1958 to December 1963]. *Voprosy geografii* 83: 173–189.

Sokolov, A.V., and E.G. Kushnarev. 1951. *Tri krugosvetnykh plavaniya M.P. Lazareva [The Three Circumnavigations by M.P. Lazarev]*. Moscow: Geografgiz.

Sokolov", A., ed. 1854. *Proekt" Lomonosova i èkspeditsiya Chichagova [Lomonosov's Proposal and the Chichagov Expedition]*. St Petersburg: Hydrographic Department.

Somov, Mikhail. 1978. *Na kupolakh zemli [On the Domes of the Earth]*. Leningrad: Lenizdat.

Sopotsko, A.A. 1978. Vakhtennye zhurnaly korablej V.I. Beringa [V.I. Bering's Logbooks]. *Izvestiya vsesoyuznogo geograficheskogo obshhestva* 110 (2): 164–170.

Sparks, John. 2016. How Russia is Engaged in a Battle for its Own History. *Sky News*, December 11. https://news.sky.com/story/how-russia-isengaged-in-a-battle-for-its-own-history-10691897.

Stackpole, Edouard A. 1955. *The Voyage of the Huron and the Huntress*. Mystic, CT: Marine Historical Association.

Starosel'skaya-Nikitina, O.A., and Others. 1949. *Istoriya estestvoznaniya 1917–1947: literatura, opublikovannaya v SSSR [Bibliography of the History of Science in the USSR 1917–1947]*. Moscow: Academy of Sciences.

———. 1955. *Istoriya estestvoznaniya 1948–1950: literatura, opublikovannaya v SSSR [Bibliography of the History of Science in the USSR 1948–1950]*. Moscow: Academy of Sciences.

Statiev, Alexander. 2012. "La Garde meurt mais ne se rend pas!": once again on the 28 Panfilov Heroes. *Kritika: Explorations in Russian and Eurasian History* 13: 769–798.

Stojkovich", Athanasij. 1813. *Nachal'nyya osnovaniya fizicheskoj geografii [Preliminary Elements of Physical Geography]*. Kharkov: University Press.

Strokov, A. 1947. O sbornike statej «Pyotr Velikij» [On the Anthology 'Peter the Great']. *Voennaya mysl'* 10: 84–87.

Strugatskij, V. 1986. *Podvig na Polyuse kholoda [Victory at the Pole of Cold]*. Leningrad: Lenizdat.

Suris, B. 1957. *I. Shhedrovskij*. Moscow: Izogiz.

Tammiksaar, E. 2016. The Russian Antarctic Expedition under the Command of Fabian Gottlieb von Bellingshausen and its Reception in Russia and the World. *Polar Record* 52 (5): 578–600.

Taylor, Prue. 2011. Common Heritage of Mankind Principle. In *The Berkshire Encyclopedia of Sustainability. Vol. 3. The Law and Politics of Sustainability*, ed. Klaus Bosselmann, Daniel Fogel, and J.B. Ruhl, 64–69. Great Barrington, MA: Berkshire.

Telitsyn, Vadim. 2013. *Gitler v Antarktike [Hitler in the Antarctic]*. Moscow: Yauza.

Thomson, John. 1817. *A New General Atlas*. Edinburgh: Self.

Tikhonov, V.V. 2012. Iz istorii ideologicheskikh kampanij v sovetskoj istoricheskoj nauke: sbornik «Pyotr Velikij» i sud'ba ego avtorov [From the History of Ideological Campaigns in Soviet Historiography: The Anthology *Peter the Great* and the Fate of its Authors]. *Istoriya i istoriki* 2009–2010: 118–133.

———. 2013. «Khudshij obraznik prekloneniya pered inostranshhinoj»: ideologicheskie kampanii «pozdnego stalinizma» i sud'ba istorika S.A. Fejginoj ['The Worst Example of Subservience before the Outside World': The Ideological Campaigns of 'late Stalinism' and the Fate of the Historian S.A. Fejgina]. *Novejshaya istoriya Rossii* 1: 199–207.

Timofeev, P. 2006. Pokorenie nichejnogo materika [The Conquest of No-man's Continent]. *Vash tainyj sovetnik* 6: 22–23.

de Traversay, Marquis. 1819. Memo, June 23 1819. St Petersburg: SARN—F–166, S–1, P–660a, fo. 42.

Tryoshnikov, A.F. 1963. *Istoriya otkrytiya i issledovaniya Antarktidy [A History of Discovery and Research in Antarctica]*. Moscow: State Geographical Press.

Tsigel'nitskij, I.I. 1988. *V morya studenye ukhodyat korabli [And Ships Move Out to Bitter Seas]*. Leningrad: Gidromet.

Tsuker, Viktor. 2015. Vydayushhijsya vklad moryakov rossii [A Notable Contribution from Russian Seamen]. *Vesti morskogo Peterburga* 1: 44–45.

Tumarkin, D.D. 1978. Materialy pervoj russkoj krugosvetnoj èkspeditsii kak istochnik po istorii i ètnografii Gavajskikh ostrovov [Materials from the First Russian Circumnavigating Expedition as a Source for the History and Ethnography of the Hawaiian Islands]. *Sovetskaya Ètnografiya* 5: 68–84.

———. 1983. Materialy èkspeditsii M.N. Vasil'eva—tsennyj istochnik po istorii i ètnografii Gavajskikh ostrovov [Materials from M.N. Vasil'ev's Expedition—A Valuable Source for the History and Ethnography of the Hawaiian Islands]. *Sovetskaya Ètnografiya* 6: 48–61.

Udintsev, G. 2002. Pervoötkryvatelyam Antarktidy [For the First Discoverers of Antarctica]. *Moskovskij zhurnal* 12: n.p. http://ruskline.ru/monitoring_smi/2002/12/01/pervootkryvatelyam_antarktidy/.

Unkovskij, S.Ya. 2004. *Zapiski moryaka 1803–1819 gg. [Memoirs of a Seaman 1803–1819]*, ed. L. Zakovorotnaya, Moscow: Sabashnikovy.

US State Department. 1948. Press Release, 28 August 1948. *Department of State Bulletin* 19 (479): 301.

Usov, Aleksej. 2010. Putin khochet sdelat' Antarktidu russkoj koloniej: $2 milliarda ot Minprirody uplyvut k pingvinam [Putin Wants to Turn Antarctica into a Russian Colony: Minprirody to Waste $2 Billion on Penguins]. *Novyj Region*, October 22. http://vlasti.net/news/107204.

Utusikov, Yu.D. 2012. *Ot Obskikh beregov do mostika «Obi» [From the Shores of the Ob' to the Bridge of the Ob']*. St Petersburg: Morskoe nasledie.

Uzin, S.V. 1950. *Zagadochnye zemli [Mysterious Lands]*. Moscow: Geografgiz.

V.I. 1881. Mikhail" Petrovich" Lazarev". *Russkij arkhiv"* 2 (2): 347–361.

Vadetskij, B. 1957. *Obretenie schast'ya [The Discovery of Fortune]*. 2nd ed. Moscow: Kirov Press.

Vaisala. 2010. Automatic Weather Reports from Antarctica. https://www.vaisala.com/sites/default/files/documents/MET%20Antarctica%20success%20story%20B210957EN-A.pdf.

Vdovin, A.I. 2008. Bor'ba s nizkopoklonstvom i kosmopolitizmom v poslevoennye 1940-e gody: prichiny, posledstviya, uroki [The Struggle Against Kowtowing and Cosmopolitanism in the Postwar 1940s: Causes, Results and Lessons]. In *Sluzhenie otechestvu: russkaya traditsiya i sovremennost'*, ed. Evgenij Troitskij, 247–258. Moscow: Granitsa.

Verne, Jules. 1871. *Vingt Mille Lieues sous les Mers [20,000 Leagues under the Sea]*. Paris: Hetzel.

Volkov, Dmitrij. 2003. Est' gory kruche Èveresta [There are Mountains Higher than Everest]. *Izvestiya*, 11 February 2003. https://iz.ru/news/272868.

Vorob'ev, V.I. 1948. Obshhij ocherk [Overview]. In *Materialy po lotsii Antarktiki*, ed. I.F. Novoselov, 1–38. Moscow: Hydrographic Department, Soviet Navy.

Vujacic, Veljko. 2007. Stalinism and Russian Nationalism: A Reconceptualization. *Post-Soviet Affairs* 23 (2): 156–183.

Vvedenskij, N. 1939. Russkaya krugosvetnaya antarkticheskaya èkspeditsiya 1819–1821 gg. [The Antarctic Circumnavigation by a Russian Expedition, 1819–1821]. *Geografiya v shkole* 3: 43–47.

———. 1940. *V poiskakh yuzhnogo materika [The Quest for a Southern Continent]*. GlavSevMorPut: Leningrad–Moscow.

———. 1941. K voprosu o russkikh otkrytiyakh v Antarktike v 1819–1821 godakh, v svete novejshikh geograficheskikh issledovanij [On the Question of Russian Discoveries in the Antarctic in 1819–1821, in the Light of Recent Geographical Research]. *Izvestiya vsesoyuznogo geograficheskogo obshhestva* 73 (1): 118–122.

Vyshinskij, A.Ya. 1950. Dokladnaya zapiska ministra inostrannykh del SSSR A.Ya. Vyshinskogo I.V. Stalinu po voprosu o memorandume ryadu pravitel'stvo rezhime Antarktiki, February 20 1950 [An Official Memo from Soviet Minister of Foreign Affairs A.Ya. Vyshinskij to I.V. Stalin on the question of the Memorandum to Several Governments about an Antarctic Regime]. Arkhiv Aleksandra N. Yakovleva, Sovetsko–Amerikanskie otnosheniya, 1949–1952. Document No.49. http://www.alexanderyakovlev.org/fond/issues-doc/71711.

Walker, Matt. 1999. Back from the Dead. *New Scientist* 2202, September 4. https://www.newscientist.com/article/mg16322020-200-back-from-the-dead/.

Weddell, James. 1825. *A Voyage Towards the South Pole, Performed in the Years 1822–24.* London: Longman and Others.

Wüllerstorf-Urbair, Bernhard, and Others. 1861–1875. *Reise der österreichischen Fregatte Novara um die Erde in den Jahren 1857, 1858, 1859 [Circumnavigation by the Austrian Frigate Novara in 1857, 1858, 1859].* 21 vols. Vienna: K. Gerold's Sohn.

Yablokov, A.V., and V.A. Zemsky, eds. 2000. *Soviet Whaling Data (1949–1979).* Moscow: Centre for Russian Environmental Policy.

Yakovlev, A. 1972. Protiv antiistorizma [Against Antihistoricism]. *Literaturnaya gazeta*, November 15: n.p. http://left.ru/2005/15/yakovlev132.phtml.

Yakovlev, V.G. 1953. O sisteme pionerskikh sborov [On the Training System for Pioneers]. *Sovetskaya pedagogika* 1: 15–30.

Yeltsin Presidential Library. 2018. Russkoj èkspeditsiej otkryta Antarktida [Antarctica Discovered by a Russian Expedition], January 28. https://www.prlib.ru/history/618985.

Yelverton, David E. 2004. *The Quest for a Phantom Strait.* Guildford: Polar Publishing.

Zhdanov, A.A. 1946. Doklad t. Zhdanova o zhurnalakh «Zvezda» i «Leningrad» [Comrade Zhdanov's Report on the Magazines *Zvezda* and *Leningrad*]. *Pravda*, 21 September 1947: 2–3.

———. 1952. *Vystuplenie na diskussii po knige G.F. Aleksandrova «Istoriya zapadnoevropejskoj filosofii», 24 iyunya 1947 [Intervention in the Discussion of the Book History of Western Philosophy by G.F. Aleksandrov, 24 June 1947].* Moscow: Politizdat.

Zotikov, Igor'. 1984a. *Za razgadkoj tajn Ledyanogo kontinenta [Unlocking the Secrets of the Icy Continent].* Moscow: Mysl'.

———. 1984b. *Ya iskal ne ptitsu kivi [I was not Looking for Kiwis].* Leningrad: Gidromet.

Zubov, N.N. 1953. Russkie moryaki—issledovateli okeanov i morej [Russian Sailors—Explorers of the Seas and Oceans]. In *Russkie moreplavateli*, ed. V.S. Lupach, iii–xxxvii. Moscow: Military Press.

Zubov, V.P. 1956. *Istoriografiya estestvennykh nauk v Rossii (XVIII—pervaya polovina XIX v.) [The Historiography of the Natural Sciences in Russia: 1700–1850].* Moscow: Academy of Sciences.

Name Index

© The Author(s), under exclusive license to Springer Nature
Switzerland AG 2021
R. Bulkeley, *The Historiography of the First Russian Antarctic
Expedition, 1819–21*,
https://doi.org/10.1007/978-3-030-59546-3

SUBJECT INDEX